Geolocation Techniques

Camillo Gentile · Nayef Alsindi
Ronald Raulefs · Carole Teolis

Geolocation Techniques

Principles and Applications

Springer

Camillo Gentile
National Institute of Standards
and Technology
Gaithersburg, Maryland
USA

Nayef Alsindi
Etisalat BT Innovation Center (EBTIC)
Khalifa University of Science, Technology
and Research (KUSTAR)
Abu Dhabi
United Arab Emirates (UAE)

Ronald Raulefs
German Aerospace Center
Wessling, Bavaria
Germany

Carole Teolis
TRX Systems
Greenbelt, Maryland
USA

ISBN 978-1-4614-1835-1 ISBN 978-1-4614-1836-8 (eBook)
DOI 10.1007/978-1-4614-1836-8
Springer New York Heidelberg Dordrecht London

Library of Congress Control Number: 2012945745

Printed on acid-free paper

Springer is part of Springer Science+Business Media (www.springer.com)

Preface

While geolocation is a relatively new topic in the multidisciplinary area of electrical, mechanical, and industrial engineering, it has grown very rapidly in the last decade due to the tremendous impact it is having on our everyday lives. Some of the most conspicuous applications, to name a few, range from location identification in mobile social networking and in automatic recognition systems, to furnishing real-time directions on the road, as well as in critical missions for precision personnel tracking in emergency situations such as firefighting.

Geolocation systems are based on a number of different technologies. For example, the Global Positioning System has existed for several decades, but only in the last couple of years has it been commercially accessible to the everyday consumer in the form of portable navigators. The pervasiveness of wireless access points for communications has, as a byproduct, provided yet another means for consumers to determine their positions using Wi-Fi technology. Fourth generation cellular systems, which are currently being rolled out, are being designed specifically—as opposed to previous generations—with location services in mind and, in turn, can deliver accuracy an order of magnitude higher. Also, the approval of the unlicensed FCC band has enabled rapid growth in the use of Ultra-Wideband technology for high-precision ranging. Finally, the maturity of inertial-based location systems coupled with their cost-effective solutions are beginning to play a central role in cheap smartphones as well as in more complex emergency responder rescue systems. All these technologies, while treated separately in the past, are coming together in the form of hybrid systems that offer robust solutions for a wide range of user communities.

This scope of this book is to provide a comprehensive overview of geolocation technologies and techniques, from radio-frequency based to inertial based—to our knowledge, the first book to do so—affording the reader a valuable resource that facilitates not only basic understanding of the subject, but also depth to serve as a reference for scholarly activities such as teaching, self-learning, or research. The book contains sufficient detail for use as a university textbook, but is broad enough to be of interest to laymen wishing to gain insight into the topic. In that capacity, it could serve as a starting point for a graduate student who wishes to conduct in-depth research on the topic. Likewise, it could be used in the industry during the first stages of product development. The audience will range from general readers who are interested to know about geolocation fundamentals to the more advanced readers such as researchers and industry engineers who will benefit from the technical depth and advanced techniques provided. The collaboration of international co-authors brings together expertise in different specific subjects to ultimately provide material that adds value to the many interested in the field of geolocation.

Acknowledgments

To Christian and Sophia. You make me so proud to be a father. And to my wife, Simona, for everything—Camillo Gentile.

I would like to thank my wife Abeer and son Nasser for their support and encouragement while working on this project—Nayef Alsindi.

Contributions to Chapter 5 and 7 have been performed in the framework of the FP7 project ICT-248894 WHERE2 (Wireless Hybrid Enhanced Mobile Radio Estimators - Phase 2) which is partly funded by the European Union. Furthermore, thanks to Armin Dammann, Christian Mensing, Siwei Zhang (all DLR), Benoit Denis (CEA) and the WHERE2 colleagues for their valuable insights and advice over the past four years. Finally, I would like to thank my family (Susanne, Bastian, Luise and Peter) for their patience and support while working on this book—Ronald Raulefs.

Contributions to Chapters 8 and 9 of research on pedestrian and robot navigation conducted by the TRX team: Benjamin Funk, Amrit Bandyopadhyay, Asher Kohn, Kamiar Kordari, Dan Hakim, John Karvounis, Jared Napora, Ruchika Verma, Chris Giles, Brian Beisel and Gilmer Blankenship are gratefully acknowledged. I would like to especially thank Professor Gilmer Blankenship who has taken the time to review several versions of these chapters. Finally, I would like to thank my family for their patience and support while working on this book—Carole Teolis.

Contents

Chapter 1
Introduction

1.1 Overview

1.1.1 Outdoor Localization

The integration of location services into our day-to-day life will grow significantly over the next decade as technologies mature and accuracy improves. The evolution of localization technologies has occurred independently for different wireless systems/standards. The Global Positioning System (GPS) was the first system to bring to light the benefits of accurate and reliable location information. Consequently, it has been incorporated into many services and applications. Currently, outdoor localization, thanks to GPS, has revolutionized navigation-based applications running on automotive GPS-enabled devices and smart phones. Applications range from location awareness, to point-by-point directions between destinations, to identifying the closest cinema or coffee shop. The basic technology behind the system is to measure the time elapsed for a signal to travel between a number of satellites orbiting the globe and a mobile device. Through a computational technique known as triangulation, the location of the mobile can be calculated from the tracked positions of the satellites and the times measured, each known as the Time-of-Arrival. The success of GPS has been due to the reliability, availability, and practical accuracy that the system can deliver; however, GPS lacks coverage indoors and in urban areas, in particular near buildings when the signal is blocked; even in the best of conditions, the accuracy is on the order of several meters.

As the growth of the number of smart devices and mobile users continues to increase without bound, the desire for new location-based services that require enhanced accuracy, including in GPS-denied areas, has emerged. To address this challenge, novel solutions attempt to integrate different wireless technologies with GPS. For example, assisted GPS (A-GPS) was developed to provide better localization information in limited coverage areas by decreasing the time necessary for GPS to obtain a position fix (Richton 2001). Specifically, in A-GPS, cellular

C. Gentile et al., *Geolocation Techniques*, DOI: 10.1007/978-1-4614-1836-8_1,

networks furnish GPS-equipped mobile devices with satellite constellation information such that they can identify the closest orbiting satellites a priori, providing a faster lock. In addition, A-GPS relieves the burden of the computationally intensive triangulation technique from the CPU-limited mobile device by forwarding the links the GPS receiver measures to the base stations (BSs), which then calculate the mobile's position and return the information to the mobile device.

Unsurprisingly, the next logical evolution of localization systems emerged from the cellular domain, where the requirement for localization was spearheaded by the Emergency-911 (E-911) mandate. Before GPS was widely available on mobile phones, cellular operators adopted and deployed varying technologies to locate mobiles within a cell radius. Time-Of-Arrival-based computational techniques, which originated from GPS systems, were adapted in order to achieve similar localization performance for the common cell phone channel sharing (multiplexing) schemes: Time Division Multiple Access (TDMA) and Code Division Multiple Access (CDMA). The use of cellular localization was limited to E-911 due to the difficulty in achieving useful accuracy especially in urban environments. The poor accuracy stemmed from clusters of buildings in urban and suburban residential areas which brought about significant signal degradation due to multipath and Non-Line-Of-Sight (NLOS) problems.

At the same time, the popular IEEE 802.11 standard emerged, enabling ubiquitous deployment of Wi-Fi hotspots which sparked enthusiasm for an alternative to cellular localization. The rapid expansion of Wi-Fi access points (AP) across the urban/indoor environments made it possible for researchers to envision alternatives to TOA-based systems. Specifically, Received Signal Strength (RSS) location fingerprinting techniques emerged. One success story for deployment in the urban environment is Skyhook Wireless, a start-up company in the Boston area. Skyhook realized the potential of exploiting Wi-Fi signals emitted from residential homes and offices (available for free!) that are continuously in use—particularly in dense urban areas. Skyhook realized they could improve localization by building databases of Wi-Fi signatures tied to locations that could be integrated to aid in the localization process. In essence, a survey is conducted by "wardriving" across a city with a Wi-Fi equipped device and a companion GPS receiver to record location. Wi-Fi RSS values and associated Medium Access Control (MAC) IDs are registered in a database for each location. During a localization query, a mobile device compares the RSSs measured from the registered APs to those in the database using a pattern matching technique. The mobile's location is then determined by the best RSS match. Skyhook Wireless's technology has attracted attention from the major players in the mobile device industry such as Apple and Google (Wortham 2009). The technique is very practical and delivers decent accuracy (tens of meters) for mobile location applications in outdoor urban environments where Wi-Fi APs are plentiful.

Of course, the aforementioned triangulation fingerprinting techniques are not just applicable to GPS, cellular, or Wi-Fi networks. They can also be readily extended to virtually any pervasive radio-frequency source, in particular to television networks. In fact, the Rosum Corporation from Redwood City, CA took advantage of the 4.5 MHz of bandwidth available in broadcast TV channels. Besides the wide bandwidth available for accurate TOA estimation, the low carrier frequency offered excellent penetration through walls to mitigate NLOS conditions. The performance of different types of RF location systems in terms of cost and accuracy will vary widely. Other system design considerations include 2D or 3D location accuracy requirements, power requirements, and whether infrastructure installation is acceptable and, if so, the density of BSs required to achieve the desired accuracy. In designing a system, there will be tradeoffs between performance and cost requirements. For example, typical RF base positioning accuracies are tens of meters accuracy at best and do not provide accurate elevation. Many solutions available to augment GPS leverage surrounding infrastructure such as cell towers, Wi-Fi hot spots, or installed RF tags. The precision of the results varies widely based on the infrastructure location and availability.

- Cellular survey-based techniques: hundreds or thousands of meters
- Cellular triangulation techniques: less than 100 m
- Television triangulation techniques: tens to hundreds of meters
- Wi-Fi survey-based techniques: tens to hundreds of meters.

The latter three techniques require signals from at least three reference stations which could lead to operational lapses indoors. It is not possible to rely on these infrastructure-based solutions in uncontrolled environments such as emergency or combat operations where the infrastructure may not exist; however, for commercial use, the accuracy and reliability provided may be adequate (Baker 2011; Mia 2011; Young 2008).

1.1.2 Indoor Localization

The lucrative business opportunities for location-enabled services are not limited to outdoors. In fact, the potential for indoor services has been projected by different sources as an untapped multibillion dollar industry (Patel 2011). The variety of indoor applications affects every aspect of our lives: from E-911 to respond to mobile emergency calls to tracking kids in day-care centers, elderly in nursing homes, inventories in warehouses, medical devices in hospitals, and personnel in emergency/first responder applications (firefighters). What is stopping or hindering the emergence of such needed—even life-saving for emergency response—applications is the difficulty in delivering the required accuracy and reliability in indoor environments. Indoor localization research has been going on for decades in the robotics field (Smith 1986; Durrant-Whyte 1988). The E-911 requirement for improved localization of cell phones indoors spurred more RF infrastructure and signals of opportunity-based research (Pahlavan et al. 1998). The fact that location research is to date a very active research area indicates that there are still many challenges left to resolve. The challenges depend on the required accuracy and reliability dictated by the application. For applications that require only coarse location information and can afford to install a significant amount of infrastructure, there are existing products, for example by the Finnish company Ekahau (EKAHAU 2012) and CISCO Wireless Location Appliance (CISCO Corporation 2012). These systems capitalize on the RSS location fingerprinting technique to deliver accuracies on the order of a few meters in the indoor environment. However, it became evident that the effectiveness and robustness of RSS-based fingerprinting techniques are limited to uncluttered environments and outdoors.

As the application domain gravitated toward the dense urban and indoor settings, where localizable assets naturally clutter together due to smaller dimensions, an alternative to legacy cellular localization and fingerprinting techniques was needed to push the accuracy boundary to sub-meter—the so-called "holy grail" of indoor localization. Many potential applications were envisioned to benefit from centimeter-level information, from inventory tracking to firefighters/soldiers tracking inside buildings. The fundamental challenge indoors is that the radio frequency environment—characterized by limited coverage, severe multipath signal fading and NLOS conditions—is not conducive to wireless propagation. Since the limitations are physical in nature, they must be dealt with by any algorithm or technique. To this end, researchers revisited TOA-based techniques—however applied Ultra-Wideband (UWB) communications which uses low power but increased bandwidth to provide protection against multipath—and NLOS mitigation algorithms to combat the effects of the propagation environment. A significant portion of this book is dedicated to addressing the challenges of harsh propagation environments.

As the form factor of mobile devices diminished in size, yet increased in complexity, a new school of thought emerged from the localization research community around the idea of collaboration using sensor networks. This area is interesting in that wireless sensor networks (WSNs) developed independently from cooperative localization and—only when applications were considered for the former—did it become obvious to the sensor network researchers that location information is indeed vital. At the same time, localization researchers analyzed the potential in collaboration between the two areas to address the propagation challenges and currently cooperative localization in WSNs is a very active research area—theory, experimental, and hardware/software development.

Since geolocation is a dynamic process, navigation and tracking techniques (similar to outdoor GPS) naturally complement "stationary" localization techniques/algorithms. The development of Microelectromechanical Systems (MEMS) technology led to the dawn of miniature inertial sensors such as accelerometers and gyroscopes, enabling smartphones and mobile/gaming devices to be equipped with navigation sensors. MEMS-based inertial navigation systems (INS) developed in parallel to RF geolocation techniques and provided another localization dimension. Inertial navigation technologies do have their own challenges and limitations—due to low-cost hardware that introduces errors/drifts/biases to the speed/acceleration estimation. The development of inertial technology integrates naturally with the evolution of "RF localization" in the sense that their complementary error properties can make possible even more accurate and robust geolocation systems.

In general, providing accurate location and navigation indoors will require extensive infrastructure or the implementation of multiple, complementary technologies (RF, gyroscopes, pressure sensors, speed sensors, etc.). In fact, the trend in indoor geolocation research seems to point towards the integration of hybrid sensor technologies. The effectiveness of different sensors can vary based on the environment of operation and the tracked subjects motion: RF propagation depends on building topology and construction material, lighting affects optical sensors, and the tracked subject's motion affects optical and inertial sensors. Inertial and RF based location sensors provide complementary location information: inertial tracking systems provide high accuracy over short durations, but suffer from significant drift over longer times in the absence of methods to mitigate their drift; in contrast, RF ranging measurements are subject to short-term outages in areas with poor RF connectivity, but can provide long-term stability when fixed references are part of a managed infrastructure. Wi-Fi only location provides an unmanaged and often changing network of APs which cannot be relied upon if a known level of accuracy is required, but may provide adequate accuracy for many consumer applications. Finally, elevation determination (floor location) presents a challenge for RF systems; here, inertial and referenced barometric pressure systems can provide support.

Developing algorithms to effectively fuse sensor data from multiple sources to produce improved localization results is a hot research topic. One popular technique is Simultaneous Localization and Mapping (SLAM). SLAM relies on data from multiple sensors to build a map of the environment that enables one to navigate for long periods of time by using the map to provide location corrections. SLAM systems use RF and inertial sensors as well as sensors that measure the environment directly such as image, Light Detection and Ranging (LIDAR), and sonar sensors to construct a geometric or topological map of the environment and then use that map for navigation. The environmental sensors help to alleviate some of the problems faced by inertial and RF techniques, but they have their own set of problems and challenges in the path to accurate mapping and localization.

1.2 Organization

The focus of this book is to provide an overview on the different types of infrastructure supported by most commercial localization systems as well as on the most popular computational techniques which these systems employ. While much of the content presented applies to outdoor systems as well, the specific concentration of the book is on robust systems which can deliver high degrees of accuracy in harsh multipath environments; these environments are most common indoors. Each chapter of this book introduces a different aspect of localization systems and describes solutions that have been developed to address specific challenges.

The organization of the book chapters follows closely the evolution of geolocation techniques for the last couple of decades. In this section we will provide a

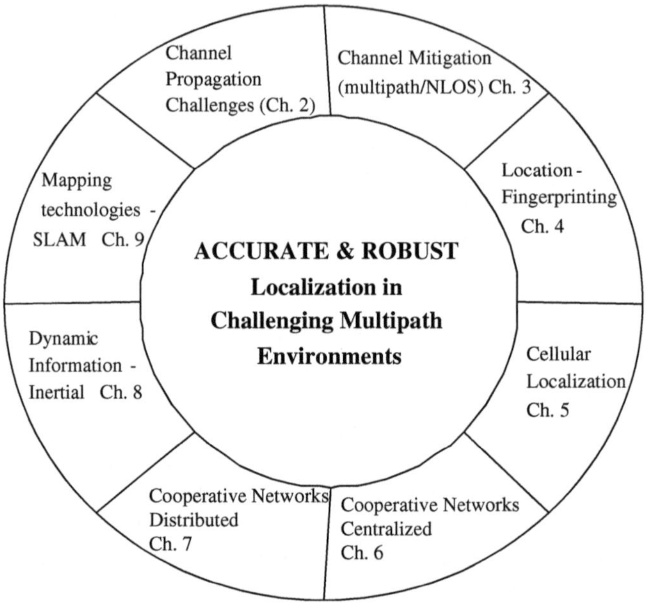

Fig. 1.1 Techniques for accurate and robust localization

detailed overview of each chapter. Figure 1.1 highlights the overall structure of the book where the focus is to introduce the fundamentals, challenges, and evolution of localization technology.

1.2.1 Chapter 2: Localization in Harsh Multipath Environments

In Chap. 2 the basics of RSS, TOA and Angle-of-Arrival (AOA) localization are first introduced. The impact of multipath and NLOS is then investigated for the two most popular ranging technologies: TOA and RSS. Finally, measurement and modeling of ranging is presented to provide an empirical analysis into the challenges of harsh multipath environments. For RSS-based systems, multipath causes the well-known fast fading phenomenon, where the received power in a given location fluctuates significantly due to constructive and destructive interference of incoming multipath signals. For TOA-based systems, the multipath impacts the distance estimation directly by adding a random bias to the estimation and it is usually a more serious problem. In low bandwidth systems, for example, the time resolution can yield significantly inaccurate distance estimates. Typically, the time domain resolution is inversely proportional to the system bandwidth. For example, the bandwidth of GSM signals is 200 kHz which translates to 5 μs or 1,500 m! This means that two paths arriving less than 1,500 m will not be resolved. For example, system bandwidth can vary between 5 and 20 MHz (UMTS/WiMAX/

LTE) in which the highest bandwidth of 20 MHz equates to $\sim$15 m of time resolution. This resolution, unfortunately, is not suitable for dense multipath environments (such as indoors) where large errors in the final localization solution can make it difficult to localize mobile devices to within even a single floor. The ambiguity resulted from poor multipath resolution is one of the major challenge facing localization technology in multipath rich environments such as dense urban or indoors.

The second major challenge facing dense urban/indoor environments is the high probability of the obstruction of the LOS between the transmitting and receiving device. This channel condition is commonly referred to as NLOS. For RSS-based systems, NLOS introduces the problem of shadow fading, where RSS is attenuated randomly as the mobile device moves from one area to the other. Since obstructions change significantly (doors, walls, elevators, etc.) the RSS changes significantly and this fluctuation makes it difficult to rely on RSS-based range estimates in NLOS. Furthermore, pathloss models that describe the distance–power relationship can be difficult to obtain for the variety of obstructions in realistic urban/indoor environments. For TOA-based systems, NLOS affects the estimation of the delay of the direct path signal. Since in most cases the direct path delay signal will not be detectible, ranging is achieved through non-direct path components which bias TOA-based estimation. This bias can range from a meter to even tens of meters depending on the propagation environment and type of obstructions.

A detailed empirical evaluation is further introduced in the last section of Chap. 2 which will shed light on the significance of the multipath and NLOS problems. The ultimate aim of the measurement and modeling of TOA- and RSS-based ranging is to be able to answer the following fundamental questions:

- How does the system bandwidth improve accuracy?
- To what extent can the increase in system bandwidth improve accuracy in LOS and NLOS environments?
- How significant are the NLOS-induced errors experienced in harsh multipath environments?
- Is the TOA-based ranging error a function of the propagation environment (e.g. building structure)?
- For a given operational multipath environment, what is the practical *ranging coverage* that can be achieved for TOA-based techniques? This question is important since a notion of ranging coverage which is analogous to communication coverage is needed in practice.
- How are RSS-based ranging techniques affected by the LOS/NLOS power variations with location?

These questions are fundamentally important to system engineers designing next-generation ranging and localization systems. In addition, channel measurement and modeling can shed light on the correlation between the channel conditions (LOS vs. NLOS) and signal metrics such as power of the first path, total signal power, etc. These relationships can be exploited in NLOS identification algorithms, which are typically required for reliable and practical ranging and localization in harsh multipath environments (NLOS identification/mitigation algorithms are introduced in Chap. 3).

1.2.2 Chapter 3: Multipath and NLOS Mitigation Algorithms

Mitigating the multipath propagation challenges is addressed in Chap. 3 and the chapter starts with describing two major techniques/technologies to mitigate the multipath problem: Super-resolution and UWB. Super-resolution techniques have shown great potential for low-bandwidth systems and the improvement in time resolution can enhance the accuracy significantly for certain scenarios. UWB is an emerging technology that utilizes very large system bandwidths and has the potential for high data rate communications (in the Gigabit range) and centimeter level TOA estimation accuracies. From a ranging/localization perspective, full usage of the designated 7.5 GHz bandwidth translates into a time domain resolution of 4 cm, which is highly desirable for accurate positioning. There are two main types of UWB systems: Single band and multiband UWB. The single band UWB is typically known as impulse radio UWB [**6**], where very narrow pulses in the time domain achieve the bandwidth that defines UWB. The latter technique is multiband in nature and the very popular multiband OFDM (MB-OFDM) implementation has been the main proponent for high-data rate and accurate localization. Results of measurements and simulation have shown that UWB has the potential to achieve sub-centimeter accuracy in LOS environments but struggles to match the accuracy in NLOS environment due to the physical obstruction problem.

The NLOS problem is addressed in the second part of Chap. 3 and it typically involves two stages: NLOS identification and NLOS mitigation. This area has received considerable attention in the research community within the last decade and it continues to provide innovation potential for researchers. NLOS identification techniques are based on estimating or identifying the condition of the channel to infer whether it is LOS or NLOS. Once the "channel" information is available, it is possible to incorporate it into an NLOS mitigation algorithm to improve the accuracy and robustness of the location estimate. NLOS identification typically operates on the physical-layer-sensed signal which can be used to extract a "metric" that can indicate the state of the channel. NLOS mitigation, however, operates at higher levels closer to the localization algorithm. As a result, identification and mitigation are typically independent; however, there are approaches that combine the identification and mitigation in one step. The robustness of NLOS mitigation algorithm depends inherently on the robustness of the NLOS identification stage. The better the detection accuracy (probability of detection for a given probability of false alarm) the more effective and useful the channel information can be for the mitigation stage and the entire localization algorithm. As a result it is no surprise that NLOS identification can be the critical element in the mitigation process.

1.2.3 Chapter 4: Fingerprinting Techniques

Survey-based localization is the focus of Chap. 4. The basic idea behind this technique is to associate physically measurable properties to discrete locations throughout a deployment area. These properties, commonly referred to as fingerprints or signatures, can then act as location identifiers. The greater the spatial variability of the signatures, the greater the capacity of the system to discriminate between locations and, in turn, to deliver finer resolution. Therefore, the same physical properties of the environment which render non-survey-based techniques more challenging—in particular multipath fading in radio frequency systems—on the contrary facilitate survey-based techniques.

Location fingerprinting techniques are categorized mainly by the type of properties which are collected. The three major radio frequency properties that have been implemented to date are: RSS, the time domain Channel Impulse Response (CIR) [or equivalent frequency domain Channel Transfer Function (CTF)], and the Frequency Channel Coherence Function (FCF). RSS is by far the most prevalent in commercially deployed wireless systems. This is due to many factors, most notably its robustness and good penetration in NLOS conditions—especially at lower carrier frequencies—its simple data structure, and the computational ease (inexpensiveness) with which it can be measured. It also stems from the fact that RSS is accessible directly from the frameware in popular wireless standards such as the IEEE 802.11. As mentioned earlier, RSS fingerprinting systems have been successfully deployed in dense urban and indoor environments by Skyhook Wireless and Ekahau, respectively. The disadvantage of using RSS as a signature—especially when only a few APs are available—is the lack of uniqueness, meaning that multiple sites in close proximity throughout a deployment area may have similar fingerprints. This translates into limited localization resolution. While CIR, CTF, and FCF provide more distinctive signatures, they also require more complex (expensive) equipment to extract and have larger data storage requirements. The latter can be prohibitive for medium to large sized databases (typical indoor environments). In addition, because the data structure is more complex, the pattern matching algorithms are more computationally intensive.

The fingerprinting technique, in a nutshell, is to construct a database of signatures from available wireless network infrastructures, such as APs. Each signature is registered at a unique location—typically at points on a uniformly spaced grid throughout a given environment (e.g. 1 m^2). This occurs in an "offline" phase, i.e., before localization is attempted. Figure 1.2 highlights the method of constructing a fingerprint at a given location. The location of a mobile device is then estimated during an "online" phase. For each query, the signature parameters are measured at the mobile device and subsequently are compared against the signatures registered in the database through a pattern matching algorithm. The location of the mobile is then designated as the location corresponding to the closest signature in the database. The role of the database in the offline and online stages is illustrated in Fig. 1.3.

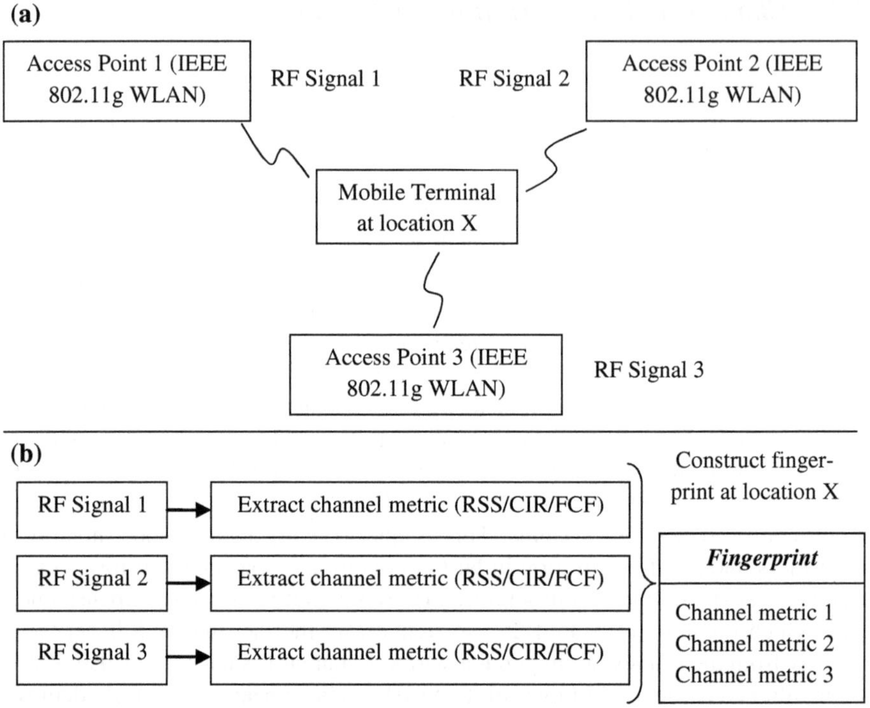

Fig. 1.2 Overview of existing fingerprint construction. **a** Mobile terminal at location *X* conducts measurements to 3 APs and captures RF signals. **b** Channel metrics are extracted from the 3 RF signals and a fingerprint is created

1.2.4 Chapter 5: Cellular Localization Systems

Cellular localization is of tremendous interest for the network operators. As mentioned before, this was publicly stimulated by the FCC requirements that were published at the end of the 1990s for E-911 calls in the US and in 2003 Europe the E-112 initiative by the European Commission. However, the communication systems, like GSM, UMTS or LTE are designed to use the well-paid spectrum efficiently for communication needs. These needs are, e.g., a robust coverage as well as high throughput—to fulfill these requirements the spectrum is used efficiently for unknown data transfer. Localization in cellular systems is performed through fingerprinting or ranging. Fingerprinting methods range between a coarse localization through the cell ID or via signal strength-based localization. Signal strength methods are based on premeasured datasets and rely on known transmitted signal strength. Common time-based ranging methods require precise synchronization between the transmitter and the receiver. Such precise synchronization is not well established in common communication systems, especially not at the mobile terminal. Furthermore, in communications a single BS is enough to

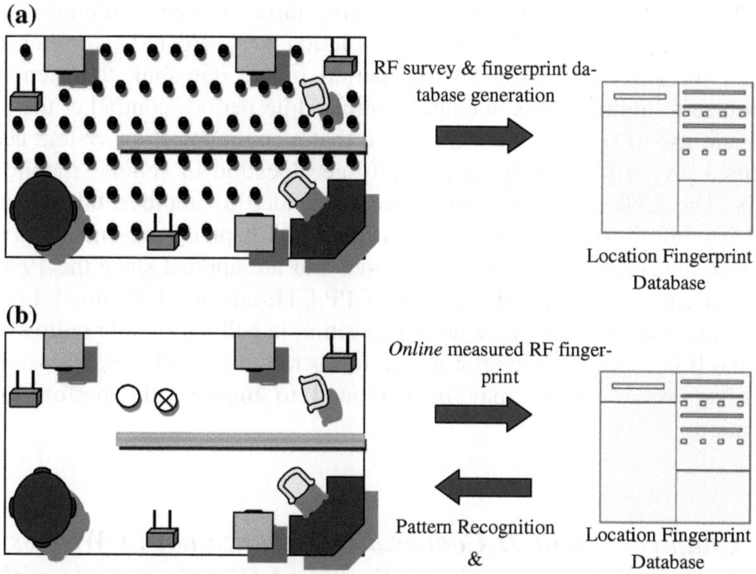

Fig. 1.3 Location fingerprinting. **a** Offline: fingerprint database generation at locations on a grid. **b** Online pattern recognition and position estimation. *Circle* is actual position and *circle/cross* is the estimated position

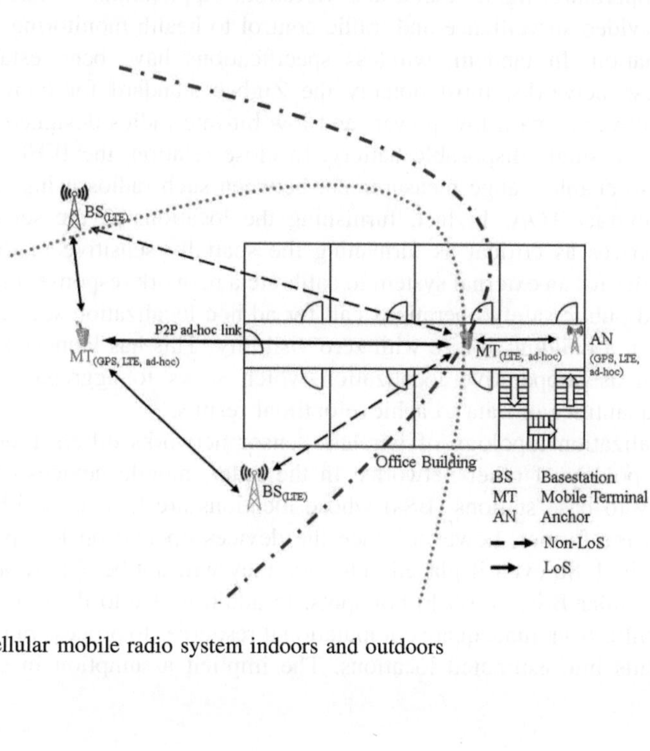

Fig. 1.4 Cellular mobile radio system indoors and outdoors

cover the basic needs. Localization requires three or more differently placed transmitters or receivers. The simplest solution to overcome interference was proposed for UMTS: Adding an idle period in the downlink to listen and to synchronize to multiple BSs one after another. Idle periods contradict the idea of an efficient use of spectrum, but it showed that a communication system needs to be defined properly to apply successfully geo-location in cellular mobile radio systems. The LTE standardization process intended to improve this, by adding special synchronization sequences for positioning. Chapter 5 presents an overview of the different methods that were proposed and are applied since the 1990s and are now discussed in standardization of 3GPP LTE-advanced. Figure 1.4 presents how the different radio links are used to position in cellular mobile radio systems. Either the BSs or the mobile terminal performs ranging. Furthermore, also indoor APs acting as an anchor may be exploited to improve the performance of localization.

1.2.5 Chapters 6 and 7: Cooperative Localization in Wireless Sensor Networks—Centralized and Distributed Algorithms

The falling price and reduced size of wireless sensors in recent years have fueled the proliferation of dense networks to gauge and relay environmental properties such as temperature, light, sound, and vibration. Applications of such networks range from video surveillance and traffic control to health monitoring and industrial automation. In tandem, wireless specifications have been established to support these networks, most notably the Zigbee standard for communication protocols between small, low-power, and low-bit-rate radios designed to operate for years on a single disposable battery. In close relation, the IEEE 802.15.4 g standard also enables range measurement between such radios using UWB technology to extract TOA. In fact, furnishing the locations of the sensors in the networks proves as critical as furnishing the spatially sensitive readings themselves in order for an external system to calibrate a network response. In particular, military and public safety operations call for ad hoc localization such as that of a man down in a building ablaze with zero visibility. This has launched a research area known as cooperative localization which seeks to aggregate potentially enormous quantities of data to achieve optimal results.

The localization topology of wireless sensor networks differs fundamentally from the topology of other networks. In the latter, mobile devices enjoy direct connectivity to base stations (BSs) whose locations are known, as illustrated in Fig. 1.5. In the former, however, since the devices operate on low power, their range is limited. So even if placed outdoors, they will not be able to access GPS satellites, cellular BSs, or Wi-Fi hot spots. In addition due to their compact size, they may suffer from inadequate computational resources to process range or angle measurements into estimated locations. The implicit assumption in cooperative

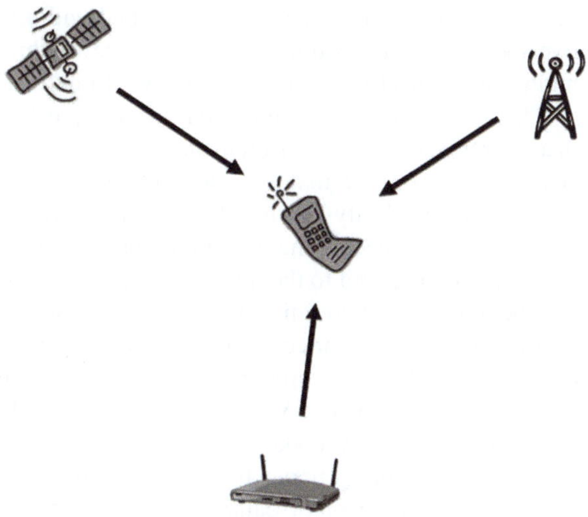

Fig. 1.5 Localization topology in GPS, cellular, and WLAN networks. The mobile device has direct connectivity to the base stations

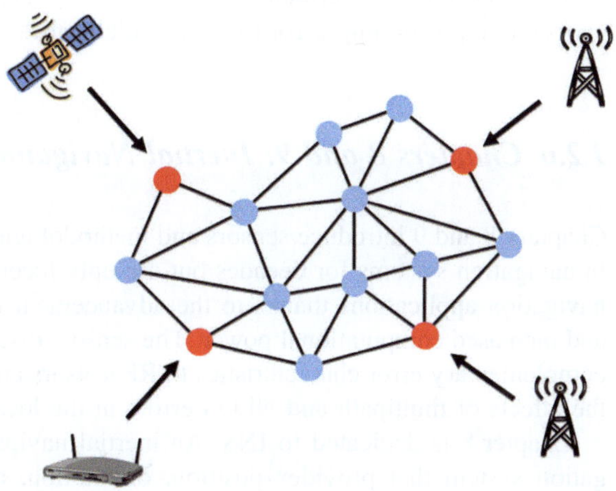

Fig. 1.6 Localization topology in wireless sensor networks. The anchors (*red*) have direct connectivity to the BSs. The sensors (*blue*) are connected to the anchors through mulit-hop links

localization is that only a small ratio of the total number of devices in the network, known as anchor nodes, are able to estimate their locations from BSs. This may be due to their favorable placement in the environment which enables connectivity, but for the most part special network devices equipped with higher power and enhanced computational resources will be required.

Hence in cooperative networks, sensors lacking direct connectivity to anchors must discern their locations through neighboring nodes whose locations are also unknown. In essence, sensors must connect to anchors through multiple hops, as illustrated Fig. 1.6. A consequence of this complex topology is that simple triangulation algorithms must be substituted with more sophisticated algorithms. And since each connection on a multi-hop link is subject to measurement error, the reliability of the composite link is diminished with respect to an otherwise direct link. However, since the number of nodes can range from the hundreds to the thousands, WSNs are often densely packed with overlapping coverage. Cooperative localization algorithms take advantage of this redundancy and, despite the multi-hop connectivity, have been shown to deliver good results.

Chapter 6 introduces centralized cooperative localization. Centralized implies that the range or angle measurements are gathered locally and then forwarded to a central processor such that they can transformed into the locations of the unknown nodes. The scalability of algorithms is a key ingredient for future wireless networks and the expected increase in the number of devices in wireless networks is exponential compared to the number of active devices today. For such networks, it may be infeasible to coordinate the devices through a centralized architecture. The positioning solutions for centralized cooperative methods are user/agent-centric. A related idea, which is common in WSNs, consists in sharing computational load onto the entire network, yet preserving reasonable complexity and low power consumption in each node. The sharing of computational load between geographically distributed nodes builds on the concept of distributed algorithms. The application requirements (scalability, energy efficiency, and accuracy) will influence the design of distributed algorithms. In Chap. 7, a variety of distributed cooperative positioning algorithms, especially message passing, is presented.

1.2.6 Chapters 8 and 9: Inertial Navigation Systems

Chapters 8 and 9 introduce sensors and methodologies that have been widely used in navigation systems for decades but are only recently applicable to commercial navigation applications thanks to the advancement in electronics miniaturization and increased computational power. The sensors discussed in Chaps. 8 and 9 have complementary error characteristics to RF sensors and so can enable mitigation of the effects of multipath and NLOS errors in the location solution.

Chapter 8 is dedicated to INS. An inertial navigation system (INS) is a navigation system that provides position, orientation, and velocity estimates based solely on measurements from inertial sensors. Inertial measurements are differential measurements in the sense that they quantify changes in speed or direction. The two primary types of inertial sensors are accelerometers and gyroscopes. Accelerometers measure instantaneous changes in speed, or equivalently force, and gyroscopes provide a fixed frame of reference with which to measure orientation or equivalently change in direction. Given its previous position and orientation as well as accelerometer and gyroscopic measurements over an elapsed period of time, an instrumented platform may calculate an estimate of its current position and orientation. Calculation of navigation information from differential measurements of speed and direction is termed dead reckoning.

Inertial navigation systems, by definition, compute their navigation solutions without the use of external references. INS were used as a prime means of navigation in the nineteenth and early twentieth century in maritime, aviation, and spaceflight applications. A main drawback of using purely inertial systems for

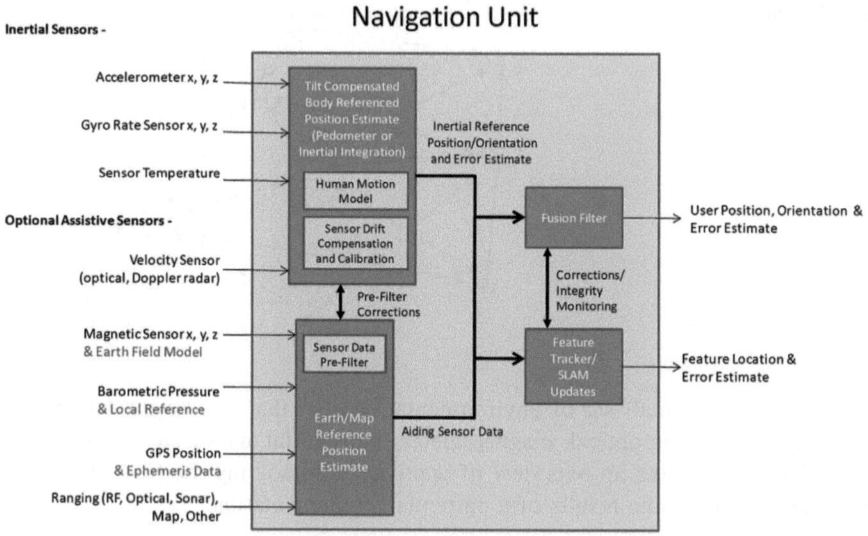

Fig. 1.7 Robust navigation solutions require input from multiple sources

navigation is that errors in the differential measurements are necessarily accumulated in the navigation solution over time. Thus, even with highly precise inertial measurements, position estimates based on them degrade over time.

It is now well accepted that a high accuracy navigation solution requires the ability to fuse input from multiple sensors making use of all available navigation information. Cross-validation allows inconsistent sensor data to be identified and suppressed in the overall navigation solution. Figure 1.7 illustrates a navigation device that takes input from multiple sources including sensors and map information.

The key to making inertial sensors part of a precision positioning system is developing methods to both ***minimize*** free inertial position error growth and ***bound*** accumulated inertial position errors. In Chap. 8 we discuss fusion of inertial sensor data with sensors and/or algorithms that provide estimates of secondary inertial state variables such as velocity, heading, and elevation.

SLAM techniques are one approach to fusing information from a variety of sensors. In Chap. 9 we introduce SLAM algorithms which incorporate past path history and derived or available map information to determine the most probable position estimates conditioned on constraints determined by map information. Figure 1.8 shows a conceptual diagram of SLAM. Both the subject's state, x_k (termed the subject pose and indicated by successive triangles), and the location of select landmarks (indicated by stars) are tracked. The basic idea of SLAM is that if the sensor and algorithms can identify a landmark and a location of that landmark relative to tracked subject, then any time that landmark is seen again, its location can be used to correct the track subject's location.

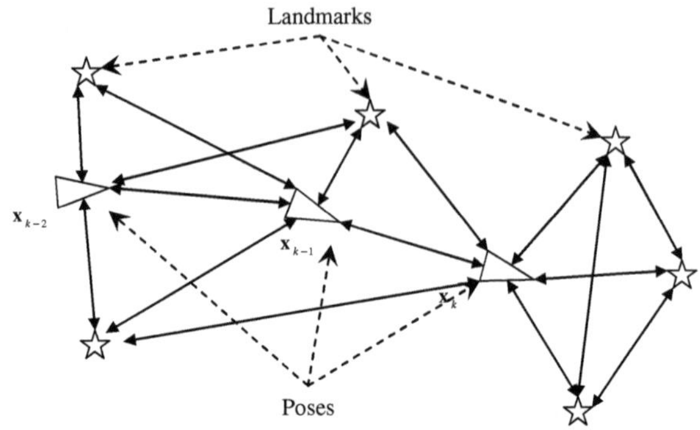

Fig. 1.8 Simultaneous localization and mapping

We discuss a small set of environmental sensors that can be used in SLAM algorithms including optical, magnetometer, and inertial and discuss how features are selected. We give an overview of approaches to solving the SLAM problem and then discuss some results of a particular implementation.

References

J. Baker, The impact of indoor cellular coverage on location accuracy, Indoor location—the enabling technologies (2011)

CISCO (2012). http://www.cisco.com

H. F. Durrant-Whyte, Uncertain geometry in robotics. IEEE J. Robot. Autom. **4**(1), 23–31 (1988). doi:10.1109/56.768. http://ieeexplore.ieee.org/stamp/stamp.jsp?tp=&arnumber=768&isnumber=50

EKAHAU (2012). http://www.ekahau.com

R.S. Mia, Indoor wireless location—the E911 perspective, Indoor location—the enabling technologies (2011)

K. Pahlavan, P. Krishnamurthy, A. Beneat, Wideband radio propagation modeling for indoor geolocation applications. Commun. Mag. **36**(4), 60–65 (1998)

N. Patel, Strategy analytics: the $10 billion rule: location, location, location, navigation: wireless media strategies. May 11, 2011. http://www.strategyanalytics.com/default.aspx?mod=reportabstractviewer&a0=6355

R.E. Richton, G.M. Djuknic, Geo-location and assisted GPS. IEEE Comput., **34**(2), 123–125 (2001)

Skyhook Wireless (2012). http://www.skyhookwireless.com

R. Smith, P. Cheeseman, On the representation of spatial uncertainty. Int. J. Robotics Res., **5**(4), 56–68 (1986)

J. Wortham, Cellphone locator system needs no satellite. New York Times (June 2009)

T. Young, TV + GPS location and timing, WPI precision indoor personnel location and tracking (2008)

Chapter 2
Ranging and Localization in Harsh Multipath Environments

In this chapter, we will first introduce the basics of geolocation techniques that are based on Time of Arrival (TOA), Time Difference of Arrival (TDOA), Angle of Arrival (AOA), and Received Signal Strength (RSS). Then we introduce the major challenges to accurate localization: multipath propagation and non-line-of-sight conditions where we will focus on the two most popular ranging techniques, TOA and RSS, and evaluate how the accuracy of localization is affected by these physical challenges. We will further highlight the relationship between the accuracy of estimation and the signal to noise ratio and bandwidth parameters through the well-known Cramer-Rao Lower Bound (CRLB) equations. Finally, we will introduce measurement and modeling of the RSS/TOA ranging that will highlight the impact of multipath and NLOS on the accuracy of ranging systems.

2.1 Basics of Geolocation

Classical geolocation techniques (non-survey based) depend on geometrical relationships between the coordinates of the reference points (satellites in GPS technology) and the associated range/angle measurements. Typically, reference points are wireless devices with known location information (e.g. x- and y-coordinates) either pre-programmed or obtained through GPS. The mobile device (seeking its own position information) exchanges RF signals with the reference points to estimate the distance or angle to each of the reference points. Equipped with the range measurements and the coordinates of the reference points, the mobile device can solve for the unknown position through a variety of techniques (geometrical, optimization, etc.). The accuracy of the location information is affected by three major factors: the accuracy of the reference points' position, the accuracy of range/angle estimates, and the geometrical configuration of the reference points and the unknown position. The non-survey geolocation techniques computes location estimates through two steps: range/angle estimation and tri-lateration/angulation. Figure 2.1 illustrates the two-step procedure.

C. Gentile et al., *Geolocation Techniques*, DOI: 10.1007/978-1-4614-1836-8_2,

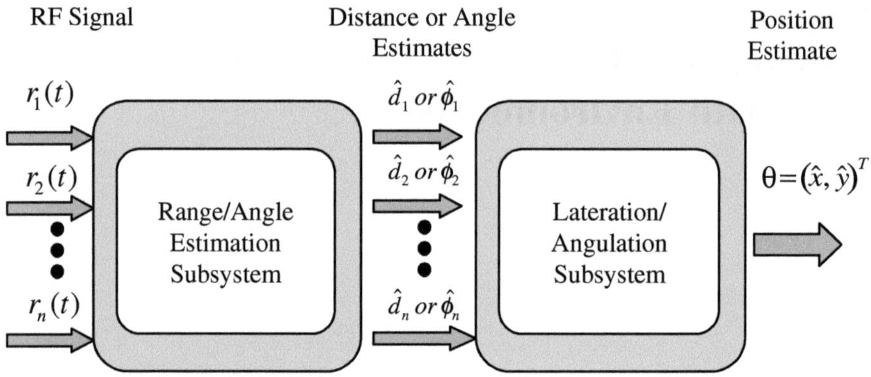

Fig. 2.1 Classical geolocation system. Range or angle information is extracted from received RF signals. Location is then estimated by lateration/angulation techniques

In this section we will introduce the most popular geolocation techniques: TOA, TDOA, AOA, and RSS and provide an evaluation of the achieved accuracy through the well-known Cramer-Rao Lower Bound (CRLB) analysis.

2.1.1 TOA-Based Techniques

Once distance/range measurements to at least 3(4) reference points are available the 2(3)-dimensional position estimate can be obtained. The set of distance measurements from the reference points to the mobile terminal forms a set of nonlinear equations that can be solved to estimate the position. Here, it is assumed that the mobile terminals exchanging range measurements are time synchronized and that they are all in LOS condition (no obstruction between the mobile device and the base stations). Figure 2.2 illustrates the basic concept of tri-lateration.

The range measurements can be used to estimate the position of a mobile device through several techniques that are generally grouped under Maximum Likelihood (ML) and Least-Squares (LS) Techniques. In ML techniques, the solution is the position that maximizes the conditional probability density function or

$$\hat{\boldsymbol{\theta}} = \arg\max_{\boldsymbol{\theta}} P\left(\hat{\mathbf{d}}|\boldsymbol{\theta}\right) \tag{2.1}$$

where $\hat{\boldsymbol{\theta}} = [\hat{x},\hat{y}]^T$ and $\boldsymbol{\theta} = [x,y]^T$ are the estimated and true position coordinates, respectively. $\hat{\mathbf{d}} = \mathbf{d} + \mathbf{w}$ is the measured/estimated distance vector to each base station or $\hat{\mathbf{d}} = \begin{bmatrix} \hat{d}_1 & \hat{d}_2 & \ldots & \hat{d}_{n_B} \end{bmatrix}^T$, $\mathbf{w}$ is zero-mean Gaussian measurement noise and n_B is the number of base stations. Assuming that the noise measurements are independent and identically distributed (i.i.d), then the conditional distribution is given by

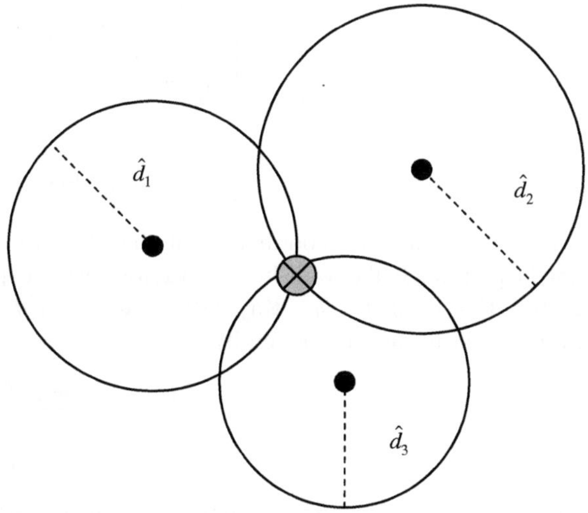

Fig. 2.2 TOA-based trilateration. Range measurements to at least three base stations make up a set of nonlinear equations that can be solved to estimate the position of a mobile device. Black points are base stations with a priori known position information while the intersection of the circles is the position estimate of the mobile device

$$P(\hat{\mathbf{d}}|\boldsymbol{\theta}) = \prod_{i=1}^{n_B} \frac{1}{\sqrt{2\pi\sigma_i^2}} \exp\left\{-\frac{(\hat{d}_i - d_i)^2}{2\sigma_i^2}\right\}.$$

where σ_i^2 is the variance of the ith measurement noise. There are two major problems with this ML approach. The first is that conditional PDF requires the knowledge of the exact distances, which is not available in practice. The second is that solving for the position using the maximization approach requires a search over all possible locations which is neither practical nor computationally efficient (Guvenc and chong 2009). There are also some variants of the original ML technique which are the two-step ML and the approximate ML (AML). The interested reader can find more details about the ML techniques in Guvenc et al. (2006), Chan and Ho (1994).

The other class of TOA-based localization algorithms is based on the LS techniques. The range measurements to the reference points form a set of nonlinear equations of which the solution is the mobile position. The LS techniques are further subdivided into nonlinear LS (NL-LS) and the linearized LS (L-LS). The NL-LS technique estimates the position by minimizing a residual error function (Caffery and Stuber 1998) or

$$\hat{\boldsymbol{\theta}} = \arg\min_{\boldsymbol{\theta}}\{R_{\text{es}}(\boldsymbol{\theta})\} = \arg\min_{\boldsymbol{\theta}}\left\{\sum_{i=1}^{n_B} \beta_i\left(\hat{d}_i - \|\boldsymbol{\theta} - \boldsymbol{\theta}_i\|\right)^2\right\}. \tag{2.2}$$

Thus the residual, $R_{\text{es}}(\boldsymbol{\theta})$, is a measure of error between the measured distances, $\hat{d}_i$, and the estimated distance obtained from computing the Euclidean distance between the reference points and the estimated position, $\|\boldsymbol{\theta} - \boldsymbol{\theta}_i\|$. β_i is a weight that can be used to emphasize range estimates which is proportional to the degree of confidence in the measurement. The L-LS solution is obtained by linearizing the nonlinear equations formed by the $n_B$distances given by

$$\mathbf{F}(\boldsymbol{\theta}) = \begin{bmatrix} \sqrt{(x-x_1)^2+(y-y_1)^2} \\ \sqrt{(x-x_2)^2+(y-y_2)^2} \\ \vdots \\ \sqrt{(x-x_{n_B})^2+(y-y_{n_B})^2} \end{bmatrix} \tag{2.3}$$

where $[x_n, y_n]$ are the coordinates of the nth base station. The linearization is obtained through the well-known Taylor series expansion around $\boldsymbol{\theta}_0$ given by $\mathbf{F}(\boldsymbol{\theta}) \approx \mathbf{F}(\boldsymbol{\theta}_0) + \mathbf{J}(\boldsymbol{\theta} - \boldsymbol{\theta}_0)$ Kay (1993) where $\mathbf{J}(\boldsymbol{\theta} - \boldsymbol{\theta}_0)$ is the Jacobian of $\mathbf{F}$ evaluated at $\boldsymbol{\theta}_0$ and it is given by

$$\mathbf{J} = \begin{bmatrix} \frac{\partial f_1}{\partial x} & \frac{\partial f_2}{\partial x} & \cdots & \frac{\partial f_{n_B}}{\partial x} \\ \frac{\partial f_1}{\partial y} & \frac{\partial f_2}{\partial y} & \cdots & \frac{\partial f_{n_B}}{\partial y} \end{bmatrix}^T_{\theta=\theta_0} \tag{2.4}$$

and the L-LS solution (mobile position estimate) is then given by Kay (1993)

$$\hat{\boldsymbol{\theta}} = \boldsymbol{\theta}_0 + \left(\mathbf{J}^H\mathbf{J}\right)^{-1}\mathbf{J}^H\left[\hat{\mathbf{d}} - \mathbf{F}(\boldsymbol{\theta}_0)\right] \tag{2.5}$$

where H is the Hermitian operation. Typically, the accuracy of localization is affected by the accuracy of the base station location; the statistics of the range measurements and the geometry of the base stations with respect to the mobile terminal. The performance of TOA-based localization can be examined by evaluating the Cramer-Rao Lower Bound (CRLB), which provides the lower bound on the variance of the estimate or Kay (1993)

$$E\left[(\hat{\boldsymbol{\theta}} - \boldsymbol{\theta})^2\right] \geq \mathbf{I}(\boldsymbol{\theta}) \tag{2.6}$$

where $\mathbf{I}(\boldsymbol{\theta})$ is the Fisher Information Matrix (FIM) and $E[\bullet]$ is the expectation operation. The FIM is given by Kay (1993)

$$\mathbf{I}(\boldsymbol{\theta}) \triangleq E\left[\left(\frac{\partial}{\partial \boldsymbol{\theta}} \ln f(\hat{\mathbf{d}}|\boldsymbol{\theta})\right)^2\right] = E\left[\frac{\partial}{\partial \boldsymbol{\theta}} \ln f(\hat{\mathbf{d}}|\boldsymbol{\theta}) \cdot \left(\frac{\partial}{\partial \boldsymbol{\theta}} \ln f(\hat{\mathbf{d}}|\boldsymbol{\theta})\right)^T\right] \tag{2.7}$$

where $f(\hat{\mathbf{d}}|\boldsymbol{\theta})$ is the joint PDF of $\hat{\mathbf{d}}$ condition on the unknown parameters. The measured distances are modeled by

$$\hat{\mathbf{d}} = \mathbf{d} + \mathbf{w} \tag{2.8}$$

where $\mathbf{d}$ is the vector containing the actual (exact) distances between the mobile device and the BS and $\mathbf{w}$ is the zero-mean Gaussian measurement noise. Since the joint PDF is a function of $\mathbf{d}$ which is a function of $\boldsymbol{\theta}$, then from the chain rule

$$\frac{\partial}{\partial \boldsymbol{\theta}} \ln f(\hat{\mathbf{d}}|\boldsymbol{\theta}) = \frac{\partial \mathbf{d}}{\partial \boldsymbol{\theta}} \cdot \frac{\partial}{\partial \mathbf{d}} \ln f(\hat{\mathbf{d}}|\mathbf{d}). \tag{2.9}$$

So (2.7) can be rewritten as

$$\begin{aligned}
\mathbf{I}(\boldsymbol{\theta}) &= E\left[\frac{\partial \mathbf{d}}{\partial \boldsymbol{\theta}} \cdot \ln f(\hat{\mathbf{d}}|\mathbf{d}) \cdot \left(\frac{\partial \mathbf{d}}{\partial \boldsymbol{\theta}} \cdot \ln f(\hat{\mathbf{d}}|\mathbf{d})\right)^T\right] \\
&= \frac{\partial \mathbf{d}}{\partial \boldsymbol{\theta}} E\left[\frac{\partial}{\partial \mathbf{d}} \ln f(\hat{\mathbf{d}}|\mathbf{d}) \left(\frac{\partial}{\partial \mathbf{d}} \ln f(\hat{\mathbf{d}}|\mathbf{d})\right)^T\right] \frac{\partial \mathbf{d}^T}{\partial \boldsymbol{\theta}} \\
\mathbf{I}(\boldsymbol{\theta}) &= \mathbf{J} \mathbf{I}_{\mathbf{d}} \mathbf{J}^T
\end{aligned} \tag{2.10}$$

where $\mathbf{J}$ is the Jacobian given in (2.4) or explicitly

$$\mathbf{J} = \begin{bmatrix} \frac{x-x_1}{\sqrt{(x_1-x)^2+(y_1-y)^2}} & \cdots & \frac{x-x_{n_B}}{\sqrt{(x_{n_B}-x)^2+(y_{n_B}-y)^2}} \\ \frac{y-y_1}{\sqrt{(x_1-x)^2+(y_1-y)^2}} & \cdots & \frac{y-y_{n_B}}{\sqrt{(x_{n_B}-x)^2+(y_{n_B}-y)^2}} \end{bmatrix}^T \tag{2.11}$$

or alternatively

$$\mathbf{J} = \begin{bmatrix} \cos\phi_1 & \dots & \cos\phi_{n_B} \\ \sin\phi_1 & \dots & \sin\phi_{n_B} \end{bmatrix} \tag{2.12}$$

where ϕ_n is the angle between the mobile device and the nth BS. The joint PDF of the distance measurements is given by

$$f(\hat{\mathbf{d}}|\mathbf{d}) = \frac{1}{(2\pi)^{n_B/2}|\boldsymbol{\Sigma}|^{1/2}} \exp\left\{-\frac{1}{2}(\hat{\mathbf{d}}-\mathbf{d})^T \boldsymbol{\Sigma}^{-1}(\hat{\mathbf{d}}-\mathbf{d})\right\} \tag{2.13}$$

where $\boldsymbol{\Sigma}$ is the covariance. $\mathbf{I}_{\mathbf{d}}$ can then be easily derived and it is given by

$$\mathbf{I}_{\mathbf{d}} = \boldsymbol{\Sigma}^{-1} = \mathrm{diag}(\sigma_1^{-2}, \sigma_2^{-2}, \ldots, \sigma_{n_B}^{-2}) \tag{2.14}$$

The CRLB for the mobile device position is then given by

$$\left[\mathbf{I}(\boldsymbol{\theta})^{-1}\right]_{2\times 2} = \left(\mathbf{J}\mathbf{I}_{\mathbf{d}}\mathbf{J}^T\right)^{-1}. \tag{2.15}$$

Another popular metric to characterize the accuracy of localization is the Geometric Dilution of Precision (GDOP) which describes the amplification of the errors in range measurements to the location error (Patwari et al. 2003) and it is given by

$$\mathrm{GDOP} = \frac{\sqrt{\sigma_x^2 + \sigma_y^2}}{\sigma_r} \tag{2.16}$$

where σ_x^2 and σ_y^2 the variances of the position estimate and σ_r is the standard deviation of the range measurement error. An alternative expression for the GDOP could be derived to emphasize the geometrical relationship between the BSs and the mobile device Spirito (2001)

$$\mathrm{GDOP}(n_B, \phi) = \sqrt{\frac{n_B}{\Sigma_i \Sigma_{j,j>i} \left|\sin(\phi_{ij})\right|^2}} \tag{2.17}$$

where ϕ_{ij} is the angle between the ith and jth BSs.

Although the CRLB derivations in this subsection assumed single-path ideal propagation (simplified zero-mean Gaussian noise model) it can provide a starting point to evaluate the performance and understand the main factors that can affect the accuracy. Different CRLB derivations that address the NLOS problem can be found in Qi et al. (2006), Dardari et al. (2006), Shen et al. (2007). The accuracy of the TOA-based techniques relies heavily on the measurement noise and the multipath condition of the channel. Thus the CRLB will only be meaningful when the models are realistic in that they reflect the actual propagation conditions. In addition it is common to assume that the BS and the mobile device are synchronized, but this is not the case in practice.

2.1.2 TDOA Techniques

Time Difference of Arrival (TDOA) technique is based on the idea that the position of the mobile device can be determined by examining the difference in time at which the signal arrives at multiple reference points (Liu et al. 2007). Adopting this technique is useful in practical scenarios where synchronization between mobile devices is not available. Each TDOA measurement constrains the location of the mobile device to be on a hyperboloid with a constant range difference between the two reference points. For two-dimensional position estimation three reference points are required. Figure 2.3 illustrates the localization technique based on TDOA measurements.

A TDOA measurement between BS1 and BS2 can be given by Sayed et al. (2005)

$$t_{21} = (t_2 - t_0) - (t_1 - t_0) = t_2 - t_1 \tag{2.18}$$

where t_0 is the clock time of the mobile device, t_1 and t_2 are the TOA between the mobile device and BS1 and BS2 respectively. The equation can be written in terms of distance through speed of light scaling or $d_{21} = (t_2 - t_1)c$. Thus the time difference (or range difference) is $d_{21} = d_2 - d_1$ where $d_2^2 = (x_2 - x)^2 + (y_2 - y)^2$ and $d_1^2 = (x)^2 + (y)^2$. Without loss of generality, the latter equation is valid with the assumption that the x- and y-coordinates of BS1 are (0,0). The range difference

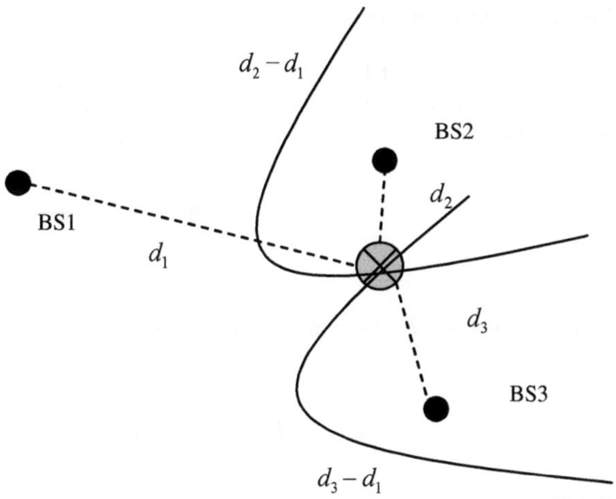

Fig. 2.3 TDOA localization. At least three BS are required for two-dimensional localization. The time (range) differences $d_2 - d_1$ and $d_3 - d_1$ form two hyperboloids of which the intersection (solution) is the estimated position

equation can be rearranged to $d_{21} + d_1 = d_2$. The TDOA equation can then be obtained by squaring both sides or

$$(d_{21} + d_1)^2 = d_2^2 = x_2^2 + x^2 - 2x_2x + y_2^2 + y^2 - 2y_2y \tag{2.19}$$

Using $K_2^2 = x_2^2 + y_2^2$ the above equation simplifies to

$$(d_{21} + d_1)^2 = K_2^2 - 2x_2x - 2y_2y + x^2 + y^2 \tag{2.20}$$

which can be further rearranged to solve for the unknowns or

$$-x_2x - y_2y = d_{21}d_1 + \frac{1}{2}\left(d_{21}^2 - K_2^2\right). \tag{2.21}$$

Two equations are required to solve for the two unknowns and the second TDOA equation between BS3 and BS1 can be similarly obtained

$$-x_3x - y_3y = d_{31}d_1 + \frac{1}{2}\left(d_{31}^2 - K_3^2\right). \tag{2.22}$$

The equations can be arranged in matrix form given by Sayed et al. (2005)

$$\mathbf{H}\boldsymbol{\theta} = d_1\mathbf{a} + \mathbf{b} \tag{2.23}$$

where $\mathbf{a} = \begin{bmatrix} -d_{21} \\ -d_{31} \end{bmatrix}$, $\mathbf{b} = \frac{1}{2}\begin{bmatrix} K_2^2 - d_{21}^2 \\ K_3^2 - d_{31}^2 \end{bmatrix}$. Solving for $\boldsymbol{\theta}$ we have

$$\boldsymbol{\theta} = d_1\mathbf{H}^{-1}\mathbf{a} + \mathbf{H}^{-1}\mathbf{b}. \tag{2.24}$$

Extension to more reference points and three dimensions is trivial and more details can be found in Sayed et al. (2005).

The performance of TDOA-based localization can be similarly examined by evaluating the CRLB. A similar derivation of the CRLB for TDOA localization

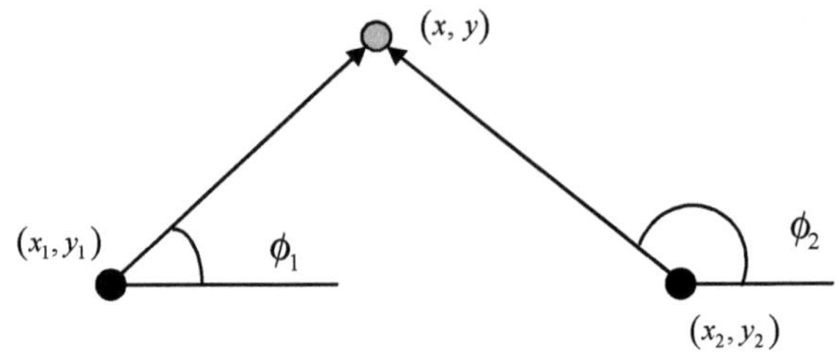

Fig. 2.4 AOA positioning (angulation). The AOA estimate from 2 base stations to the mobile terminal can be used to estimate the position

follows from (2.7). In fact it can be shown that the TDOA CRLB is given by Qi et al. (2006)

$$\left[\mathbf{I}_{\mathrm{TDOA}}(\boldsymbol{\theta})^{-1}\right]_{2\times 2} = \left(\mathbf{J}_{\mathrm{TDOA}}\mathbf{I}_{\mathrm{TDOA}}\mathbf{J}_{\mathrm{TDOA}}^{T}\right)^{-1} \tag{2.25}$$

where

$$\mathbf{J}_{\mathrm{TDOA}} = \begin{pmatrix} \cos\phi_1 & \cos\phi_2 & \cdots & \cos\phi_{n_B} \\ \sin\phi_1 & \sin\phi_1 & \cdots & \sin\phi_{n_B} \\ 1 & 1 & \cdots & 1 \end{pmatrix} \tag{2.26}$$

$$\mathbf{I}_{\mathrm{TDOA}} = \mathbf{I}_{\mathrm{TOA}}. \tag{2.27}$$

where ϕ_n is the angle between the mobile device and the nth BS.

2.1.3 AOA-Based Techniques

Localization using angle-of-arrival is simpler than time-based techniques in that only two angle measurements are required, as opposed to three range measurements, in order to estimate the two-dimensional position. However the challenge is presented when obtaining accurate angle of arrival estimation using wireless devices. In typical scenarios, the base stations are equipped with K antenna array elements spaced by Δ which are capable of estimating the angle of arrival which is then used to locate the mobile device. Figure 2.4 illustrates the basic concept of AOA localization.

The relationship between the coordinates and the angles is given by

$$\frac{y - y_1}{x - x_1} = \tan(\phi_1) \quad \frac{y - y_2}{x - x_2} = \tan(\phi_2) \tag{2.28}$$

These equations can be combined to estimate the position of the mobile terminal as Dempster (2006)

$$\boldsymbol{\theta} = \begin{bmatrix} \tan\phi_1 & -1 \\ \tan\phi_2 & -1 \end{bmatrix}^{-1} \begin{bmatrix} x_1\tan\phi_1 & -y_1 \\ x_2\tan\phi_2 & -y_2 \end{bmatrix} \tag{2.29}$$

The CRLB for AOA can be similarly obtained from the formulation in (2.6) and (2.7), but with specific models for the angle measurements. In practice, the antenna array is capable of measuring a function of the angle or Qi et al. (2006)

$$\hat{\varphi}_n = \varphi_n(\phi_n) + w_n \tag{2.30}$$

where n is the index identifying the BS and w_n is a zero mean Gaussian noise with a variance given by Qi et al. (2006)

$$\sigma_w^2 = \left(2\Upsilon \cdot \frac{d\mathbf{a}_n^H(\varphi_n)}{d\varphi_n} \cdot \frac{d\mathbf{a}_n(\varphi_n)}{d\varphi_n}\right)^{-1} \tag{2.31}$$

where $\mathbf{a}_n(\varphi_n)$ is the steering vector for a specific antenna array configuration and Υ is the Signal to Noise Ratio (SNR). For an antenna array with K elements spaced by Δ the steering vector is given by

$$\mathbf{a}_n(\varphi_n) = [1 \quad \exp(i\varphi_n) \quad \ldots \quad \exp(i(K-1)\varphi_n)]^T \tag{2.32}$$

where $\varphi_n = 2\pi\Delta\cos\phi_n$. The variance of the estimation error is then given by Qi et al. (2006)

$$\sigma_w^2 = \frac{3}{K(K+1)(2K+1)\Upsilon}. \tag{2.33}$$

Given the above model parameters of the AOA localization system the CRLB can be given by Qi et al. (2006)

$$\left[\mathbf{I}_{\text{AOA}}(\boldsymbol{\theta})^{-1}\right]_{2\times 2} = \left(\mathbf{J}_{\text{AOA}}\mathbf{I}_{\text{AOA}}\mathbf{J}_{\text{AOA}}^T\right)^{-1} \tag{2.34}$$

where

$$\mathbf{I}_{\text{AOA}} = \frac{K(K+1)(2K+1)}{3}\text{diag}(\Upsilon_1 \quad \Upsilon_2 \quad \ldots \quad \Upsilon_{n_B}) \tag{2.35}$$

and

$$\mathbf{J}_{\text{AOA}} = 2\pi c\Delta \cdot \begin{pmatrix} \frac{1}{d_1}(\sin\phi_1)^2 & \frac{1}{d_2}(\sin\phi_2)^2 & \cdots & \frac{1}{d_{n_B}}(\sin\phi_{n_B})^2 \\ -\frac{1}{d_1}\cos\phi_1\sin\phi_1 & -\frac{1}{d_2}\cos\phi_2\sin\phi_2 & \cdots & -\frac{1}{d_{n_B}}\cos\phi_{n_B}\sin\phi_{n_B} \end{pmatrix} \tag{2.36}$$

The performance of AOA positioning techniques in LOS conditions is satisfactory. However, in severe NLOS multipath conditions the reliability and accuracy of AOA techniques suffers considerably. As a result in these unfavorable propagation conditions, TOA- or RSS-based techniques are preferred. Furthermore, hybrid positioning techniques can be used to incorporate the advantages of two different techniques which usually outperform the individual techniques.

2.1.4 Received Signal Strength Localization

Localization using Received Signal Strength (RSS) is very similar to TOA-based technique in that the distances to n_B base stations are used in a tri-lateration approach to estimate the position. The difference is the method in which the distance is estimated. For a mobile device and n_B base stations, the unknown location can be estimated using the LS method similar to that of the TOA presented in (2.5) or

$$\hat{\boldsymbol{\theta}} = \boldsymbol{\theta}_0 + \left(\mathbf{J}^H\mathbf{J}\right)^{-1}\mathbf{J}^H\left[\hat{\mathbf{d}}_{\text{RSS}} - \mathbf{F}(\boldsymbol{\theta}_0)\right] \tag{2.37}$$

The difference between (2.37) and (2.5) is the estimated distance vector. For RSS-based localization the distance can be estimated through the power–distance relationship that is very well known for wireless propagation in different environments. The RSS between the mobile device and the nth base station is modeled by

$$P_r^{\text{dBm}} = -10\gamma\log_{10} d_n + S_n \tag{2.38}$$

where γ is the pathloss exponent (governing the rate of power decay with distance), S_n is the log-normal shadow fading component with variance $\sigma_{S_n}^2$ and d_n is the distance between the mobile devices and the nth base station. The ML estimate of the distance is given by $\hat{d}_n = 10^{(-P_r)/(10\gamma)}$ Patwari et al. (2003). Then the distance vector in (2.37) is given by $\hat{\mathbf{d}}_{\text{RSS}} = \left[\hat{d}_1, \hat{d}_2, \ldots, \hat{d}_{n_B}\right]^T$. The CRLB for RSS-based localization can be similarly derived from (2.6) to (2.7) (Qi et al. 2006)

$$\left[\mathbf{I}_{\text{RSS}}(\boldsymbol{\theta})^{-1}\right]_{2\times 2} = \left(\mathbf{J}_{\text{RSS}}\mathbf{I}_{\text{RSS}}\mathbf{J}_{\text{RSS}}^T\right)^{-1} \tag{2.39}$$

where

$$\mathbf{I}_{\text{RSS}} = \text{diag}\left(\,\sigma_{S_1}^{-2} \quad \sigma_{S_2}^{-2} \quad \cdots \quad \sigma_{S_{n_B}}^{-2}\,\right) \tag{2.40}$$

and

$$\mathbf{J}_{\text{RSS}} = \frac{10\gamma c}{\ln 10}\cdot\begin{pmatrix} \frac{\cos\phi_1}{d_1} & \frac{\cos\phi_2}{d_2} & \cdots & \frac{\cos\phi_{n_B}}{d_{n_B}} \\ \frac{\sin\phi_1}{d_1} & \frac{\sin\phi_2}{d_2} & \cdots & \frac{\sin\phi_{n_B}}{d_{n_B}} \end{pmatrix}. \tag{2.41}$$

2.2 The Multipath Problem

The presence of multipath fading in harsh propagation environments can have a significant impact on the performance of TOA-, RSS-, or AOA-based ranging and localization systems. Multipath is the reception of multiple copies of the transmitted signal—each arriving from different propagation paths—which combine in

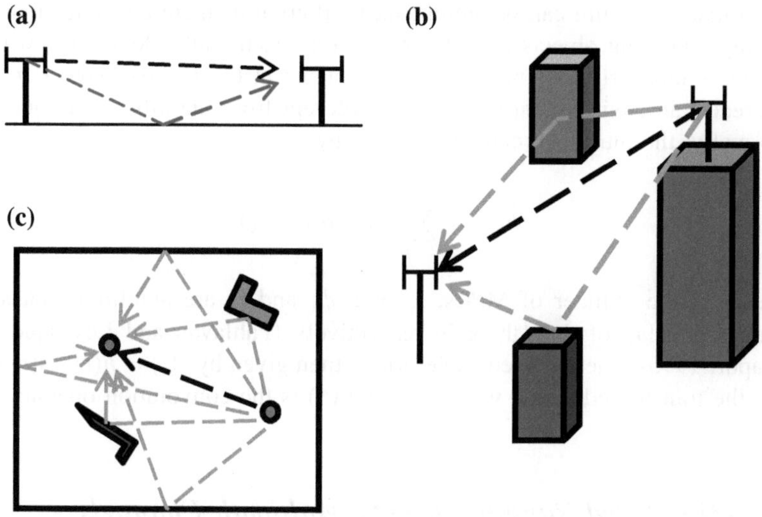

Fig. 2.5 LOS multipath channels. **a** Outdoor open space—single bounce model, **b** urban LOS, **c** indoor LOS

either a constructive or destructive manner that distorts the received signal. The transmitted signal undergoes reflections and diffractions along different propagation paths to the receiver. At the receiver, replicas of the transmitted signal arrive attenuated, phase-shifted, and time-delayed. For RSS-based systems, multipath causes the well-known fast fading phenomenon, where the received power in a given location fluctuates significantly due to constructive and destructive interference of incoming multipath signals. For TOA-based systems, the multipath impacts the distance estimation directly by adding a random bias to the estimation. In this section, we will introduce the multipath problem and highlight its impact on RSS- and TOA-based ranging/localization systems.

In order to appreciate the impact of multipath, it is important to analyze it in LOS environments, since multipath is the major error contributor. LOS propagation can behave drastically different based on the environment. For example, performance in outdoor open-field LOS, outdoor urban LOS, and indoor LOS can exhibit different TOA estimation behavior. In outdoor open-field LOS, the direct path between the transmitter and receiver is unobstructed and there is at least a single ground reflection at the receiver. In urban LOS or indoor LOS, there may be many signals arriving at the receiver that were reflected or diffracted from the surrounding buildings or objects. Figure 2.5 illustrates different possible LOS multipath scenarios.

In outdoor open space, the multipath structure is mainly composed of the direct path signal and a single-bounce ground reflection (see Fig. 2.5a). In urban LOS, reflections from the surrounding buildings make up the multipath environment (see Fig. 2.5b). The density of the buildings and surrounding obstacles will dictate the structure of the multipath environment. Finally, in indoor LOS environments, the multipath structure can be significantly different as there are reflections from the many cluttering objects and also reflections from walls, doors, and windows with closer interarrival of multipath components at the receiver (see Fig. 2.5c). This creates an environment that is very different from the urban environment.

Formally, the multipath can be modeled by

$$h(\tau) = \sum_{k=1}^{L_p} \alpha_k e^{j\phi_k} \delta(t - \tau_k) \tag{2.42}$$

where L_p is the number of MPCs, α_k and, ϕ_k and τ_k are amplitude, phase and propagation delay of the kth path, respectively (Pahlavan and Levesque 2005; Rappaport 1996). The received waveform is then given by $r(t) = h(t) * s(t)$ where $s(t)$ is the transmitted signal waveform and (*) is the convolution operator.

2.2.1 TOA-Based Ranging in LOS Multipath Channels

The basic idea behind TOA-based ranging is to estimate the distance between a transmitter and a receiver through measuring the signal propagation delay. For a transmitter at location (x_1, y_1) and a receiver at location (x_2, y_2) the Euclidean distance is given by $d = \sqrt{(x_1 - x_2)^2 + (y_1 - y_2)^2}$. In practice the distance can be calculated from the speed of light/propagation delay relationship given by $c = d/\tau$, where c is the speed of signal propagation (in free space $c = 3e8$ m/s) and τ is the propagation delay. But in realistic applications, the propagation delay estimates are always corrupted by noise—additive white Gaussian noise (AWGN). Thus, the measured distance can be written as $\hat{d} = c \times \tau + w = d + w$. Here w is a zero-mean Gaussian noise. In practice, the delay can be estimated using two methods: one-way TOA ranging or two-way TOA ranging. The latter requires no synchronization and it is the basic ranging technique proposed in IEEE 802.15.4a (IEEE 802.15.TG4a). The former requires strict synchronization since the distance is estimated from the received waveform. This is practically challenging for two reasons. The first is that extracting the TOA of the first path arrival is difficult (Lee and Scholtz 2002; Guvenc and sahinoglu 2005). The second is that synchronization of wireless devices in multipath environments is very difficult to achieve and is in fact an open research area. The main challenges are due to the clock drift over time and the effect of temperature and humidity on the accuracy of clock frequency (Sundararaman et al. 2005). Two-way TOA ranging techniques are the most popular due to the fact that they do not require synchronization and the protocols are very simple. For treatment of one-way TOA ranging further details can be found in (Guvenc and sahinoglu 2005). Two-way TOA ranging is achieved by noting the time that the ranging reference signal is sent out with the time it takes to receive it. Figure 2.6 illustrates an example where Device 1 is attempting to estimate the distance to Device 2.

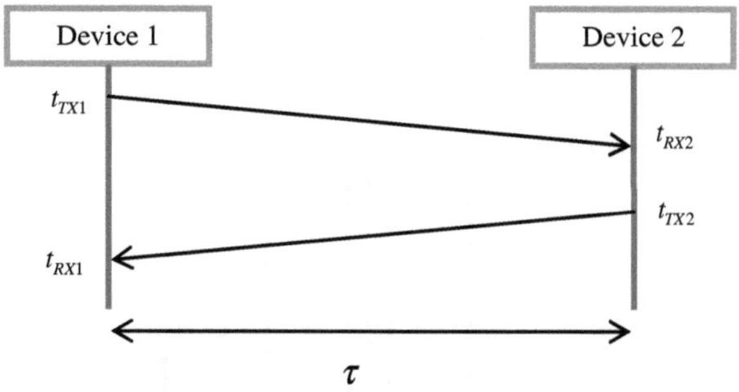

Fig. 2.6 Two-way TOA ranging. Devices 1 and 2 exchange transmit and receive time information. With these four time stamps, the propagation delay (distance) between the devices can be estimated

Device 1 initiates the two-way ranging by sending a ranging packet (signal) to Device 2 and noting the time as t_{TX1}. Device 2 receives the signal at t_{RX2} and prepares its own ranging signal (after a processing delay) and sends out a response ranging signal at time t_{TX2}. Finally Device 1 receives the response at t_{RX1}. Given that Device 2 shares the time stamp information t_{RX2} and t_{TX2} with Device 1 it is now possible to estimate the propagation delay (distance) between the two devices by

$$\tau = \frac{t_{\text{total}} - t_{\text{round-trip}}}{2} = \frac{(t_{RX1} - t_{TX1}) - (t_{TX2} - t_{RX2})}{2} \tag{2.43}$$

where $t_{\text{total}} = 2\tau + t_{\text{round-trip}}$ is the total time it takes for the two-way ranging and $t_{\text{round-trip}}$ is the round-trip delay at Device 2. The assumptions regarding this two-way TOA ranging are overly simplistic and not valid in practice. In reality, the clocks of the two devices are not synchronized and not perfect. This means that with time the clocks will drift and the delay estimation will not be accurate (biased). Recently, researchers have investigated this problem and proposed some practical techniques to estimate the delay in non-ideal scenarios (clock drift and bias) (Zheng and Wu 2010; Wu et al. 2011).

The performance of TOA estimation in single path AWGN ideal scenario is usually analyzed using the Cramer Rao Lower Bound, which is a statistical approach to quantifying the variance of TOA estimation. Essentially any algorithm, in theory, can achieve the CRLB given that both the CRLB and algorithm follow the same assumptions (for example LOS single path model and same noise variance). The variance of TOA estimation σ^2_{TOA} is bounded by the CRLB given by Gezici et al. (2005),

$$\sigma^2_{\text{TOA}} \geq \frac{1}{8\pi^2 \Upsilon T B f_0^2 \left(1 + \frac{B^2}{12 f_0^2}\right)} \tag{2.44}$$

where T is the signal observation time, Υ is the SNR, f_0 is the frequency of operation, and B is the system bandwidth. This relationship highlights that the accuracy of TOA estimation can be improved by either increasing the SNR—since higher signal level will enable the estimation of the direct path signal with greater accuracy—or increasing the system bandwidth—since higher system bandwidth

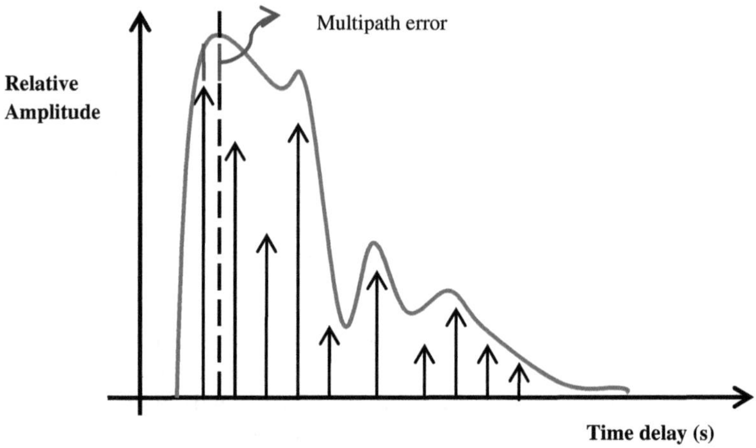

Fig. 2.7 Power delay profile highlighting the multipath error corrupting TOA-based range estimates

means higher time–domain resolution leading to better range estimates. The increase in time–domain resolution of the channel can be attributed to narrower time–domain signals/pulses. This makes it possible to discriminate or *resolve* the different multipath arrivals and improve the TOA estimation. Multipath signals (especially in dense cluttered environment) tend to arrive fairly close to the direct path. If the interarrival time between the multipath components is much smaller than the time–domain resolution of the system (low bandwidth systems) then at the receiver those multiple signals will *combine* to create a new cluster. The TOA estimate (from the receiver's point of view) will then be the peak of the cluster. In order to clarify this phenomenon, Fig. 2.7 illustrates a power delay profile example and the resulting envelope. A power delay profile is a representation of the channel impulse response where the power from different arrival paths can be measured and analyzed. In the figure there are ten multipath components where the first multipath component is the strongest and, in this case, is the LOS or *direct* path. The multipath components arriving after the direct path fall in close proximity to each other (because of the nature of the propagation environment). For this narrowband system, the multipath components arrive and combine (due to low time–domain resolution) and *appear* at the receiver as four multipath components (the peaks of the blue envelope). As a result, the peaks will ultimately be detected as path arrivals. The first path arrival will be estimated as the LOS path and thus used for distance estimation. It is clear in this case that the actual TOA is not equal to the estimated TOA. This difference in estimation is the multipath error.

For higher system bandwidths, the multipath error in LOS environments is usually smaller. For example, Fig. 2.8 illustrates a measured channel impulse response (measurement systems will be described in detail later in the chapter) for 200 MHz bandwidth in a typical LOS office environment.

Fig. 2.8 Measured power delay profile highlighting TOA estimation in indoor LOS office at 200 MHz bandwidth

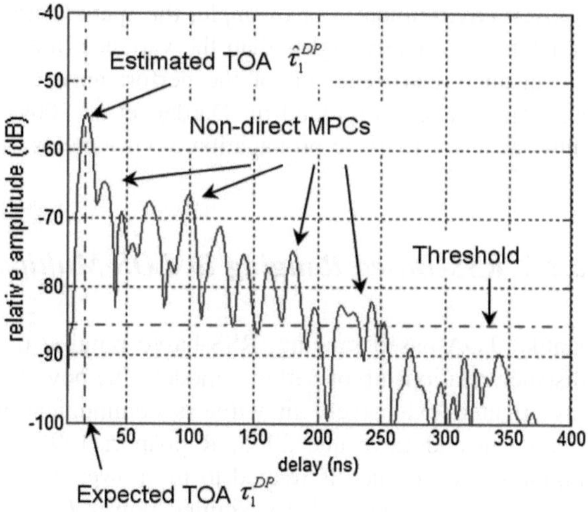

As can be seen from the figure, the actual/expected TOA is very close to the peak of the measured TOA of the first path. Also, note that there are about 15 multipath components for a noise threshold of –85 dBm. These non-direct multipath components can originate from wall reflections, furniture diffractions, and scattering from other objects in the office.

One way to assess the performance of TOA-based ranging is to analyze the ranging error. In LOS environments, the ranging error could be attributed to both multipath and measurement noise. Let α_1^{DP} and τ_1^{DP}denote the DP amplitude and propagation delay, respectively. The distance between the transmitter and the receiver is $d^{\mathrm{DP}} = v \times \tau_1^{\mathrm{DP}}$, where v is the speed of signal propagation. Then ranging error which is defined as the difference between the estimated and the actual distance or,

$$\varepsilon = \hat{d} - d_{\mathrm{DP}} \tag{2.45}$$

In a general LOS multipath environment, the ranging device will experience varying error behavior depending on the structure of the propagation environment and the system bandwidth. In LOS, the distance estimate can be modeled by

$$\hat{d}_{\mathrm{DP}} = d_{\mathrm{DP}} + \varepsilon_{\mathrm{DP}}(B) + w \tag{2.46}$$

where $\varepsilon_{\mathrm{DP}} = \tilde{b}_m(B)$ is a bias induced by the multipath and it is a function of the system bandwidth and w is a zero-mean additive measurement noise. As we will later discuss, the statistics of the multipath bias can be modeled differently. One popular approach is to model it spatially as a zero-mean Gaussian (Alavi and Pahlavan 2003). This means that an ensemble of LOS measurements in a given LOS environment will generally result in a Gaussian distribution. The variance of the distribution will be directly related to the variation in the multipath structure in a given environment. For example, the spatial variance in an indoor office environment is typically higher than the variance in an outdoor, flat terrain.

An analytical treatment of the performance of TOA estimation in multipath environments can be found in (Dardari et al. 2009) where Ziv-Zakai Bounds are introduced for realistic propagation environments.

2.2.2 RSS-Based Ranging in LOS Multipath Environments

Unlike TOA-based ranging, RSS-based ranging depends on an a priori power–distance relationship or pathloss model. The power–distance relationship has been investigated extensively in wireless communications for different technologies (Pahlavan and Levesque 2005; Rappaport 1996). In many of the experimental findings, the distance is related to the power law. For a narrowband transmitted signal in free-space with transmitted power P_t, the received signal power P_r is given by Pahlavan and Levesque (2005)

$$P_r = P_t G_t G_r \left(\frac{\lambda}{4\pi d} \right)^2 \tag{2.47}$$

where G_t and G_r are the transmitter and receiver antenna gains, respectively. λ is the wavelength of the transmitted signal and d is the distance between the transmitter and receiver. A reference received power at distance $d = 1$ m is usually defined as $P_0 = P_t G_t G_r (\lambda/4\pi)^2$ then the distance–power relationship in free space can be given by

$$P_r = \frac{P_0}{d^2}. \tag{2.48}$$

RSS ranging is based on models which assume an a priori relationship between the distance and the received power (or pathloss of the signal). A popular model in LOS channels relates the received power to the transmitted power by the following equation

$$\log_{10} P_r = \log_{10} P_0 - 10\gamma \log_{10} d \tag{2.49}$$

where γ is the pathloss exponent that determines the rate of power loss with increasing distance. Note that this is equivalent to (2.48) for $\gamma = 2$. If we define pathloss to be the ratio of received power to transmitted power then the above power–distance relationship can be rewritten in terms of pathloss L given by

$$L = L_0 + 10\gamma \log_{10} d \tag{2.50}$$

where $L_0 = 10 \log_{10} P_t - 10 \log_{10} P_0$ and $L = 10 \log_{10} P_t - 10 \log_{10} P_r$. In order to model the power–distance relationship more accurately, a random component that models the shadow (slow) fading is included or

$$L = L_0 + 10\gamma \log_{10} d + S \tag{2.51}$$

where S is a normally distributed random variable in the log domain and it models the fluctuation of the signal away from the median pathloss. This fluctuation stems from the presence of different obstructions between the transmitter and receiver which "shadow" the signal. RSS-ranging is mainly affected, however, by fast-fading (Pahlavan and Levesque 2005; Rappaport 1996). At the receiver, the attenuated and phase shifted replicas of the transmitted signal combine either constructively or destructively. The effect is a fast fluctuation of power at a given distance. One way to deal with this fast fading problem is to collect more RSS measurements and "average out" the fluctuations by taking the mean of the measurements. Then, for a given pathloss exponent and P_0, the Maximum Likelihood Estimate (MLE) of the distance between a transmitter and a receiver can be estimated from the measured received power as Patwari et al. (2003)

$$\hat{d}_{\mathrm{MLE}} = 10^{(P_0 - P_r)/(10\gamma)} \tag{2.52}$$

The major weakness with RSS-based distance estimation is the assumption that the pathloss exponent (pathloss model) is known a priori when in fact the exponent changes between multipath environments—and even within the same environment. Furthermore, the accuracy of the range estimate cannot be improved by averaging the received signal power alone. Averaging of the RSS prior to estimating the distance will only remove the small-scale fading (fast fading) due to the multipath but not the shadow fading (which is more common in NLOS environments). Typical values for the pathloss exponent in LOS multipath environments range between 1 and 2 (Pahlavan and Levesque 2005). There are approaches that attempt to estimate the pathloss exponent prior to the localization stage, but that approach presents some challenges as well (Li 2006). The statistical performance of RSS ranging can be analyzed through the well-known CRLB given by Qi and Kobayashi (2003)

$$\sigma_{\mathrm{RSS}}^2 \geq \frac{(\ln 10)^2 \sigma_S^2 d^2}{100\gamma} \tag{2.53}$$

where σ_S^2 is the variance of the shadow fading term. This relationship indicates that RSS-based ranging estimation is affected by the pathloss exponent and the variance of the shadow-fading in addition to the distance. As the distance increases, RSS estimation degrades. More importantly, as the variance of the shadow fading increases, the variance of RSS ranging also increases. This basic, yet powerful relationship highlights the challenges of RSS-based ranging. In typical multipath environments, the shadow fading variance is significant and thus reliable estimation of the distance can be difficult. In addition, the inverse dependency on the pathloss exponent indicates that performance of RSS ranging in LOS environments (lower pathloss exponent $\sim$1–2) is expected to be much better than NLOS environments (typical pathloss exponents $\sim$3–5). These challenges to RSS-based ranging make it a more practical, but inaccurate option for localization.

2.3 The NLOS Problem

This section introduces the NLOS problem and describes the impact of NLOS channels on TOA- and RSS-based ranging. For the former, NLOS affects the estimation of the direct path signal. Since in most cases the direct path will not be detectible, ranging is achieved through non-direct path components which bias TOA-based estimation. For the latter, NLOS introduces the problem of shadow fading, where RSS is attenuated randomly as the mobile device moves from one area to the other.

2.3.1 TOA-Based Ranging in NLOS Multipath Environments

In the previous section, the basics of TOA-based ranging in LOS environments were introduced. A natural extension of the LOS case is a more challenging and complex situation where the transmitter and receiver experience an NLOS multipath channel. Specifically, when considering NLOS cases, there is an obstruction in the path of the transmitter and receiver. Depending on the type of obstruction and the relative distances of the transmitter/receiver to the obstruction, the channel behavior can vary significantly. There are two specific NLOS cases that occur in typical obstructed environments. The first is when the direct path (DP) signal is attenuated but detected (albeit weak SNR). This situation can arise naturally when the transmitter and receiver are separated by "light" obstructions such as a glass or a wooden door. Indeed, in this scenario TOA estimates can be obtained with good accuracy due to the detection of the DP signal. The second NLOS case is when there is a "heavy" or severe obstruction between the transmitter and receiver, where the direct path is severely attenuated and "buried" under the noise floor of the receiver, making it undetectable. The first non-Direct path (NDP) component is then used for TOA estimation. This results in a significant bias that corrupts the TOA estimation and ultimately the position estimate. In this severe NLOS condition, the variance of TOA estimation with time is usually large due to the fact that the estimated first arrival path varies significantly due to the shadowing problem. For a quasi-static channel, the first path can be detected. However, when some perturbation is introduced to the multipath structure (another person moves around/close to the TX-RX path), then the estimation of the first path arrival will fluctuate significantly. It is clear, then, that NLOS does not only introduce a bias, but also introduces significant TOA estimation perturbations that can degrade the real-time distance estimation.

Formally stated, in the absence of the DP, ranging is achieved using the amplitude and propagation delay of the first Non-Direct Path (NDP) component—denoted as α_1^{NDP} and τ_1^{NDP} respectively—resulting in a longer distance $d^{\mathrm{NDP}} = v \times \tau_1^{\mathrm{NDP}}$, where $d^{\mathrm{NDP}} > d^{\mathrm{DP}}$. In order for the receiver to successfully identify the DP, the ratio of the strongest multipath component to that of the DP, given by

$$\kappa_1 = \frac{\max\left(|\alpha_i|_{i=1}^{L_p}\right)}{\alpha^{\mathrm{DP}}}, \tag{2.54}$$

must be less than the receiver dynamic range, κ, and the power of the DP must be greater than the receiver sensitivity, φ. These constraints are given by

$$\kappa_1 \le \kappa \tag{2.55}$$

$$P_{\mathrm{DP}} > \varphi \tag{2.56}$$

where $P_{\mathrm{DP}} = 20\log_{10}\left(\alpha_1^{\mathrm{DP}}\right)$.

In an indoor environment the mobile device will experience varying error behavior depending on the availability of the DP and, in the case of its absence, on the characteristics of the DP blockage. It is possible to categorize the error based on the following ranging states (Alsindi et al. 2009). In the presence of the DP, both the constraints above are met and the distance estimate is accurate, yielding

$$\hat{d}_{\mathrm{DP}}^{\mathrm{NLOS}} = d_{\mathrm{DP}} + \varepsilon_{\mathrm{DP}}^{\mathrm{NLOS}} + w \tag{2.57}$$

$$\varepsilon_{\mathrm{DP}}^{\mathrm{NLOS}} = b_{\mathrm{pd}} + \tilde{b}_m(B) \tag{2.58}$$

where $\tilde{b}_m$ is the zero-mean random bias induced by the multipath, b_{pd} is the bias corresponding to the propagation delay caused by NLOS conditions and w is a zero-mean additive measurement noise. It has been shown that $\tilde{b}_m$ is indeed a function of the bandwidth and signal to noise ratio (SNR) (Pahlavan et al. 1998), while b_{pd} is dependent on the medium of the obstacles (Gentile and Kik 2007). In the more severe case, the DP is completely attenuated and the requirement that $\kappa_1 \le \kappa$ is not met because the DP is shadowed by some obstacle, burying its power under the dynamic range of the receiver. In this situation, the ranging estimate experiences a larger error compared to the LOS condition. Emphasizing that ranging is achieved through the first arriving NDP component, the estimate is then given by

$$\hat{d}_{\mathrm{NDP}}^{\mathrm{NLOS}} = d_{\mathrm{DP}} + \varepsilon_{\mathrm{NDP}}^{\mathrm{NLOS}} + w \tag{2.59}$$

$$\varepsilon_{\mathrm{NDP}}^{\mathrm{NLOS}} = \tilde{b}_m(B) + b_{\mathrm{pd}} + b_{\mathrm{NDP}} \tag{2.60}$$

where b_{NDP} is a deterministic additive bias representing the nature of the blockage. Unlike the multipath biases, and similar to biases induced by propagation delay, the dependence of b_{NDP} on the system bandwidth and SNR has its own limitations, as reported in Pahlavan et al. (1998). Figure 2.9 illustrates the two specific conditions occurring in NLOS environments.

An example of the measured channel profiles in the NLOS conditions is shown in Figs. 2.10 and 2.11.

It is clear from the figures that in NLOS channel conditions large ranging errors are possible, highlighting the major limitation to deploying accurate geolocation

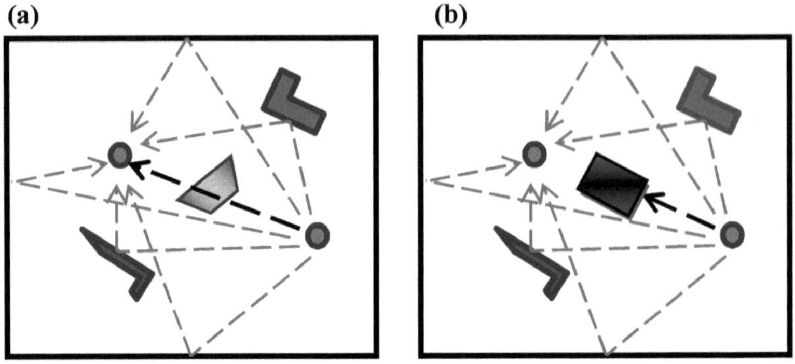

Fig. 2.9 Indoor NLOS multipath channels. **a** "Light" NLOS—the DP is attenuated but can be detected **b** Severe NLOS—the DP is not detected

Fig. 2.10 Measurement of a "light" NLOS channel—the DP is attenuated but can be detected

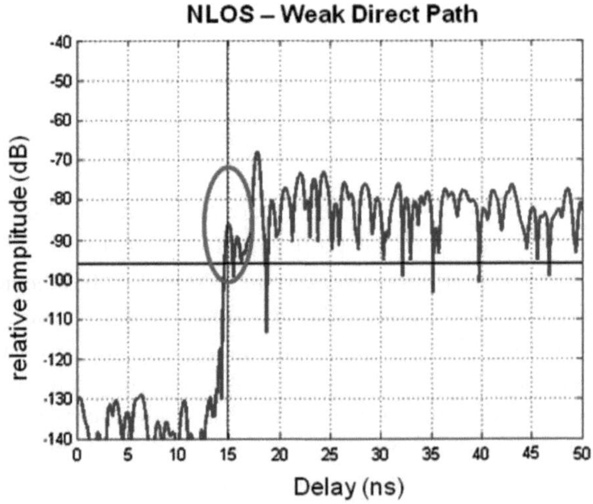

systems in urban and indoor environments. The impact of NLOS range measurements on the localization performance can be evaluated through CRLB-type bounds. Given that the statistics of the NLOS biases are available then it is possible to derive the Generalized-CRLB which integrates the statistical information. The analytical treatment of this problem can be found in Qi et al. (2006).

2.3.2 RSS-Based Ranging in NLOS Multipath Environments

In the previous section, RSS-based ranging in LOS multipath environment was introduced and it was illustrated how a simple pathloss model can be used to estimate the distance. Besides the limitation due to the unknown parameters of the

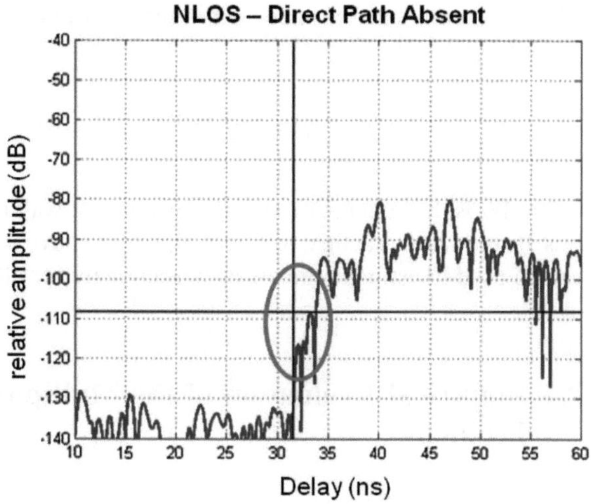

Fig. 2.11 Sample measurement of a Severe NLOS multipath channel—the DP is not detected

pathloss model, the challenge of RSS ranging in NLOS is exacerbated by the fact that obstructions between the transmitter and receiver can further complicate the distance–power relationship, making it difficult to directly estimate the distance accurately. For example, consider a mobile station moving away from a base station in a typical LOS environment. The pathloss model for this scenario is a typical LOS propagation model with pathloss exponent around 1–2 and minimal shadowing variance. However, as the mobile moves behind a wall, cabinet, or even an elevator, the power suddenly fluctuates and severe attenuation perturbs the LOS distance–power relationship. It then becomes very difficult to achieve accurate distance estimation in light of this problem. Although Li (2006) proposed a technique to estimate the pathloss exponent in real-time, the limitations still affect the accuracy and practicality of this approach. As a result, numerous research efforts have focused instead on an alternative RSS-based localization technique, namely fingerprinting-based localization, an approach to which Chap. 4 of this book is completely dedicated.

In NLOS environments the pathloss model introduced earlier for LOS environments can be further extended

$$L = L_0 + 10\gamma^{\mathrm{NLOS}} \log_{10} d + S^{\mathrm{NLOS}} \tag{2.61}$$

where γ^{NLOS} and S^{NLOS} is the pathloss exponent and shadow fading parameters for NLOS. Usually $\gamma^{\mathrm{NLOS}} > \gamma^{\mathrm{LOS}}$, with γ^{NLOS} ranging between 3 and 6 (Pahlavan and Levesque 2005). The NLOS pathloss model will be significantly different when considering the type and number of obstructions separating the transmitter and receiver. For example in indoor NLOS environments, the number of walls between the transmitter and receiver can significantly change the pathloss behavior. An additional parameter to incorporate the wall effect has been modeled in the literature as

$$L = L_0 + 10\gamma^{\text{NLOS}} \log_{10} d + S^{\text{NLOS}} + \sum_{n=1}^{N} W_n \tag{2.62}$$

where W_n is the attenuation specific to a type of wall (Durantini and Cassioli 2005). It is clear that, in practice, it is very difficult to have an accurate pathloss model that can be used to estimate the distance accurately for all the environments.

2.4 Empirical Evaluation of the Multipath and NLOS Problems

In order to understand the impact of the propagation channel on the effectiveness of existing TOA-based and RSS-based algorithms and to appreciate the limitations that they face, it is necessary to empirically characterize the radio propagation channel for the ranging- or geolocation-specific application. The TOA- and RSS-specific propagation studies help to shed light on the fundamental aspects of the ranging technique and the parameters that control its performance. In this section, we will provide an overview of the measurement techniques, results, and modeling efforts that have been carried out for TOA- and RSS-based ranging. The aim of this section is to introduce the reader to the methodologies used to measure and characterize the wireless channel for geolocation applications. This will serve as a foundation through which it is possible to understand the limitations facing some of the popular ranging and localization techniques that will be introduced in the later chapters.

2.4.1 *Channel Measurement Systems*

In order to characterize the behavior of TOA- or RSS-based ranging in multipath environments, the channel impulse response (CIR) or the power delay profile of the channel must be measured. The CIR is the time-delay characterization of the multipath and it provides the amplitude/delay relationship of the arriving multipath components. In practice the CIR can be measured directly by either using a time–domain measurement system or indirectly by using a frequency–domain measurement system. For geolocation-specific measurements and modeling, either systems can be used to extract relevant information for TOA-based ranging. Specifically, the measurement systems can be used to measure the large-scale, spatial characteristics of the direct path, mainly the $\hat{\alpha}_1^{\text{DP}}$ and the $\hat{\tau}_1^{\text{DP}}$, which can be used to examine the ranging coverage (pathloss characterization) and accuracy, respectively. In the absence of the DP, it is possible to measure the first detected

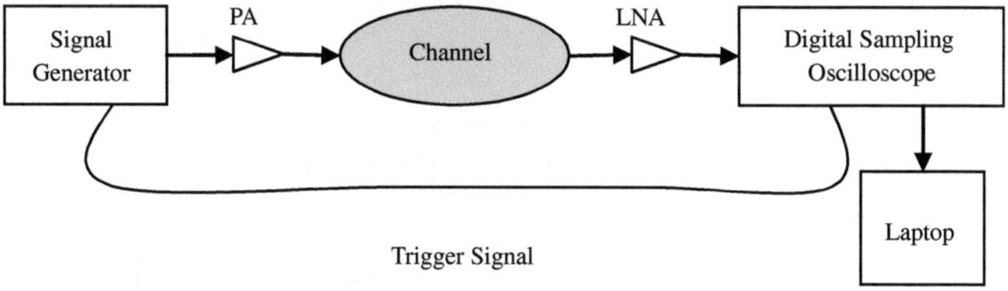

Fig. 2.12 Time domain measurement system block diagram

path, $\hat{\tau}_1^{\mathrm{NDP}}$, and analyze the probability of blockage and the error statistics in this condition. These TOA-based parameters can be extracted directly from the measured CIR.

2.4.1.1 Time Domain Systems

One way to capture the channel multipath profile is through the well-known time domain measurement system. The channel is captured by transmitting a known waveform (with special autocorrelation properties) and post-processing the received waveform by cross-correlation with the known template. Since the arriving waveform will be a superposition of shifted and attenuated replicas of the original signal, then the output of the cross-correlation will contain "peaks" at the delay values of the multipath components. A typical time domain measurement system is depicted in Fig. 2.12.

Typically, the template waveform can be either pulses or PN-sequences, employed in direct-sequence spread spectrum systems (Ciccognani et al. 2005). After amplification, the received waveform is captured by a digital sampling oscilloscope and stored for post-processing (Cassioli et al. 2002). Depending on the waveform type, the multipath profile can be extracted from the received waveform. In the case of the PN-sequence waveform, the received signal is correlated (after demodulation) with a replica of the transmitted sequence (Janssen and Vriens 1991). Note that for this measurement system the signal generator must be "synchronized" with the digitally sampling oscilloscope. That is a trigger signal is typically used to trigger the events for correlation purposes.

2.4.1.2 Frequency Domain Systems

One of the most popular and practical methods to measure the wireless channel is through the use of the frequency-domain measurement system. For such measurement systems a generic vector network analyzer (VNA) can be used. Frequency-domain measurement techniques have been previously employed to characterize the channel impulse response (Ghassemzadeh et al. 2004); Chong and

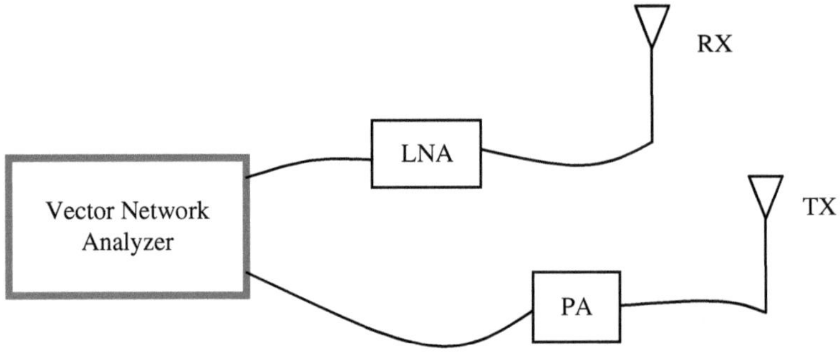

Fig. 2.13 Frequency-domain measurement system block diagram

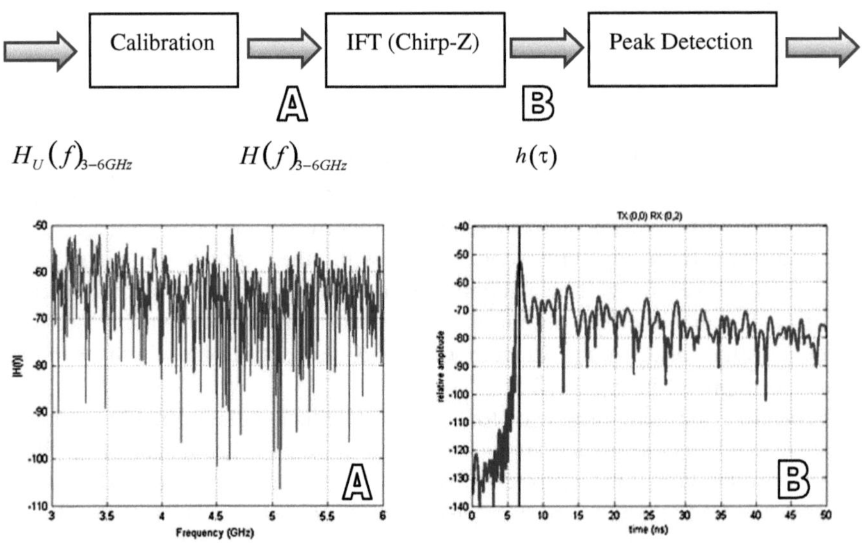

Fig. 2.14 Measurement post-processing and CIR generation

Yong (2005), Pahlavan and Levesque (2005), Howard and Pahlavan (1990) but for modeling the communication channel—characterizing RMS delay spread and power-distance relationships. The frequency measurement system captures the channel transfer function (CTF) and the time domain CIR can then be obtained by the inverse Fourier Transform (IFT).

The core of the measurement system is the VNA, which is used to sweep the frequency spectrum of a desirable system bandwidth with a certain sampling interval. The CTF can be captured by measuring the S21 S-parameter on the VNA which are samples of the frequency domain of the channel. Figure 2.13 illustrates an example measurement system setup. Further details of the measurement system can be found in Ghassemzadeh et al. (2004), Pahlavan and Levesque (2005) and Alsindi et al. (2009).

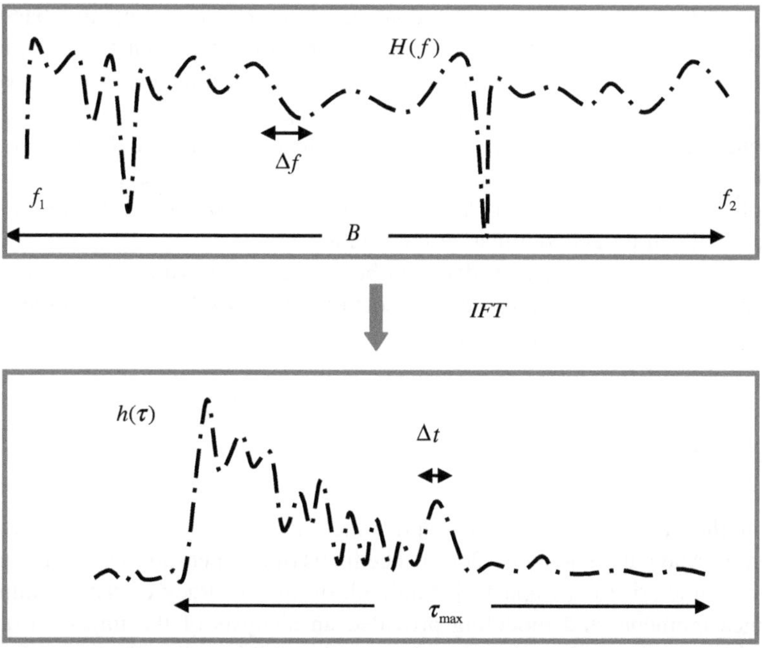

Fig. 2.15 Frequency domain measurement system—parameter relationships between the frequency and time domain signals

Table 2.1 Common frequency/time domain definitions and relationships

$B = f_2 - f_1 = N_f \Delta f$
$\tau_{\max} = t_{N_t} - t_1 = N_t \Delta t$
$B \propto \frac{1}{\Delta t}$ and $\tau_{\max} \propto \frac{1}{\Delta f}$

The CIR is then obtained by an IFT process and Fig. 2.14 highlights the system block diagram of the post-processing stage.

The uncalibrated measured CTF from the VNA is passed through a post-measurement calibration process that removes the channel response of the cables, LNA and PA. The CIR is then obtained by the IFT or the Chirp-Z transform which has a signal processing "zooming" capability. The time and amplitude of the multipath delays are then extracted by passing the raw estimated CIR through a peak detection algorithm, that essentially identifies the peaks in the profile that are greater than a certain noise threshold (typically -120 to -110 dBm).

The frequency domain measurement parameters are related to the time domain channel impulse response. The parameters that can be controlled in the VNA when measuring the frequency response are the swept frequencies (bandwidth), the number of samples, and the transmitted power. The frequency spacing is determined by the number of samples in a given bandwidth. For a CTF measurement

$H(f) = H(f_1, f_2)$ the VNA can be configured to measure a certain bandwidth between f_1 and f_2, or $B = f_2 - f_1$. Selecting the number of points will dictate the frequency spacing Δf. The relationship between the number of measured frequency samples, N_f, and the frequency spacing, Δf is $N_f = (f_2 - f_1)/\Delta f$. The frequency samples on the VNA directly affect the time domain CIR. The measured bandwidth B controls the time domain resolution Δt and the frequency spacing Δf controls the maximum time delay, τ_{max}, that can be measured. Figure 2.15 and Table 2.1 illustrate and summarize the relationship.

The collected measurement data can be then used to extract the TOA or RSS parameters for analysis. In the next subsectionss we introduce some of the models developed for the indoor environment.

2.4.2 Alavi Models

One of the earliest TOA-based ranging measurements and modeling was conducted by Alavi and Pahlavan (2006). The focus of the measurement and modeling was to characterize the impact of multipath on the accuracy of range estimation. The measurements and modeling provided an analysis of the impact of system bandwidth on the multipath-induced error. In addition, the TOA-specific measurements errors were analyzed under different NLOS conditions. Specifically, in this work, ranging error was referred to as Distance Measurement Error (DME) and it is given by

$$\varepsilon_B(d) = \hat{d}_B - d \tag{2.63}$$

where d is the ground-truth distance, $\hat{d}_B$ is the measured distance, and its dependence on system bandwidth is explicitly given by the subscript B. As a result, the error is a function of the distance between the transmitter and receiver and the bandwidth. Furthermore, depending on the condition of the indoor channel, the error can be significantly different: in LOS environments, multipath is the dominant source of error while in NLOS the absence of the DP—also known as Undetected Direct Path (UDP)—dominates the error. UDP is essentially severe NLOS where the DP cannot be detected due to a large obstruction between the transmitter and receiver which causes the DP path to be buried under the receiver noise floor. The models were obtained by conducting frequency domain measurements using the VNA described in the previous subsections. Figures 2.8 and 2.11 illustrate LOS versus NLOS with undetected DP.

By comparing the two measured profiles, it is clear that the error in UDP conditions contains a combination of the multipath error and a "UDP" error, which is essentially a bias in the time delay estimation. Note from the figure that the direct path is severely attenuated and lies below the noise threshold, which makes its detection very difficult. Based on the measurements in an indoor environment, Alavi introduced a model that incorporates the different ranging conditions. Specifically, the error is modeled as

$$\varepsilon_B(d) = \varepsilon + \varepsilon_{M,B}(d) + \xi(B)\varepsilon_{U,B}(d) \tag{2.64}$$

where $\varepsilon_{M,B}(d)$ is the multipath error, $\varepsilon_{U,B}(d)$ is the UDP error or bias, and $\xi_B(B)$ is a random variable that takes the value of "1" when a UDP condition occurs and "0" otherwise. The model also includes ε, which is an error that models the inaccuracies occurring during measurement of the actual distance between the transmitter and receiver. Typically, this error can be assumed zero-mean Gaussian with a variance that depends on the accuracy of the measurement error. Since ε cannot be separated from the multipath error, it is assumed that $\varepsilon + \varepsilon_{M,B}(d) \approx \varepsilon_{M,B}(d)$, which simplifies the model to

$$\varepsilon_B(d) = \varepsilon_{M,B}(d) + \xi_B(B)\varepsilon_{U,B}(d). \tag{2.65}$$

The multipath error $\varepsilon_{M,B}(d)$ can been modeled by

$$\varepsilon_{M,\,B}(d) = X(m_{M,\,B}, \sigma_{M,\,B})\log(1 + d) \tag{2.66}$$

where $X(m_{M,\,B}, \sigma_{M,\,B})$ is a Gaussian random variable with mean $m_{M,\,B}$ and standard deviation $\sigma_{M,\,B}$. The UDP error component was similarly modeled as Gaussian $X(m_{U,B}, \sigma_{U,B})$. As a result, the overall model is given by

$$\begin{aligned}\hat{d} &= d + \mathrm{MDME} + \xi_B(d)\mathrm{UDME} \\ &= d + X(m_{M,B}, \sigma_{M,B})\log(1 + d) + \xi_B(d)X(m_{U,B}, \sigma_{U,B})\end{aligned} \tag{2.67}$$

The random variable $\xi_B(d)$ can be modeled as

$$f_{\xi_W}(y) = \left(1 - p_{U,B}(d)\right)\delta(y) + p_{U,B}(d)\delta(y - 1). \tag{2.68}$$

The proposed models have been verified to fit actual data in Alavi and Pahlavan (2006).

The work in Alavi and Pahlavan (2006) also investigated the impact of the system bandwidth on the DME. Basically, as the system bandwidth increases, the error decreases due to enhanced time resolution. The finding further supports the idea that one way to mitigate the multipath problem is to increase the system bandwidth. This observation was also highlighted in Gentile and Kik (2007).

2.4.3 Alsindi Models

As stated earlier, the Alavi models were the first models developed for TOA-based ranging that analyzed the impact of LOS/NLOS and system bandwidth on the accuracy. The results of these models highlighted the fundamental limitations and challenges facing TOA-based ranging in harsh multipath environments. The

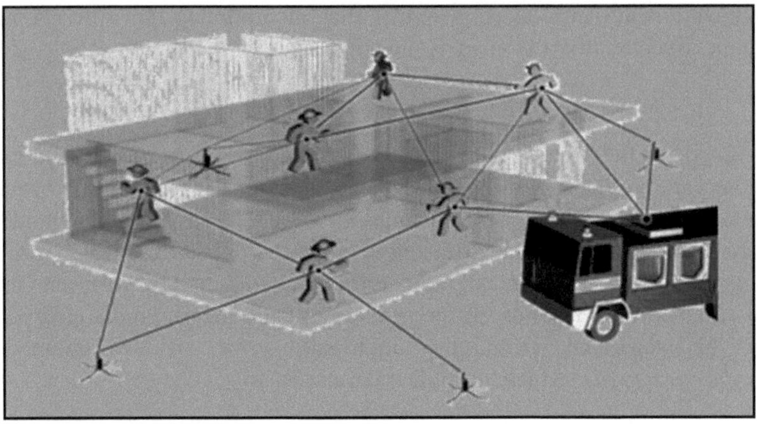

Fig. 2.16 Firefighter/soldier localization scenario in hostile indoor environments

measurements and models, however, were limited in scope since they were based on a single floor/office of an indoor environment and thus lacked comprehensive analysis in different buildings/environments. In addition, the models do not provide any indication of the coverage aspect of the ranging systems. As a continuation of the modeling efforts, Alsindi's work focused on developing models for Ultra-Wideband (UWB) TOA-based systems that characterize in detail some of the fundamental parameters such as ranging coverage, ranging error in LOS, NLOS-presence of DP, and NLOS-absence of DP (Alsindi et al. 2009). UWB is defined as any system operating with a bandwidth of 500 MHz or with a bandwidth exceeding 20 % of the center frequency.

The objective of the measurement campaign was to develop models for firefighter/soldier TOA-based ranging/localization in hostile indoor environments. In such scenarios, beacons or anchors were placed surrounding a given building in order to aid firefighters/soldiers to localize and navigate themselves in an indoor environment through cooperative localization using wireless sensor networks (WSN). Cooperative localization is dealt with in Chaps. 6 and 7 where centralized and distributed techniques will be discussed in more detail. Figure 2.16 illustrates the localization scenario that was considered for the measurement campaign.

In order to develop reliable systems operating in these challenging environments, it is necessary to understand the propagation characteristics that impact ranging and localization accuracy. It is clear from the figure that three distinct ranging scenarios are possible: Indoor-to-Indoor (ITI), Outdoor-to-Indoor (OTI) and Roof-to-Indoor (RTI). In addition four different building types were investigated: old office (Atwater Kent—AK), new office (Fuller Labs), residential (Schussler) and manufacturing floor (Norton). All the buildings are in Worcester, MA, USA. From this application point of view, it is then interesting to investigate the following issues:

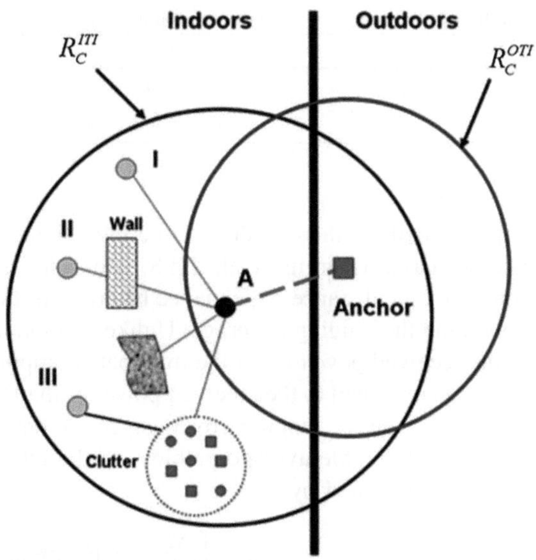

Fig. 2.17 NLOS challenges facing the firefighter localization application

- For the outdoor beacons (OTI & RTI), how far can the devices reliably provide TOA-based ranging estimates? What is the *ranging coverage*?
- What is the probability of DP blockage in NLOS environments?
- What are the ranging error characteristics in ITI, OTI and RTI?
- How is the ranging/localization performance impacted for different building types: residential, office, etc.?

For the firefighter/soldier localization scenario, the multipath and NLOS problems can be difficult challenges that will impact the accuracy of the localization directly. Figure 2.17 highlights the NLOS challenges facing OTI/RTI and ITI scenarios.

For OTI/RTI scenarios, the signal propagating through the external walls of the building typically undergoes significant attenuation because the walls are usually thick in construction and are composed of brick and steel material. As a result, the ranging coverage can be limited significantly and, in most cases, is much less than the ITI scenarios. For ITI scenarios the ranging coverage, although higher than OTI/RTI, is significantly different for LOS and NLOS scenarios.

Alsindi's models focused on characterizing the ranging coverage and ranging error in these different scenarios and environments. For the former the distance–power relationship of the Direct Path (DP) signal provides an empirical evaluation of the *ranging coverage* which is the maximum distance where the DP can be detected. For the latter the spatial distribution of the ranging error in different scenarios and environments provides an empirical evaluation of the physical limitation facing indoor geolocation.

In indoor environments, the distance-dependence of the received power, which can be used to determine the communication coverage, is usually predicted from

Table 2.2 Summary of TOA-based ranging error conditions

LOS	NLOS-DP	NLOS-NDP/UDP
$\hat{d}_{DP} = d_{DP} + \varepsilon_{DP}(B) + w$ $\varepsilon_{DP}(\omega) = b_m(B)$	$\hat{d}_{DP}^{NLOS} = d_{DP} + \varepsilon_{DP}^{NLOS} + w$ $\varepsilon_{DP}^{NLOS} = b_{pd} + b_m(B)$	$\hat{d}_{NDP}^{NLOS} = d_{DP} + \varepsilon_{NDP}^{NLOS} + w$ $\varepsilon_{NDP}^{NLOS} = b_m(B) + b_{pd} + b_{NDP}$

experimental pathloss models of the total signal energy in different environments and scenarios (Durgin et al. 1998; Molisch 2005; Ghassemzadeh et al. 2004). Similarly, the distance-dependence behavior of the power of the DP can be used to determine the ranging coverage. Unlike communication coverage which is related to the received power of all the multipath components at a given distance, ranging coverage is related to the received power of the DP component. For a given system dynamic range, κ, ranging coverage, R_r, is defined as the distance in which the maximum tolerable average pathloss of the DP is within κ (Alsindi et al. 2009). This is represented by

$$\max\left\{\overline{L_p^{DP}}\right\} = 10\gamma\log_{10}(R_r) \leq \kappa \tag{2.69}$$

where $\overline{L_p^{DP}}$ is the average pathloss of the DP and γ is the pathloss exponent. The pathloss behavior of the DP is distance-dependant, but because of the attenuation and energy removed by scattering, its intensity decreases more rapidly with distance compared with the total signal energy (Siwiak et al. 2003). This means that for typical indoor multipath scattering environment, communication coverage is greater than ranging coverage, $R_c > R_r$. Operating out of ranging coverage causes large TOA estimation errors and performance degradation (always in NLOS-NDP condition). The characterization of ranging error in different scenarios has been introduced earlier in the chapter and it is summarized in Table 2.2 for convenience.

2.4.3.1 Modeling the Pathloss: Ranging Coverage

Using the same established pathloss modeling approach used in the literature, (Ghassemzadeh et al. 2004; Pahlavan and Levesque 2005), Alsindi characterized the distance–power dependence of the measured DP (Alsindi et al. 2009) and compared it to the distance–power relationship of the total received power (RSS). The pathloss exponent is determined from measurement data through least-square (LS) linear regression. The pathloss relationship is provided in (2.61) but an additional factor attributed to the power loss due to penetration through walls can be incorporated as L_X (which is depending on the ranging scenario OTI, RTI, etc.). Thus the modified expression is given by

$$L(d) = L_0 + L_X + 10\gamma\log_{10}(d/d_0) + S, \quad d \geq d_0. \tag{2.70}$$

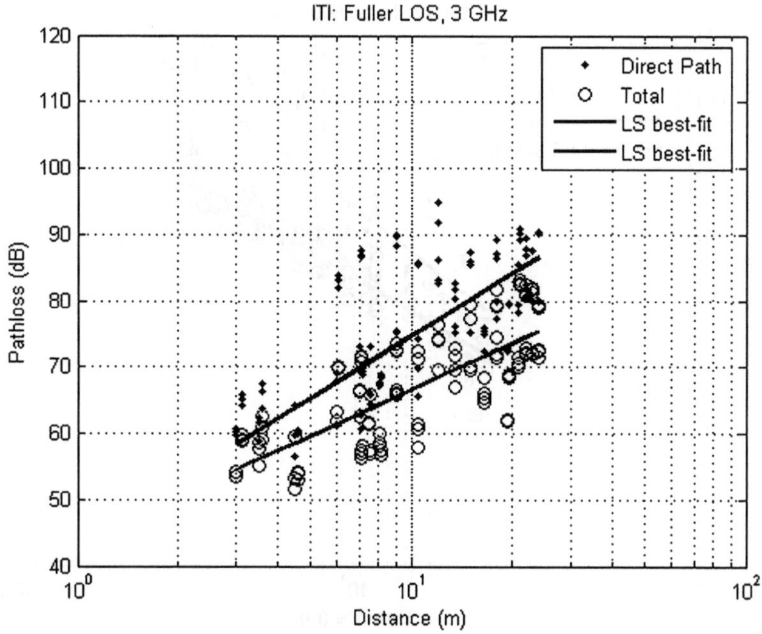

Fig. 2.18 Pathloss scatter plots in Fuller ITI LOS at 3 GHz bandwidth

All the parameters of the model in (2.70) are a function of the building type/propagation environment. Figures 2.18, 2.19 and 2.20 show sample measured scatter plots of the pathloss as a function of TX-RX separation for different buildings and ranging scenarios.

The pathloss model parameters are summarized in Table 2.3.

Several observations can be made from the table and the figures. The first is that for all the measurement data the pathloss exponent is higher for the DP relative to the total signal power, which is consistent with the modeling approach. Second, the DP power experiences greater fluctuations around the mean pathloss as compared with the total signal counterpart. This observation makes sense because small variations on the transmitter location affect the DP power more than the total power. Third, L_X changes for the different penetration scenarios. In ITI scenarios Schussler NLOS suffers 6 dB penetration loss due to the walls compared to 7.5 in AK. Norton ITI measurements are a mixture of LOS/NLOS because the manufacturing floor contained scattered machines. The impact can be clearly seen on the pathloss exponent when the bandwidth increases, hence higher attenuation. Results of OTI measurements show that Fuller and AK exhibit the largest penetration loss mainly because the signal had to penetrate a thicker building construction when compared with Norton and Schussler. In addition, the pathloss exponents in AK are large mainly because the measurement locations were conducted inside a metal shop on the edge of the building and between concrete corridors and rooms. AK in general imposes a very challenging environment for ranging because of the

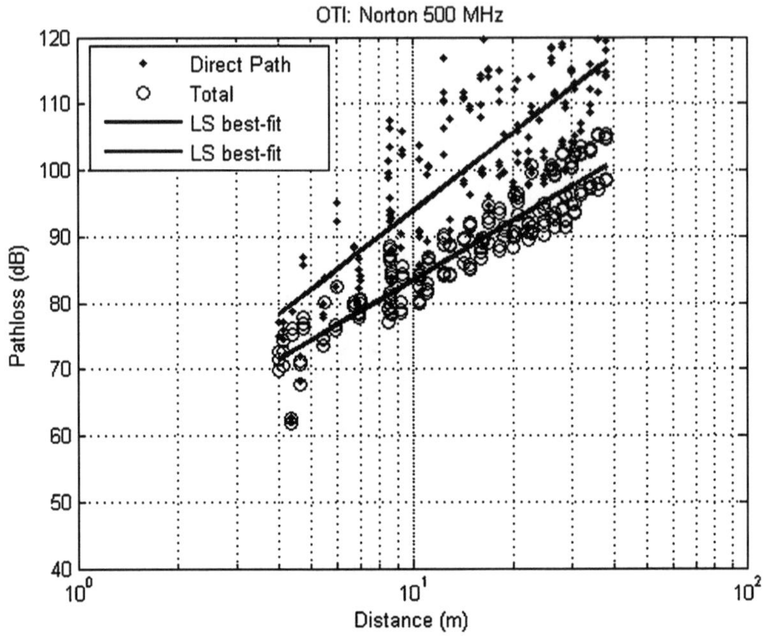

Fig. 2.19 Pathloss scatter plots in Norton OTI at 500 MHz bandwidth

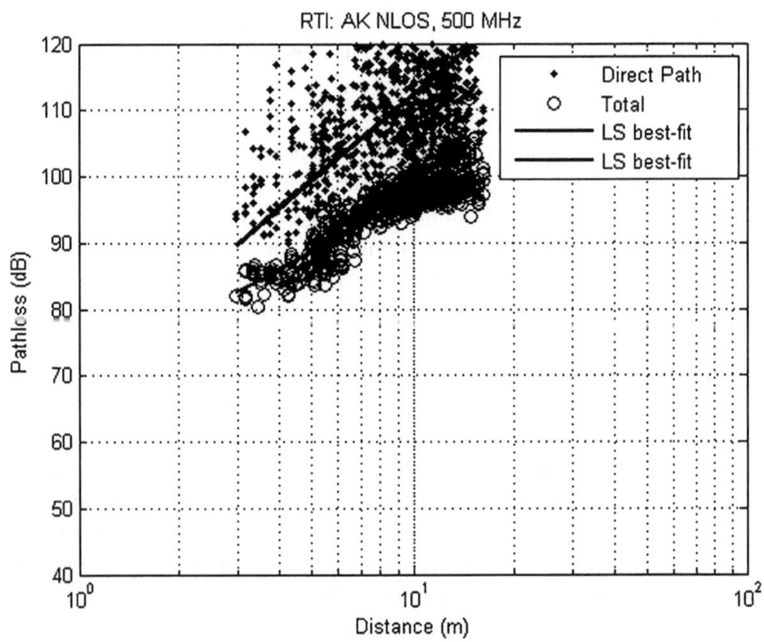

Fig. 2.20 Pathloss scatter plots in AK RTI at 500 MHz

Table 2.3 Pathloss modeling parameters

Scenario	Environment	L_X(dB)	Direct Path				Total signal	
			500 MHz		3 GHz			
			γ	S (dB)	γ	S (dB)	γ	S (dB)
ITI	Fuller (LOS)	0	3.2	8.9	3.3	7.1	2.4	5.5
	Norton (Mixed)	0	3.5	8.5	4.5	9.1	2.6	3.4
	Schussler (NLOS)	6	3.4	7.9	4.0	8.4	3.0	4.6
	AK (NLOS)	7.5	5.4	6.2	5.6	8.5	3.6	6.2
OTI	Fuller	14.3	3.4	13.7	3.7	14.1	2.2	7.7
	Norton	8.7	3.9	7.8	5.0	10.1	3.3	4.4
	Schussler	7.6	4.1	10.5	4.2	11.1	3.2	6.1
	AK	10	4.6	8.7	5.1	8.9	3.1	3.2
RTI	AK	24.5	4.3	7.6	5.3	8.8	2.9	1.7

building material and dense cluttering. RTI measurements experienced the largest penetration loss and high pathloss exponent. Finally, note that the harsher the indoor environment, the higher the pathloss exponent difference when moving to a higher system bandwidth. This is mainly due to the fact that larger system bandwidths provide better time domain resolution at the cost of reduced power per multipath component. This implies that the advantage of higher time domain resolution comes at a cost of shorter ranging coverage.

2.4.3.2 Modeling the Ranging Error

The spatial characteristics of the ranging errors are determined through the behavior of the biases, which are random due to the unknown structure of the indoor environment and the relative location of the user to them. Since the errors are highly dependent on the absence or the presence of the DP, the models introduced by Alsindi are based on the classification in Table 2.2. Further, in order to model and compare the behavior in different building environments and scenarios, the normalized ranging error was modeled instead as

$$\psi = \frac{\varepsilon}{d} = \frac{(\hat{d} - d)}{d}. \tag{2.71}$$

The range error observed in an indoor environment can then be modeled by combining the conditions in Table 2.2 through the following expression

$$\psi = \psi_m + G\left(\psi_{\text{pd}} + X\psi_{\text{NDP}}\right) \tag{2.72}$$

where ψ_m is the normalized multipath error that exists in both the presence and absence of the DP. ψ_{pd} is the normalized propagation delay-induced error, and ψ_{NDP} is the normalized error due to DP blockage. In order to distinguish between

Table 2.4 Probabilities of the presence and absence of the DP

Scenario	Environment	500 MHz		3 GHz	
		$p(\zeta_1)$	$p(\zeta_2)$	$p(\zeta_1)$	$p(\zeta_2)$
ITI	Fuller	0.1	0.90	0.2	0.98
	Norton	0.96	0.4	0.83	0.17
	Schussler	0.89	0.11	0.87	0.13
	AK	0.39	0.61	0.32	0.68
OTI	Fuller	0.42	0.58	0.39	0.61
	Norton	0.57	0.43	0.24	0.76
	Schussler	0.77	0.23	0.60	0.40
	AK	0.40	0.60	0.22	0.78
RTI	AK	0.58	0.42	0.37	0.63

the error behavior in LOS and NLOS, a Bernoulli random variable, G was used. That is,

$$G = \begin{cases} 0, & \text{LOS} \\ 1, & \text{NLOS} \end{cases} \tag{2.73}$$

where $p(G = 0) = p(\text{LOS})$ is the probability of being in LOS and $p(G = 1) = p(\text{NLOS})$ is the probability of being in NLOS. Similarly, X is a Bernoulli random variable that models the occurrence of DP blockage and is given by

$$X = \begin{cases} 0, & \zeta_1 \\ 1, & \zeta_2 \end{cases} \tag{2.74}$$

where $p(X = 0) = p(\zeta_1)$ denote the probability of detecting a DP, while $p(X = 1) = p(\zeta_2)$ denotes the probability of the occurrence of blockage. It is important to emphasize that Alsindi's modeling approach focuses on the DP and not the traditional definition of NLOS used for communications. This means that a mobile station and a base station separated by a wall, for instance, is considered NLOS, but does not necessarily imply the absence of the DP. In the remainder of the chapter, ranging error, bias, and normalized error will be used interchangeably.

The results of the measurement and modeling also revealed a significant difference in the probability of DP blockage among the different environments, which is highlighted in Table 2.4.

Several observations can be concluded. First, a positive correlation between the system bandwidth and the blockage probability $p(\zeta_2)$ exists due to lower energy per MPCs in higher system bandwidths. Second, as expected, DP blockage increases from ITI, to OTI, and RTI. Attenuation due to penetration from exterior walls and ceiling results in higher $p(\zeta_2)$. Third, blockage is highly correlated with the building type. In residential environments, blockage probability is low since the interior is composed of wooden structures with few metallic objects (e.g. a fridge, laundry room, etc.). Office buildings, however, pose harsher conditions with thicker walls, metallic beams, vending machines, metallic cabinets, shelves,

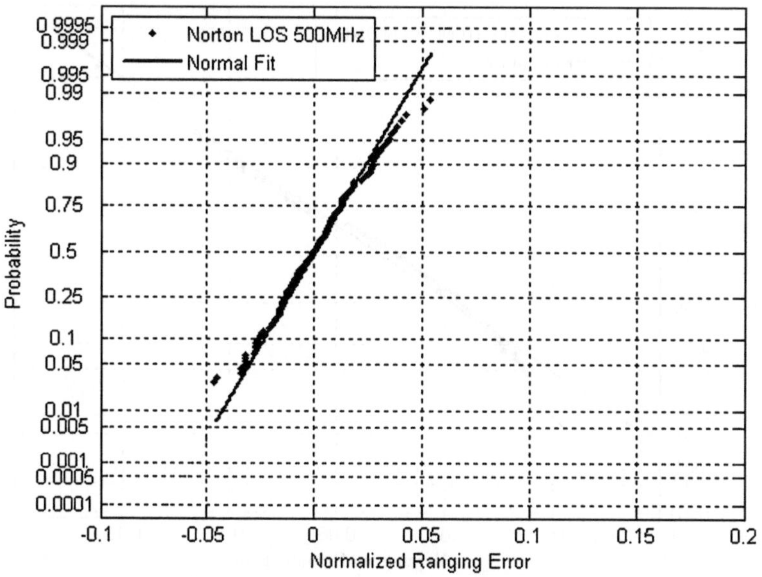

Fig. 2.21 Norton ITI at 500 MHz bandwidth: confirming the normality of the biases in LOS conditions

and elevator shafts, resulting in a substantial blockage up to 90 %, see Fuller and AK (ITI/OTI). Also, ITI measurements on the manufacturing floor highlight the impact of occasional clutter of machineries. Finally, it is worth mentioning that these results were measured using a 120 dB dynamic range provided by the external amplifiers and LNA extending the measured range. In realistic UWB systems, unfortunately, this would be prohibitively high in terms of implementation expense, which means that the results here can be seen as a lower bound.

The models also analyze the behavior of ranging error in the presence and in the absence of the DP. The measurement results of the ranging error in LOS scenarios revealed that the impact of the multipath can be modeled through a normal distribution since the DP is available and the error deviates in both directions relative to the actual distance. In addition, normality of the ranging error in this condition has been reported in Alavi and Pahlavan (2003, 2006). The error distribution can then be explicitly modeled as,

$$f(\psi|G=0) = \frac{1}{\sqrt{2\pi\sigma_{\mathrm{LOS}}^2}}\exp\left[-\frac{(\psi-\mu_{\mathrm{LOS}})^2}{2\sigma_{\mathrm{LOS}}^2}\right] \tag{2.75}$$

with mean μ_{LOS} and standard deviation σ_{LOS} specific to the LOS multipath-induced errors. Figure 2.21 further confirms the normality of errors in this condition.

A similar observation of the multipath effect in indoor LOS environments has been reported through measurements (Alavi and Pahlavan 2006). In NLOS

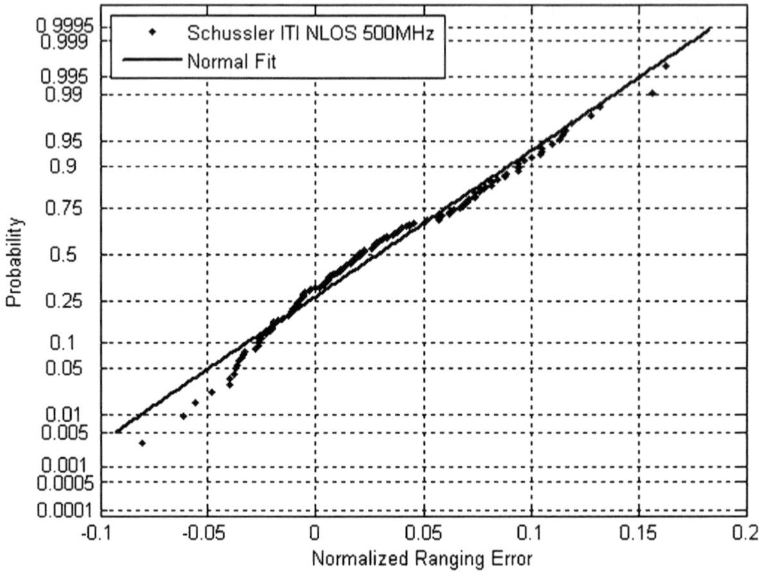

Fig. 2.22 Schussler ITI NLOS—mean of biases is larger than LOS

scenarios, when the DP is present, the amount of propagation delay and multipath due to obstructing objects such as wooden walls causes the biases to be more positive. The results show (see Fig. 2.22) that the spatial characteristics retain the statistics of the LOS counterpart but with a higher mean and standard deviation.

According to these results, the normalized ranging error is modeled similar to (2.75), but with emphasis on the condition. This is given by,

$$f(\psi|G=0,X=0)=\frac{1}{\sqrt{2\pi\sigma^2_{\mathrm{NLOS-DP}}}}\exp\left[-\frac{(\psi-\mu_{\mathrm{NLOS-DP}})^2}{2\sigma^2_{\mathrm{NLOS-DP}}}\right] \tag{2.76}$$

The subscripts in (2.76) specify the contributing error factors. Table 2.5 provides the modeling parameters of all the scenarios and environments in the presence of the DP.

The results show a positive correlation between the statistics of the normal distribution with the complexity of environment and/or ranging scenario. Negative correlation can be seen between the statistics and the system bandwidth due to reduction of multipath error in higher bandwidths.

The ranging error behavior in the absence of the DP is significantly different. The shadowing of the DP impacts the error behavior in several ways. First, only positive errors occur, since the blockage induces a higher positive bias that dominates compared to the multipath counterpart. Second, there are occasionally large positive range errors that occur due to heavier indoor constructions such as elevator shafts, clustering of cabinets, or even metallic doors. Third, the diversity of blocking material in indoor environments means that the spatial distribution of

Table 2.5 DP normal distribution modeling parameters for normalized ranging error

Scenario	Environment	500 MHz		3 GHz	
		μ_{LOS}	σ_{LOS}	μ_{LOS}	σ_{LOS}
ITI	Fuller (LOS)	0	0.028	0	0.006
	Norton (LOS)	0	0.022	0	0.007
		$\mu_{NLOS-DP}$	$\sigma_{NLOS-DP}$	$\mu_{NLOS-DP}$	$\sigma_{NLOS-DP}$
	Fuller (NLOS)	0.058	0.028	0.003	0.01
	Schussler	0.029	0.047	0.014	0.016
	AK (NLOS)	0.023	0.020	0.009	0.004
OTI	Fuller	0.015	0.017	0.002	0.011
	Norton	0.019	0.029	0.002	0.015
	Schussler	0.041	0.045	0.011	0.013
	AK	0.034	0.023	0.012	0.004
RTI	AK	0.029	0.041	0.012	0.012

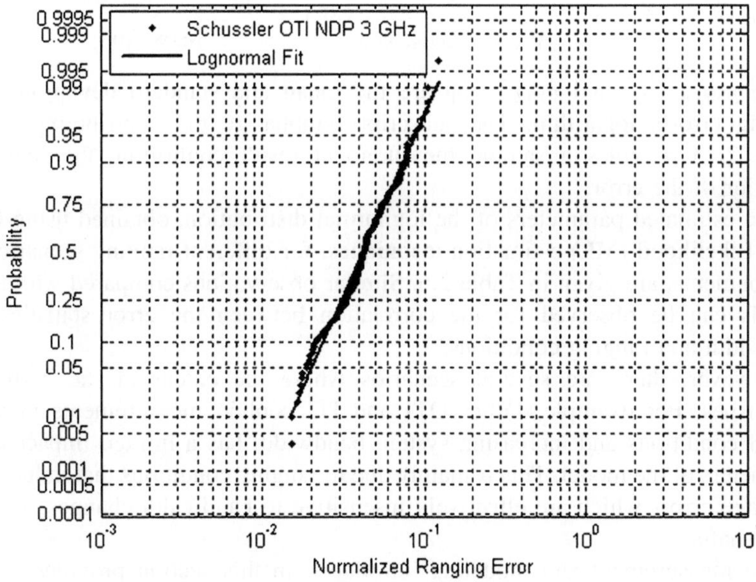

Fig. 2.23 Schussler OTI at 3 GHz bandwidth—confirming the lognormality of the measured normalized ranging error

errors will in general exhibit a heavier positive tail. By examining the PDF of the errors in this condition, it is observed that different subsets of the data showed varying tail behavior. The "heaviness" of the tail depended on the ranging environment and scenario. Thus harsher blockage conditions, i.e., higher number of blocked MPCs, exhibited heavier tails. As a result, the ranging error in this condition was modeled as log-normally distributed. The lognormal model is then given by,

Table 2.6 Lognormal distribution modeling parameters of the normalized ranging error in the absence of the direct path

Scenario	Environment	500 MHz		3 GHz	
		$\mu_{NLOS-NDP}$	$\sigma_{NLOS-NDP}$	$\mu_{NLOS-NDP}$	$\sigma_{NLOS-NDP}$
ITI	Norton (NLOS)	−3.13	0.62	−4.29	0.45
	Fuller (NLOS)	−1.68	0.88	−1.90	1.13
	Schussler	−1.59	0.49	−2.72	0.53
	AK (NLOS)	−2.17	0.45	−2.89	0.81
OTI	Fuller	−2.33	0.75	−2.99	1.17
	Norton	−2.78	0.65	−3.82	0.52
	Schussler	−2.03	0.58	−3.16	0.45
	AK	−2.32	0.51	−3.11	0.77
RTI	AK	−1.99	0.54	−3.01	0.61

$$f(\psi|G=1,X=1)=\frac{1}{\psi\sqrt{2\pi\sigma^2_{\mathrm{NLOS-NDP}}}}\exp\left[-\frac{(\ln\psi-\mu_{\mathrm{NLOS-NDP}})^2}{2\sigma^2_{\mathrm{NLOS-NDP}}}\right] \tag{2.77}$$

where $\mu_{NLOS-NDP}$ and $\sigma_{NLOS-NDP}$ are the mean and standard deviation of the ranging error's logarithm. The subscripts emphasize the contributing factors. Figure 2.23 provides a sample measurement result confirming the lognormal behavior of the error.

The estimated parameters of the lognormal distribution, obtained using Maximum Likelihood (ML) estimation techniques, for different ranging scenarios and environments, are given in Table 2.6. Similar observations compared with earlier models can be observed for the correlation between the error statistics with bandwidth and ranging conditions.

However, there are several scenarios where the extent of the correlation diminishes. For example, Fuller OTI and ITI contain measurements in severe NLOS conditions and increasing system bandwidth has a limited impact on the parameters of the model. This is mainly due to ranging conditions that induce large blockage errors which are effectively insensitive to bandwidth changes, e.g., elevator shafts.

The measurement and modeling introduced in this section provides realistic insight into these challenges, which is necessary for performance evaluation through CRLB and algorithm design and development.

2.5 Conclusion

The development of location-enabled services is mainly hindered by the realities of harsh propagation in environments where the devices are to be deployed—typically the dense urban and indoor environments. These environments pose serious challenges to system designers and engineers developing next generation location enabled devices. Specifically, multipath and NLOS are the two main physical limitations that need to be resolved in order to enable accurate and reliable localization. In this chapter we have first introduced the basics of geolocation techniques such as TOA, TDOA, AOA, and RSS. Then the multipath and NLOS problems for TOA- and RSS-based ranging techniques were presented.

Through channel measurements and modeling, the impact of multipath on TOA-based ranging as a function of bandwidth was investigated. It was shown that an increase in system bandwidth can reduce the multipath error significantly. For RSS-based ranging systems, however, the bandwidth does not play a major role in mitigating the multipath problem. Instead, averaging can remove the fast-fading variations of power due to multipath, yielding better distance estimation. With regard to the NLOS problem, both RSS- and TOA-based ranging suffer from the physical limitations. For the former, large power variations (shadow fading) affect the power–distance relationship and make it difficult to accurately estimate the distance; for the latter, NLOS introduces biases that corrupt the distance estimation and cause large errors that can affect the accuracy of any localization algorithm. In the next chapter, we will investigate popular techniques to mitigate the multipath and NLOS problems.

References

A. Durantini, D. Cassioli, A multi-wall pathloss model for indoor UWB propagation. in *Proceedings of Vehicular Technology Conference (VTC)*, pp. 30–34, May 2005

B. Alavi, K. Pahlavan, Modeling of the TOA-based distance measurement error using UWB indoor radio measurements. IEEE Commun. Lett. **10**(4), 275–277 (2006)

N. Alsindi, B. Alavi, K. Pahlavan, Measurement and modeling of ultrawideband TOA-based ranging in indoor multipath environments. IEEE Trans. Veh. Technol. **58**(3), 1046–1058 (2009)

B. Alavi, K. Pahlavan, Modeling of the distance error for indoor geolocation. in *Proceedings of IEEE Wireless Communications and Networking (WCNC)*, vol 1 New Orleans, LA, USA, (2003), pp. 668–672

B. Alavi, K. Pahlavan, Studying the effect of bandwidth on performance of UWB positioning systems. in *Proceedings of IEEE Wireless Communications and Networking Conference (WCNC)* (Las Vegas, NV, USA, 2006), pp. 884–889

C. Gentile, A. Kik, A comprehensive evaluation of indoor ranging using ultra-wideband technology. EURASIP J. Wirel. Commun. Netw. **2007**, Article ID 86031 (2007)

J. J. Caffery, G.L. Stuber, Overview of radiolocation in CDMA cellular systems. IEEE Commun. Mag. **36**(4), 38–45 (1998)

D. Cassioli, M.Z. Win, A.F. Molisch, The UWB indoor channel: from statistical model to simulations. IEEE J. Select. Areas Commun. **20**(6), 1247–1257 (2002)

C.-C. Chong, S.K. Yong, A generic statistical-based UWB channel mode for high-rise apartments. IEEE Trans. Antennas Propag. **53**(8), 2389–2399 (2005)

Y.T. Chan, K.C. Ho, A simple and efficient estimator for hyperbolic location. IEEE Trans. Signal Process. **42**(8), 1905–1915 (1994)

W. Ciccognani, A. Durantini, D. Cassioli, Time domain propagation measurements of the UWB indoor channel using PN-sequence in the FCC-compliant band 3.6–6 GHz. IEEE Trans. Antennas Propag. **53**(4), 1542–1549 (2005)

D. Dardari, C.-C. Chong, M.Z. Win, Improved lower bounds on time of arrival estimation error in realistic UWB channels. in *Proceedings of IEEE 2006 Conference On Ultra-Wideband,* pp. 531–537, 2006

D. Dardari, A. Conti, U. Ferner, A. Giorgetti, M.Z. Win, Ranging with ultrawide bandwidth signals in multipath environments. Proc. IEEE **97**(2), 404–426 (2009)

A.G. Dempster, Dilution of precision in angle-of-arrival positioning systems. Electron. Lett. **42**(5), 291–292 (2006)

G. Durgin, T.S. Rappaport, H. Xu, Measurements and models for radio pathloss and penetration loss in and around homes and trees at 5.85 GHz. IEEE Trans. Commun. **46**(11), 1484–1496 (1998)

G. Janssen, J. Vriens, High resolution coherent radiochannel measurements using direct sequence spread spectrum modulation. in *Proceedings of 6th Mediterranean IEEE Electrotechnical Conference,* vol 1, 1991, pp. 720–727

S. Gezici, Z. Tian, G.B. Giannakis, H. Kobayashi, A.F. Molisch, H.V. Poor, Z. Sahinoglu, Localization via ultra-wideband radios. IEEE Signal Process. Mag. (Special Issue on Signal Processing for Positioning and Navigation with Applications to Communications) **22**(4), 70–84 (2005)

S.S. Ghassemzadeh, R. Jana, C.W. Rice, W. Turin, V. Tarokh, Measurement and modeling of an ultra-wide bandwidth indoor channel. IEEE Trans. Commun. **52**(10), 1786–1796 (2004)

I. Guvenc, C.-C Chong, A survey on TOA based wireless localization and NLOS mitigation techniques. IEEE Commun. Surv. Tutor. **11**(3), 3rd Quarter (2009)

I. Guvenc, Y.T. Chan, H.Y.C. Hang, P.C. Ching, Exact and approximate maximum likelihood localization algorithms. IEEE Trans. Veh. Technol. **55**(1), 10–16 (2006)

H. Liu, H. Darabi, P. Banerjee, J. Liu, Survey of wireless indoor positioning techniques and systems. IEEE Trans. Sys. Man Cybern. Part C Appl Rev **37**(6), 1067–1080 (2007)

S.J. Howard, K. Pahlavan, Measurement and analysis of the indoor radio channel in the frequency domain. IEEE Trans. Instrum. Meas. **39**(5), 751–755 (1990)

I. Guvenc, Z. Sahinoglu, Threshold-based TOA estimation for impulse radio UWB systems. in *Proceedings of International Conference on Ultra-Wideband,* (2005)

J. Zheng, Y.-C Wu, Joint time synchronization and localization of an unknown node in wireless sensor networks. IEEE Trans. Signal Proces. **58**(3), 1309–1320 (2010)

J.Y. Lee, R.A. Scholtz, Ranging in a dense multipath environment using an UWB radio link. IEEE Trans. Select. Areas Commun. **20**(9), 1677–1683 (2002)

S.M. Kay, *Fundementals of Statistical Signal Processing: Estimation Theory* (Prentice Hall, Upper Saddle River, 1993)

X. Li, RSS-based location estimation with unknown pathloss model. IEEE Trans. Wirel. Commun. **5**(12), 3626–3633 (2006)

A.F. Molisch, Ultrawideband propagation channel-theory, measurement and modelling. IEEE Trans. Veh. Technol. **54**(5), 1528–1545 (2005)

K. Pahlavan, P. Krishnamurthy, J. Beneat, Wideband radio channel modeling for indoor geolocation applications. IEEE Commun. Mag. **36**(4), 60–65 (1998)

K. Pahlavan, A. Levesque, *Wireless Information Networks,* 2nd edn. Wiley (2005)

N. Patwari, A.O. Hero, M. Perkins, N.S. Correal, R.J. O'Dea, Relative location estimation in wireless sensor network. IEEE Trans. Signal Process. **51**(8), 2137–2148 (2003)

Y. Qi, H. Kobayashi, H. Suda, Analysis of wireless geolocation in a non-line-of-sight environment. IEEE Trans. Wirel. Commun. **5**(3), 672–681 (2006)

A.H. Sayed, A. Tarighat, N. Khajehnouri, Network-based wireless location: challenges faced in developing techniques for accurate wireless location information. IEEE Signal Process. Mag. **22**(4), 24–40 (2005)

K. Siwiak, H. Bertoni, S.M. Yano, Relation between multipath and wave propagation attenuation. IEE Electron. Lett. **39**(1), 142–143 (2003)

M.A. Spirito, On the accuracy of cellular mobile station location estimation. IEEE Trans. Veh.r Tech. **50**(3), 674–685 (2001)

B. Sundararaman, U. Buy, A. Kshemkalyani, Clock synchronization for wireless sensor networks: a survey. Ad Hoc Netw. (Elsevier) **3**(3), 281–323 (2005)

T.S. Rappaport, *Wireless Communications: Principles and Practice* (Prentice-Hall 1996)

IEEE 802.15.TG4a official web page, http://www.ieee802.org/15/pub/TG4a.html

Y. Qi, H. Kobayashi, On relation among time delay and signal strength based geolocation methods. in *Proceedings of IEEE Global Telecommunications Conference (GLOBECOM03)*, vol 7, (San Francisco, CA, 2003), pp. 40794083

Y. Shen, M.Z. Win, Fundamental limits of wideband localization accuracy via Fisher Information. in *Proceedings of IEEE Wireless Communications and Networking Conference (WCNC)*, pp. 3046–3051, March 2007

Y.-C Wu, Q. Chaudhari, E. Serpedin, Clock synchronization of wireless sensor networks. IEEE Signal Proces. Mag., **28**(1), 124–138 (2011)

Chapter 3
Multipath and NLOS Mitigation Algorithms

In Chap. 2, the multipath and NLOS problems were introduced and the degrading impact on distance estimation was highlighted through channel measurements and modeling. In this chapter, we will first introduce popular multipath mitigation techniques and then highlight the major approaches to dealing with the NLOS problem. For the multipath problem two mitigation techniques will be introduced: Super-resolution algorithms and Ultra Wideband (UWB) technology. The former is a spectral estimation technique that improves the TOA estimation through enhancing the time-domain resolution. The latter approach is an emerging technology that transmits very narrow pulses in time (very large bandwidths) and thus has the benefit of improved time-domain resolution which results in higher TOA estimation accuracy. The second part of the chapter is dedicated to NLOS identification and mitigation algorithms. An important pre-requisite to NLOS mitigation is channel identification. The effectiveness of the mitigation algorithms will rely mainly on the accuracy of NLOS channel identification. Thus, we will first introduce popular approaches to NLOS identification and then conclude the chapter with NLOS mitigation algorithms.

3.1 Multipath Mitigation

In Chap. 2, the degrading impact of multipath fading was introduced and illustrated through channel measurements and modeling. In order to improve ranging and localization accuracy in harsh multipath environments, as we shall see here, it is necessary to apply multipath mitigation techniques. Multipath for TOA-based ranging is a more pressing problem compared to RSS-based ranging. This is mainly because narrowband RSS-based systems suffer from multipath fast fading but a simple, yet effective, time-averaging can be performed as practical mitigation technique. Also for RSS-based ranging, there are more pressing issues than multipath such as the estimation of the path loss exponent and the shadow fading problem that makes the distance-power relationship difficult to know for different

C. Gentile et al., *Geolocation Techniques*, DOI: 10.1007/978-1-4614-1836-8_3,

environments. As a result the focus of this section will be on multipath mitigation for TOA-based systems.

To appreciate the magnitude of the multipath problem for TOA-based ranging, consider geolocation systems operating with low bandwidths. The time resolution of TOA systems is roughly inversely proportional to the system bandwidth. That is the ability to resolve two successive multipath components arriving after each other is dictated by the system bandwidth. For example, the bandwidth of GSM signals is 200 kHz, which translates to 5 μs or equivalently to a resolution of 1,500 m! This means that two paths arriving less than 1,500 m apart will not be resolved. Multipath also affects next generation wideband systems such as Digital TV (DVB) signals, UMTS 3G/4G, WIMAX and WiFi. What is common among those systems is that the operational bandwidth is suitable for communication, but not for accurate ranging/localization. For example, system bandwidth can vary between 5 and 20 MHz (UMTS/WiMAX); however, even the highest bandwidth of 20 MHz equates to only $\sim$15 m of time resolution. This resolution, unfortunately, is not suitable for dense multipath environments (such as indoors). Thus, for many of the existing systems it is necessary to investigate multipath mitigation techniques.

In this subsection, we will introduce two popular approaches to mitigate the impact of multipath fading on TOA estimation in cluttered multipath environments. The first is known as super-resolution, which is a spectral estimation technique that can deliver higher time resolution to compensate for low-bandwidth systems. The algorithm can be easily integrated with wideband systems such as DVB, 3G, and WiFi. The second approach is based on deploying Ultra-Wideband (UWB) localization systems. UWB is an emerging technology that utilizes very large system bandwidths and has the potential for high data rate communications (in the Gigabit range) and centimeter level TOA estimation accuracies (in LOS). Thus the latter approach is more of an alternative system that can be deployed alongside the existing communication systems, while the former is an algorithm that can be integrated with existing wireless systems. As we shall see in the following sections UWB can have a much better TOA estimation capabilities mainly because of the very large bandwidth (high time-resolution).

3.1.1 Super-Resolution Technique: MUSIC Algorithm

Estimating the time-domain delays of a multipath signal/channel is essentially a spectral estimation problem that can be applied in either the time domain or the frequency domain. The super-resolution algorithm takes advantage of the underlying multipath model to solve this problem. The model as was first introduced in (2.42) is described as a train of multipath arrivals, each with discrete delays and varying amplitudes, such as the tapped-delay line model for the indoor environment (Hashemi 1993). The low-pass impulse response of the multipath channel is given by

$$h(t) = \sum_{k=0}^{L_p-1} \alpha_k e^{j\phi_k} \delta(t - \tau_k) \tag{3.1}$$

which was introduced in (2.42) but presented here for convenience. The Fourier transform of (3.1) is the frequency domain channel impulse response which is given by

$$H(f) = \sum_{k=0}^{L_p-1} \beta_k e^{-j2\pi f \tau_k} \tag{3.2}$$

where $\beta_k = \alpha_k e^{j\phi_k}$. A harmonic signal model can be created by exchanging the role of time and frequency variables in (3.2) which yields,

$$H(\tau) = \sum_{k=0}^{L_p-1} \beta_k e^{-j2\pi f_k \tau} \tag{3.3}$$

This model is well-known in the spectral estimation field (Manolakis et al. 2000). As so, spectral estimation techniques that are suitable for a harmonic model can be applied to the frequency response of the indoor radio channel such that time-domain analysis can be performed. A popular high resolution parametric spectral estimation technique is the multiple signal classification (MUSIC), which was originally proposed by Schmidt in 1977 in the context of sensor arrays. MUSIC can be used to accurately estimate the time delays by converting the channel frequency response to a channel impulse response in the time domain. In essence, the spectral estimation technique estimates the spectral components (the time-delays) in the time domain. Once the multipath delays are estimated, the TOA of the direct path (DP) can be estimated. The discrete measurement data is obtained by sampling the channel frequency response $H(f)$ from (3.3) at L frequencies equally spaced by Δ. Considering additive white noise in the measurement, the sampled discrete frequency domain channel response is given by

$$x(l) = H(f_l) + w(l) = \sum_{k=0}^{L_p-1} \beta_k e^{-j2\pi(f_0 + l\Delta f)\tau_k} + w(l) \tag{3.4}$$

where $l = 0, 1, \ldots, L-1$ and $w(l)$ denotes additive white measurement noise with zero mean and variance σ_w^2.

The signal model can be rewritten in vector form as

$$\mathbf{x} = \mathbf{H} + \mathbf{w} = \mathbf{V}\beta + \mathbf{w}, \tag{3.5}$$

By defining $\mathbf{V} = [\mathbf{v}(\tau_0) \quad \mathbf{v}(\tau_1) \quad \ldots \quad \mathbf{v}(\tau_{L_p-1})]^T$ and $\mathbf{v} = [1 \quad e^{-j2\pi\Delta f \tau_k} \quad \ldots \quad e^{-j2\pi(L-1)\Delta f \tau_k}]^T$. The MUSIC super-resolution algorithm is based on the eigendecomposition of the autocorrelation matrix of the signal model in (3.5). The autocorrelation matrix is given by

$$\mathbf{R}_{xx} = E\{\mathbf{x}\mathbf{x}^H\} = \mathbf{V}\mathbf{B}\mathbf{V}^H + \sigma_w^2\mathbf{I} \tag{3.6}$$

where $\mathbf{B} = E\{\beta\beta^H\}$ and superscript H is the Hermitian, conjugate transpose, of a matrix. Therefore, the L-dimensional subspace that contains the measurement vector $\mathbf{x}$ is split into two orthogonal subspaces, known as signal subspace and noise subspace. The spaces are spanned respectively by the signal eigenvectors (EVs) and noise EVs. From (3.5), we know that the signal vector $\mathbf{v}(\tau_k)$, $0 \leq k \leq L_p - 1$ by definition lies in the signal subspace; hence it must be orthogonal to the noise subspace. This implies that

$$\mathbf{P}_w\mathbf{v}(\tau_k) = 0 \tag{3.7}$$

where $\mathbf{P}_w\mathbf{v}(\tau_k)$ is the projection matrix of the noise subspace. Thus the multipath delays τ_k, $0 \leq k \leq L_p - 1$, can be determined by finding the delay values, at which the condition in (3.7) is met, or equivalently the following MUSIC pseudospectrum achieves peak values:

$$S_{\text{MUSIC}}(\tau) = \frac{1}{\|\mathbf{P}_w\mathbf{v}(\tau)\|^2} = \frac{1}{\sum_{k=L_p}^{L-1} |\mathbf{q}_k\mathbf{v}(\tau)|^2} \tag{3.8}$$

where $\mathbf{q}_\mathbf{k}$ are the noise EVs. In practical implementations, when only one snapshot of length N is available, the data sequence is divided into M consecutive segments of length L. The estimate of the correlation matrix can be further improved by using the *forward–backward correlation matrix* which serves to decorrelate the signals—more details are described in Li and Pahlavan (2004). In theory, the decomposition results in signal eigenvalues and noise eigenvalues (corresponding to the signal and noise subspaces, respectively) for which the noise eigenvalues are all equal to the variance of the noise. In practice, however, this is most often not the case. As a result, a slight variation on the MUSIC algorithm is used, which is the EV method. The pseudospectrum is defined as

$$S_{\text{EV}}(\tau) = \frac{1}{\sum_{k=L_p}^{L-1} \frac{1}{\lambda_k} |\mathbf{q}_w^H\mathbf{v}(\iota)|^2} \tag{3.9}$$

where λ_k $L_p \leq k \leq L - 1$ are the noise eigenvalues. Effectively, the pseudospectrum of each EV is normalized by its corresponding eigenvalue, giving a greater weight to the smaller eigenvalues. Ideally, the signal EVs are associated with the L_p largest eigenvalues and the $L - L_p$ are associated with the smallest eigenvalues. In practice, however, there may be overlap between the two sets, in particular when the noise variance is high. So in the linear combination containing the noise EVs in (3.9), the EVs which are associated with the smaller eigenvalues are given a greater weight. This is because the smaller the eigenvalue, the greater the confidence that the associated EV is indeed a noise EV and not mistakenly a signal EV. The performance of the EV method is less sensitive to inaccurate estimate of the

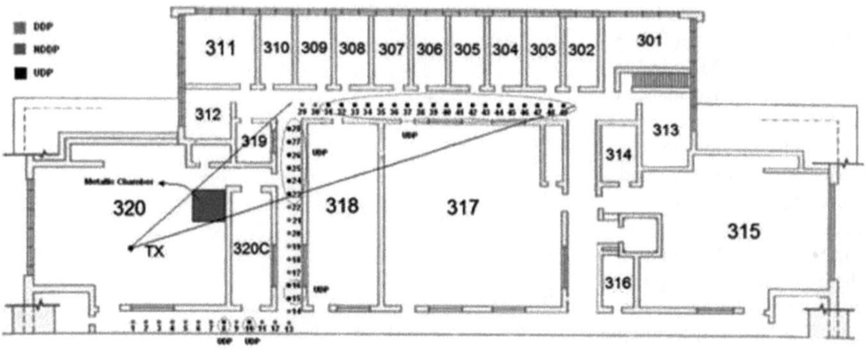

Fig. 3.1 Measurement locations at 3rd floor of Atwater Kent building in WPI. Measurements were used to test the effectiveness of super-resolution algorithm (MUSIC) in mitigating multipath-induced ranging errors

parameter L_p, which is highly desirable in practical implementation (Manolakis et al. 2000).

As a practical illustration to the performance of super-resolution algorithm, MUSIC algorithm (using the EV with FBCM) was tested on measurement data collected in a typical indoor office environment. The measurement campaign was conducted on the 3rd floor of Atwater Kent building at Worcester Polytechnic Institute where the locations of the measurements are provided in Fig. 3.1 for illustration.

Different measurement scenarios and system bandwidth were considered. Specifically, LOS/NLOS channels were measured and the system bandwidth upto 200 MHz was used. Figure 3.2 highlights a typical super-resolution estimation result in LOS channels at 40 MHz bandwidth. Note that the MUSIC algorithm has the capability to accurately resolve and mitigate the multipath and enhance the TOA estimation accuracy. To compare the performance, the Inverse Fourier Transform (IFT) and Direct Sequence Spread Spectrum (DSSS) techniques are. Basically for the IFT technique the measured data in the frequency domain is transformed into time domain without any additional signal processing. This can be considered as the most basic channel estimation. The DSSS technique is simulated by convolving the measured data with a raised cosine filter prior to the IFT operation. This effectively emulates the cross-correlation method using DSSS signals.

The multipath mitigation capabilities of MUSIC can be further highlighted by examining measurement results for NLOS channels. Recall from Chap. 2 that in NLOS propagation two conditions might arise. The first is when the DP signal is detected and the other when it is undetected. Figures 3.3 and 3.4 provide a clear indication of the advantage of using super-resolution algorithms to mitigate multipath error in low-bandwidth systems.

Super-resolution algorithms can resolve multipath components whose signal-to-noise ratio is greater than one (i.e., the signal eigenvalues are greater than the noise eigenvalues). In the extreme cases of NLOS-NDP, super-resolution algorithms do

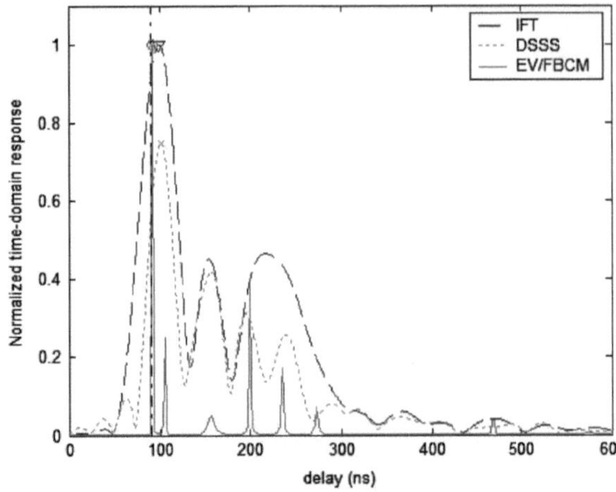

Fig. 3.2 Super-resolution MUSIC algorithm multipath mitigation at 40 MHz in LOS channels. The performance of MUSIC algorithm is compared with inverse fourier transform (*IFT*) and direct sequence spread spectrum (*DSSS*). The former is the simple IFT technique where no additional signal processing techniques were used. The later is a technique that uses the traditional cross-correlation method with DSSS signals

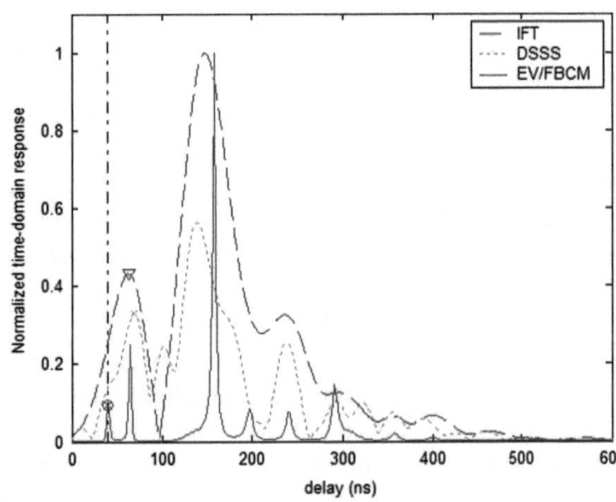

Fig. 3.3 Super-resolution MUSIC algorithm multipath mitigation at 40 MHz in NLOS-DP channels. Note that the DP is only detected by MUSIC while it is unresolvable by traditional techniques

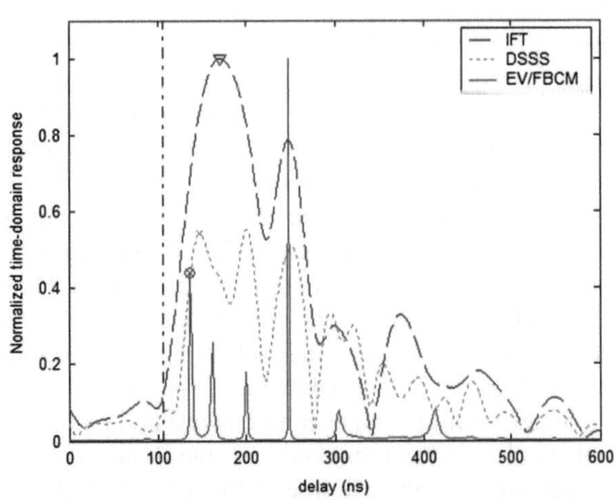

Fig. 3.4 Super-resolution MUSIC algorithm multipath mitigation at 40 MHz in NLOS-NDP channels. Note that the DP is not detected by all the algorithms. MUSIC is superior in resolving/ mitigating multipath but cannot "revive" a lost path

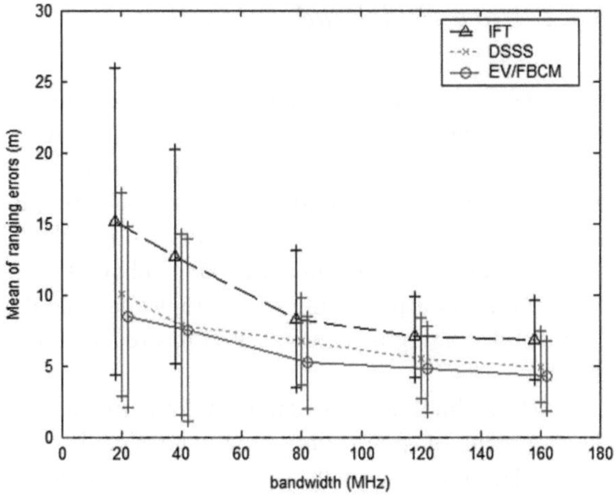

Fig. 3.5 Mean and STD of ranging errors for NLOS-NDP channel condition using different TOA estimation algorithms. Note that the Super-resolution TOA estimation error is around ~4–5 m—a significant value which is due to the absence of the direct path

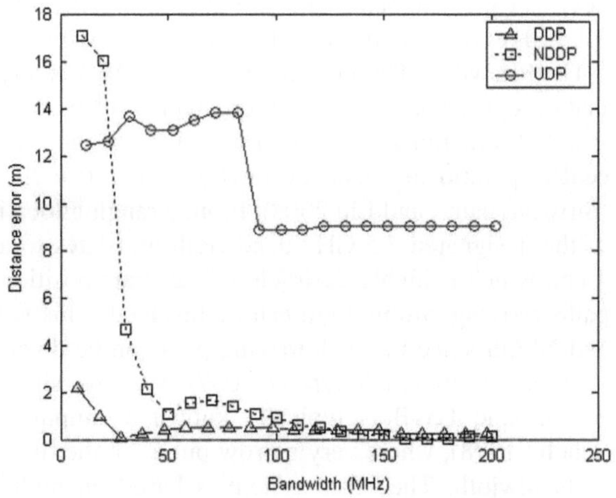

Fig. 3.6 Measured TOA estimation error (distance estimation error) versus different system bandwidths for the system under different multipath conditions (LOS/DDP, NLOS-DP/NDDP, NLOS-NDP/UDP)

not have the capability to recover a "lost" path. In other words, if the DP path is so severely attenuated and it is buried under the receiver noise, then it will not be possible for super-resolution algorithms to recover that path. Super-resolution algorithms can, however, mitigate multipath error in those scenarios by enhancing the estimation of the first arrival path. Figure 3.4 and 3.5 highlight the performance under an extreme NLOS-NDP condition. Note that MUSIC (EV/FBCM) algorithm has a marginal improvement compared to DSSS technique. This is due to the fact that in such environments the DP is not available and thus multipath mitigation can only improve the detection of the first arrival paths.

Figure 3.6 illustrates the behavior of distance estimation with bandwidth under the three channel condition. Note that MUSIC's ability to mitigate multipath depends on whether the DP is detected or not. In the latter case significant ranging error cannot be alleviated with system bandwidth and/or super-resolution algorithms.

3.1.2 Ultra Wideband Technology

It has been established in the previous sections that multipath fading presents a serious challenge to accurate ranging and localization in rich multipath environments. For band-limited systems (narrowband/wideband), multipath can be mitigated through the use of advanced spectral estimation techniques such as MUSIC and other super-resolution algorithms. Although effective for different scenarios, super-resolution algorithms still have inherent limitations due to limited system bandwidth. As a promising technology that has the potential for delivering sub-centimeter ranging accuracy, UWB has received considerable attention in the past decade.

UWB technology first came to light in the 1960s when it was used for the development of short-pulse radar and stealth communication systems (Hussain 1998). The recent growth of UWB is attributed to the Federal Communications Commission (FCC) ruling in 2003 that introduced an unlicensed band for UWB devices (Siriwongpairat and Liu 2008). Specifically, they have been assigned the $3.1-10.6$ GHz spectrum for legal operation with a power spectral density limit of -41 dBm/MHz. The FCC definition of UWB is any transmission system/scheme that occupies a bandwidth of more than 500 MHz or a fractional bandwidth greater than 0.2. Fractional bandwidth is defined as the system bandwidth divided by the center operational frequency, or B/f_c, where $B = f_H - f_L$ is the -10 dB bandwidth (Siriwongpairat and Liu 2008). From a ranging/localization perspective, full usage of the designated 7.5 GHz bandwidth translates to a time-of-arrival resolution of 4 cm, which is highly desirable for accurate positioning. This means that any two paths arriving within 4 cm can be resolved. This is highly desirable in both LOS and NLOS since the first arriving path can be detected with great accuracy.

There are two main types of UWB systems: single-band and multi-band UWB. Single-band UWB is typically known as impulse radio (IR) UWB (Win and Scholtz 1998), where very narrow pulses in the time-domain results in GHz range of bandwidth. The other system is based on multi-band Orthogonal Frequency Division Multiplexing (MB-OFDM) which has been the main proponent for high-data rate and accurate localization. In the sequel, we introduce both system implementations.

3.1.2.1 Impulse Radio Ultra Wideband

Impulse Radio Ultra Wideband (IR-UWB) is the traditional implementation that uses a single-band approach in which the transmitted signal does not employ any carrier (it is also known as carrier-free communications) (Siriwongpairat and Liu 2008). The time-domain pulses usually have duration on the order of nanoseconds and the waveform shapes are typically Gaussian, Laplacian, Hermitian, and Rayleigh (Fontana 2004). The basic Gaussian pulse is the most popular for UWB systems and it is given by Sheng et al. (2003)

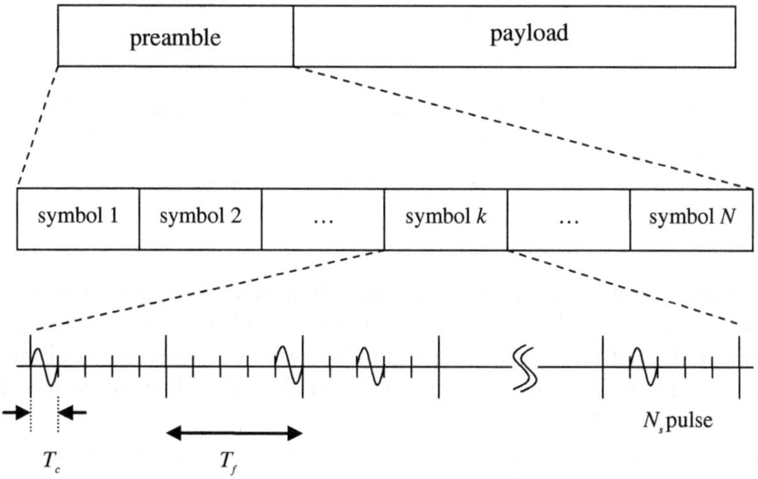

Fig. 3.7 IR-UWB preamble structure used for TOA estimation. The narrow UWB pulses within a symbol occupy a time slot according to a user-defined pseudorandom TH sequence

$$p_G(t) = \exp\left(\frac{-2\pi t^2}{\sigma_p^2}\right) \tag{3.10}$$

Alternatively, the *n*th derivative of the basic Gaussian pulse can also be adopted for UWB systems and it is given by

$$p(t) = p_G^n(t)\sqrt{\frac{(n-1)!}{(2n-1)!\pi^n\sigma_p^{(1-2n)}}}, \quad n > 0 \tag{3.11}$$

where $p_G^n(t)$ is the nth order derivative of $p^n(t)$.

In system implementations, the narrow Gaussian pulses, each with duration T_p, are transmitted in the preamble of a data packet–packet communications is most common for multiple access architecture. The preamble is subdivided into N symbols and each symbol is further subdivided into N_f frames, each of duration T_f. Within each frame the UWB pulse can occupy one of the N_c time slots/chips (of duration T_c) according to a user-specific pseudorandom TH sequence $\{c_k^u\}$ (Win and Scholtz 2000). Figure 3.7 illustrates the transmitted preamble concept.

Typical approaches for IR-UWB TOA estimation include the Stored Reference (SR) and the Energy Detection (ED). SR technique is based on correlating the received signal with a reference template and integrating the results to estimate the first arrival path (Guvenc et al. 2006). Cross correlation techniques are similarly used in DSSS systems to extract the multipath arrivals through RAKE receiver architecture. On the other hand, the ED technique is based on the concept of detecting the energy of the first arrival path. This is achieved through squaring the incoming signal (by a square-law device) and then integrating and sampling it.

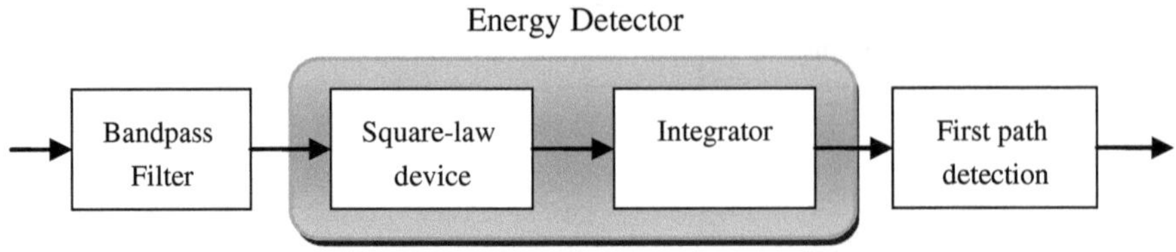

Fig. 3.8 ED TOA estimation in IR-UWB systems

The ED has a simpler architecture, but the SR technique is more robust to noise since a noise-free template is used for correlation with incoming received signals (Guvenc et al. 2006). For details regarding the system architecture of different IR-UWB based TOA estimation refer to Guvenc et al. (2006); Lee and Scholtz (2002) and Stoica et al. (2006).

One way to estimate the TOA is to implement Maximum Likelihood (ML) estimators, but they are not practical due to the high sampling rate required for systems with such large bandwidths. Instead, a popular and practical technique to estimate the TOA of the first arriving path is to process the preamble signal using ED (Dardari et al. 2009). Figure 3.8 illustrates the system diagram of ED-based TOA estimator.

The filtered signal can be expressed as

$$r(t) = s(t) + w(t) \tag{3.12}$$

where the transmitted signal is $s(t)$ and $w(t)$ is the AWGN. The transmitted signal is given by Dardari et al. (2009)

$$s(t) = \sum_{n=0}^{Nt-1} x\left(t - c_n T_c - nT_f\right) \tag{3.13}$$

where

$$x(t) = \sqrt{\frac{E_s}{N_s}} \sum_{k-1}^{L_p} \alpha_k p(t - \tau_k) \tag{3.14}$$

The output of the ED detector is then samples that contain the multipath arrival information and thus estimating the TOA of the first path involves searching for the first arrival that is above the noise threshold.

Several techniques have been proposed in literature and they vary according to the method in which they search for the TOA of the first path. The following is a brief summary of the techniques and the interested reader can find more details in Dardari et al. (2009) and the references therein.

1. *Max Technique*: The simplest TOA search is the *Max* technique where the strongest sample is chosen as the first arrival signal. This approach works in LOS environments but fails in NLOS environments where the strongest signal is not always the first arriving signal.

2. *P-Max Technique*: The second approach is the *P-Max* technique and it is based on the concept that the *P* largest arrivals are stored and then the earliest path arrival is chosen as the TOA estimate. Naturally the performance of this technique depends on the parameter *P*.
3. *Simple Thresholding*: An alternative technique is called the *Simple Thresholding* where for a given threshold; the first arrival sample that crosses the threshold is taken to be the TOA of the first path. The design of the threshold is the main factor that will affect the performance and that depends on the operating conditions and the channel statistics.
4. *Jump Back and Search Forward Technique*: When the receiver is synchronized to the strongest path it is possible to implement the *Jump Back and Search Forward* technique which is based on the detection of the strongest sample and then carry out a forward search algorithm. Note that a forward search in the TOA context means increasing delay and a backward search implies decrease in delay.
5. *Serial Backward Search Technique*: This technique detects the strongest path and then performs a backward search to find the TOA of the first path. The search ends when the tested sample falls below a certain threshold.
6. *Serial Backward Search for Multiple Clusters*: Finally this algorithm similarly detects the strongest sample and then performs a backward search. However the difference here is in the termination point; this algorithm acknowledges that for UWB channels, multipath signals arrive in clusters and thus the backward search should not end when a sample falls under the noise threshold; instead, the search continues backward until subsequent samples are below the threshold indicating that there are no further multipath clusters and thus the sample is the TOA of the first arrival path.

3.1.2.2 Multi-Band Orthogonal Frequency Division Multiplexing

The other popular UWB technique is based on MB-OFDM. Instead of sending very narrow pulses in the time domain (very large signal bandwidth), MB-OFDM divides the UWB bandwidth into subbands, each in excess of 500 MHz, and employs the OFDM modulation technique in each subband (Ghavami et al. 2007). The advantage of the multi-band approach is that parallel transmission over each subband eliminates the pressing requirements of transmitting very large bandwidth signals. Thus information is processed over a much smaller bandwidth, which reduces the overall system design complexity and improves spectral flexibility (Siriwongpairat and Liu 2008). The use of OFDM technique ensures high-data rates combined with spectral efficiency. MB-OFDM is a leading proposal for the IEEE 802.15.3a standard. One frequency allocation scheme proposed in the standard is illustrated in Fig. 3.9, where the 7.5 GHz UWB band is subdivided into 14 bands of 528 MHz each.

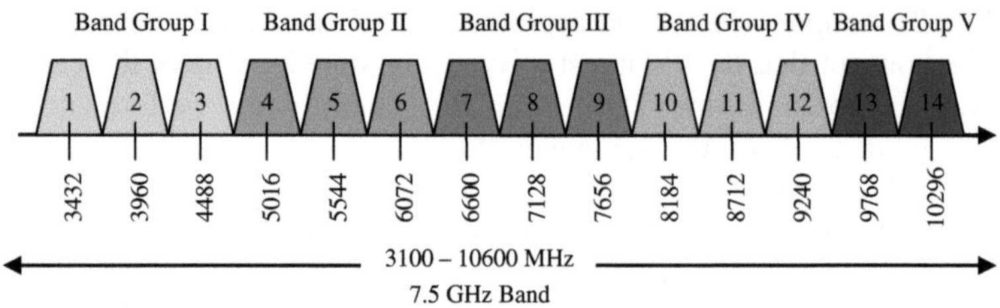

Fig. 3.9 WiMedia specification for UWB spectrum allocation. The 7.5 GHz band is sub-divided into 14 bands of 528 MHz bandwidth each

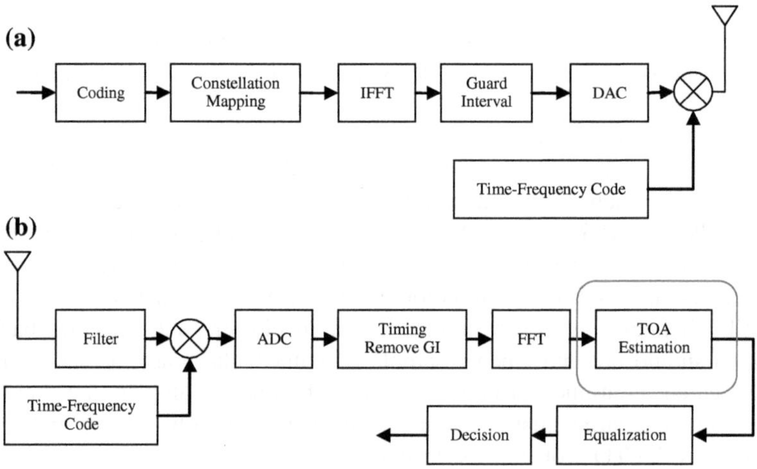

Fig. 3.10 MB-OFDM Transciever architecture. (**a**) Transmitter (**b**) receiver. Note that the TOA estimation typically occurs at the receiver after the FFT operation

Figure 3.10 illustrates a block diagram of a MB-OFDM transceiver architecture highlighting where TOA estimation typically occurs (Xu et al. 2008).

In MB-OFDM, the TOA estimation is typically achieved by processing the estimated frequency domain received signal (which is the output of the FFT block in Fig. 3.10). For a given subband, the frequency domain received signal $R(n)$ on the nth subcarrier can be give by Xu et al. (2008)

$$R(n) = H(n)S(n) + W(n), \quad \forall n \in [1, N] \tag{3.15}$$

where N is the number of sub-carriers in a subband, $S(n)$ is the transmitted pilot signal, $W(n)$ is the AWGN and $H(n)$ is the channel frequency response coefficient that can be described as

$$H(n) = \sum_{k=1}^{L} h_k \exp(-j2\pi f_n \tau_k) \tag{3.16}$$

where f_n is the nth subcarrier, h_k and τ_k are the amplitude and delay of the kth path. The least-squares (LS) channel estimate is then

$$Y(n) = R(n)S^*(n) = H(n) + W(n)S^*(n) \tag{3.17}$$

assuming that $S^*(n)S(n) = 1$.

From the Least Squares formulation, the TOA estimate (or the channel estimate in time domain) can be achieved through different approaches. One approach is to estimate the CIR using the space-alternating generalized EM (SAGE) algorithm (Fleury et al. 1999). EM is a well-known Expectation and Maximization algorithm for the ML estimation. The problem with LS estimation is that it will cause the energy leakage problem that arises when the impulse response is mis-sampled (the sampling interval does not fall on the location of the time of arrival of a path). Mis-sampling the impulse response causes the energy of one channel path to disperse to all the other taps in the recovered channel estimate which increases the TOA estimation error. An alternative technique that suppresses leakage of multipath components due to sampling was proposed in Xu et al. (2008) where the CIR is first recovered using a simple tap-spaced model given by

$$\bar{h}(t) = \sum_{k=1}^{L_M} \bar{h}_k \delta(t - \bar{\tau}_k) \tag{3.18}$$

where $\bar{\tau}_k = (k-1)T_p + \bar{\tau}_1$ and $T_p = T_h/L_M$ is the tap interval (inverse of the bandwidth), T_h is the multipath channel length and L_M is the number of taps. The taps are equally spaced and distributed in $[\bar{\tau}_1, \bar{\tau}_1 + T_h]$. $\bar{h}_k$ is the amplitude value of the CIR at each tap (bin). The tap-spaced model divides the delay into taps (bins) of equal length that is related to the time-resolution which is inverse of the system bandwidth, B. The CIR can then be estimated using the frequency domain observations on the nth subcarrier or

$$Y(n) = \sum_{k=1}^{L_M} \bar{h}_k \exp(-j2\pi f_n \bar{\tau}_k) \tag{3.19}$$

For all the subcarriers (3.18) can be re-written in matrix form or

$$\mathbf{y} = \mathbf{Fh} \tag{3.20}$$

where $\bar{\mathbf{h}} = [\bar{h}_1, \bar{h}_2, \ldots, \bar{h}_{L_M}]^T$, $\mathbf{F}$ is an $N \times L_M$ Fourier transform matrix with elements $\exp(-j2\pi f_n \bar{\tau}_k)$. The LS estimate can then be given as (Xu et al. 2008)

$$\bar{\mathbf{h}} = \left(\mathbf{F}^H\mathbf{F}\right)^{-1}\mathbf{F}^H\mathbf{y} \tag{3.21}$$

The TOA is then obtained as the estimated first arrival path from (3.20).

In general for MB-OFDM systems there are two major approaches to channel and TOA estimation. The first is to estimate TOA in individual subbands and then combine/average the TOA estimates across all subbands (Berger et al. 2006;

Dabak et al. 2005). This approach is typically non-coherent in nature. The alternative, a coherent approach, is to concatenate all the subbands in the frequency domain and then perform the LS estimation procedure outlined above. The performance results in Xu et al. (2008); Saberinia and Tewfik (2004) verified the expected higher resolution which is due to the fact that all the 7.5 GHz UWB bandwidth imply very high time-domain resolution compared to 528 MHz. In the next sub-section the mitigation capabilities of UWB will be investigated through experimental measurements of the CIR in typical indoor environments. The focus will be on the effectiveness of multipath mitigation from a channel propagation point of view.

3.1.2.3 Evaluation of UWB Multipath Mitigation

The effectiveness of UWB systems in resolving multipath arrivals can be evaluated by examining results of channel measurements versus different bandwidths. Using the frequency domain measurement system described in Chap. 2, it is possible to measure the entire FCC-allocated UWB band. The impact of system bandwidth on TOA estimation can be subsequently examined by selecting different system bandwidths prior to taking the IFT. The relationship between the multipath error and system bandwidth of UWB was experimentally illustrated by Alavi and Pahlavan (2006). In LOS environments, for 20 MHz bandwidth an RMSE of 10 m was observed. As the bandwidth increases by an order of magnitude, the error drops to 2 m. At 2 GHz the RMSE is less than 0.1 m. It is therefore possible to see why the two OFDM approaches have gained considerable attention for accurate TOA-based ranging in harsh multipath environments. In NLOS environments, the RMSE error does not follow the improvements in LOS. This observation further emphasizes the need for effective NLOS identification and mitigation algorithms since bandwidth alone cannot solve this specific problem.

3.2 NLOS Identification and Mitigation

Chapter 2 has provided sufficient motivation to appreciate the NLOS problem that localization systems face in harsh multipath environments. Although the indoor environment has been used as an example, the problem applies to any propagation environment where there is high probability of NLOS. Naturally, in order to enable effective and accurate localization in such environments, it is necessary to deal with the NLOS problem since it causes bias in range estimates. A popular research area that has grown significantly during the last decade is NLOS identification and mitigation. NLOS identification techniques are based on estimating or identifying the condition of the channel to infer whether it is LOS or NLOS. Once the "channel" information is available, it is possible to incorporate it into a NLOS mitigation algorithm to obtain a better location estimate. Figure 3.11 illustrates an example of the effectiveness of NLOS identification and mitigation techniques.

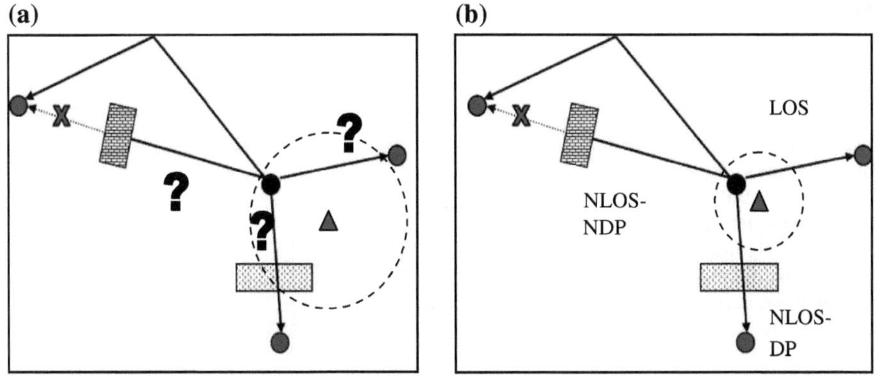

Fig. 3.11 The impact of NLOS identification and mitigation on localization accuracy. **a** Traditional "blind" approaches assume all range measurements as LOS. **b** Incorporating NLOS identification and mitigation can reduce the impact of bias and decrease the uncertainty of location estimation significantly. The black circle is the actual mobile terminal location. The triangle is the estimated location

NLOS identification typically operates on the physical-layer-sensed signal which can be used to extract a "metric" that can indicate the state of the channel. NLOS mitigation, however, operates at higher levels closer to the localization algorithm. As a result, identification and mitigation are typically independent but there are algorithms that mitigate NLOS and adopt an implicit NLOS identification as we shall see later. Figure 3.12 illustrates the general localization process with NLOS identification and mitigation.

An important clarification on the definition of the channel conditions is necessary before introducing the NLOS identification and mitigation techniques. LOS and NLOS have been used traditionally in communications terminology to describe the absence and presence of an obstruction between the transmitter and receiver, respectively. For localization, the terminology can be confusing, especially for TOA-based systems. In TOA-based systems, NLOS can still be used to describe the existence of an obstruction between the transmitter and receiver. However, the performance of TOA estimation algorithms can vary significantly for different NLOS channels. For example, in practice there are situations where a transmitter and receiver might be in NLOS but the channel "exhibits" LOS properties. This can happen, for example, when the transmitter and receiver are separated by light obstruction that attenuates the DP. Since the DP is detected, the range estimate will be very similar to that of a LOS channel (since the bias is negligible). In extreme NLOS cases, the DP path cannot be detected due to thicker obstructions which cause severe bias errors. Most of the NLOS identification and mitigation techniques in literature are based on the two channel conditions (LOS/NLOS) where the NLOS-DP is usually ignored (since it cannot be grouped with the LOS conditions). The three-channel condition classification is even more important when considering channel (CIR)-based NLOS identification algorithms

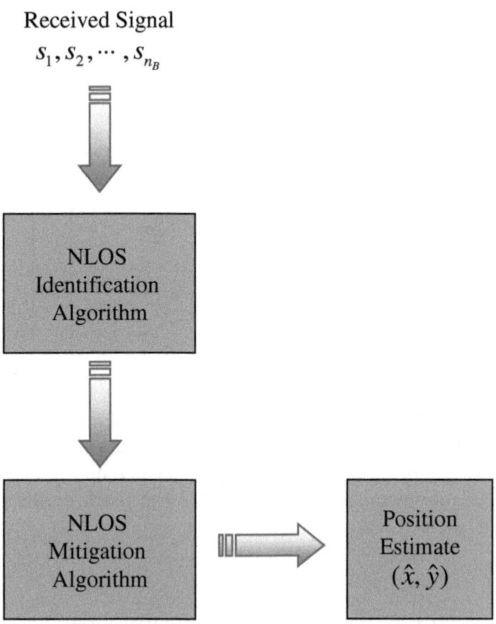

Fig. 3.12 In NLOS identification/mitigation enabled systems, the range measurements are first passed through a NLOS identification algorithm. Once the "bad links" are identified, that information is incorporated in a NLOS mitigation algorithm prior to position estimation, which can improve accuracy substantially

since the performance is directly related to the behavior of the metrics in the presence and absence of the DP, see Table 2.2.

3.2.1 NLOS Identification Techniques

The basic idea of NLOS identification is to infer the state of the channel by examining a certain metric of the received RF signal. For example by examining the received signal power, it is possible to identify the channel condition by evaluating the variance of the power with time. High power fluctuations generally indicate that the channel is in NLOS. However this "crude" approach is not optimal and more complex metrics can provide more robust performance. The existing NLOS identification techniques are generally divided into two main approaches: channel (CIR)–based and non-channel (CIR) based identification techniques. Figure 3.13 highlights the major techniques under each approach.

In non-CIR based techniques the identification is achieved *without* estimating the CIR. Instead, identification is achieved by either examining some characteristic of the received signal or by assessing the impact of NLOS on the position estimation (that usually combine identification and mitigation in one step). In CIR-based techniques, the channel is first estimated and a metric is devised to distinguish between the channel conditions. In either approach NLOS identification involves designing a hypothesis test which requires the availability of some a *priori* information about the statistics of the metrics that is used in the identification.

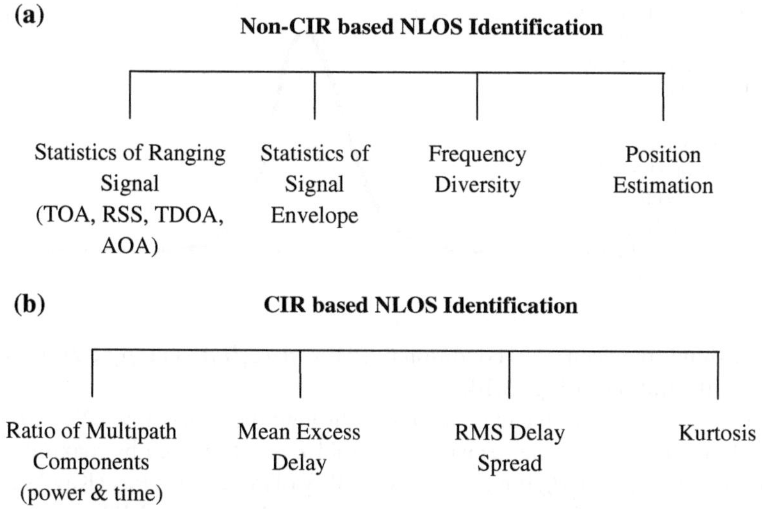

Fig. 3.13 Overview of NLOS identification literature. **a** Non-CIR based NLOS identification **b** CIR-based NLOS identification techniques

For a generic metric, ξ, the conditional PDFs under each hypothesis are used in the NLOS identification algorithms. The formulation of the hypothesis test will depend on the available data and also the assumptions of the channel. Thus, two main approaches are used: LOS/NLOS versus LOS/NLOS-DP/NLOS-NDP which is a binary versus multiple (ternary) hypothesis tests.

In the "traditional" NLOS identification approach a binary hypothesis test is used to distinguish between LOS (H_0) and NLOS (H_1) or

$$\begin{cases} H_0 : \hat{d} = d + b_m + w & \text{LOS} \\ H : \begin{cases} \hat{d} = d + b_m + b_{\text{pd}} + w & \text{NLOS} - \text{DP} \\ \hat{d} = d + b_m + b_{\text{pd}} + b_{\text{NDP}} + w & \text{NLOS} - \text{NDP} \end{cases} \end{cases} \tag{3.22}$$

Note that in almost all the literature on NLOS identification, the two conditions NLOS-DP and NLOS-NDP are usually either combined into one or the former is ignored (see Table 2.2 for description of the ranging conditions). This approach has limitations since this "black" or "white" view point is not an accurate reflection of reality and thus any identification algorithm devised with these assumptions will lack in detection accuracy and robustness. However, in many cases the a *priori* information of the three different conditions might not be available (such as existing IEEE channel models where only LOS/NLOS classification exists—mainly due to the fact that the models were developed with communication perspective without specific attention to the geolocation problem). In this case, it is possible to use the binary formulation in (3.21). A binary hypothesis test can be devised where we are particularly interested in

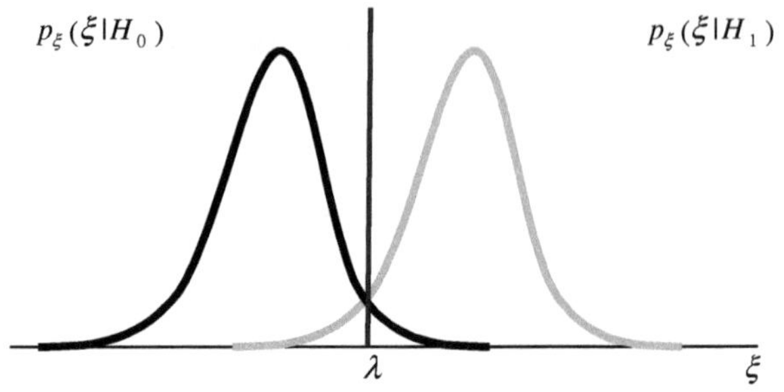

Fig. 3.14 Traditional NLOS identification based on a binary hypothesis approach. The null hypothesis is the LOS condition and the alternative hypothesis is the NLOS condition (both the –DP and –NDP sub-conditions are usually grouped under the alternative hypothesis

distinguishing between the conditional PDFs $p_\xi(\xi|H_0)$ and $p_\xi(\xi|H_1)$, see the example illustration in Fig. 3.14.

The optimum detection can be achieved through the well-known Neyman-Pearson (NP) theorem where the decision threshold λ is determined by maximizing the probability of detection P_D for a given probability of false alarm P_{FA} (Kay 1998; Van Trees 2001). As a result for a given P_{FA} a likelihood ratio test (LRT) is given by

$$L(\xi) = \frac{p_\xi(\xi|H_1)}{p_\xi(\xi|H_0)} \underset{H_0}{\overset{H_1}{\gtrless}} \lambda \tag{3.23}$$

where the threshold can be determined based on a certain P_{FA} given by

$$P_{\mathrm{FA}} = \int_{\lambda}^{\infty} p_\xi(\xi|H_0)d\xi \tag{3.24}$$

Similarly the achieved probability of detection P_D is given by

$$P_{\mathrm{D}} = \int_{\lambda}^{\infty} p_\xi(\xi|H_1)d\xi \tag{3.25}$$

Figure 3.15 illustrates the block diagram for a NLOS identification system under a binary hypothesis test approach.

If the a *priori* statistical characterization of the metric in the three channel conditions LOS, NLOS-DP, and NLOS-NDP is available then another approach is to construct a multiple (ternary in this case) hypothesis test or

$$\begin{cases} H_0 : \hat{d} = d + b_m + w & \text{LOS} \\ H_1 : \hat{d} = d + b_m + b_{\mathrm{pd}} + w & \text{NLOS} - \text{DP} \\ H_2 : \hat{d} = d + b_m + b_{\mathrm{pd}} + b_{\mathrm{NDP}} + w & \text{NLOS} - \text{NDP} \end{cases} \tag{3.26}$$

where the H_0, H_1 and H_2 represent LOS, NLOS-DP, and NLOS-NDP, respectively. Note that this classification provides a platform for a more robust identification since H_2 typically is the cause of significant ranging errors in TOA-based geolocation systems. For a multiple hypothesis problem a NP approach can be extended

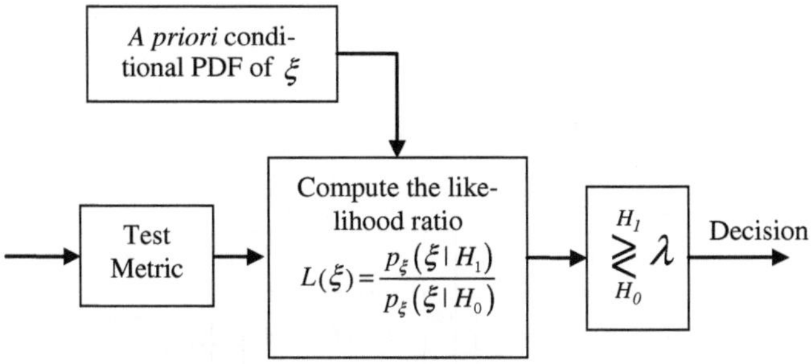

Fig. 3.15 NLOS identification using a binary hypothesis test. The optimum detector can be achieved using the NP theorem for a given PFA. Note that in most CIR-based identification techniques the a priori information that characterizes the identification metric is required. This is typically available in literature through numerous channel models and channel measurement for different wireless systems

from the binary case, however, a Bayesian approach is more popular in literature due to the practical formulation of the problem (Kay 1998; Van Trees 2001). The basic idea behind the Bayesian approach to detection and identification is to reach a decision that minimizes the Bayesian Risk given by Kay (1998), Van Trees (2001)

$$\Re = \sum_{i}^{2}\sum_{j}^{2} C_{ij}P(H_i|H_j)P(H_j) \tag{3.27}$$

where C_{ij} is the cost assigned to the decision to choose H_i when H_j is true. Typically the following particular cost assignment is assumed

$$C_{ij} = \begin{cases} 0 & i = j \\ 1 & i \neq j \end{cases} \tag{3.28}$$

This implies that the cost of making an error is higher than the cost of making the correct decision. In addition the cost in (3.27) can be modified to emphasize that the cost of making an error in DP estimation can be higher. For example, $C_{02} = C_{20} > C_{12} = C_{21} > C_{01} = C_{10}$. Using (3.27) the decision rule to minimize (3.26) is given by Kay (1998)

$$C_i(\xi) = \sum_{\substack{j=0 \\ j\neq i}}^{2} P(H_j|\xi) = \sum_{j=0}^{2} P(H_j|\xi) - P(H_i|\xi) \tag{3.29}$$

$C_i(\xi)$ is minimized by maximizing $P(H_i|\xi)$ which yields the following decision rule to decide for H_k (Kay 1998)

$$P(H_k|\xi) > P(H_i|\xi) \quad i \neq k \tag{3.30}$$

where $P(H_i|\xi) = p(\xi|Hi)P(H_i)/p(\xi)$ is the a posteriori probability. Since (3.29) is a threshold comparison between a posteriori probabilities this is often referred to

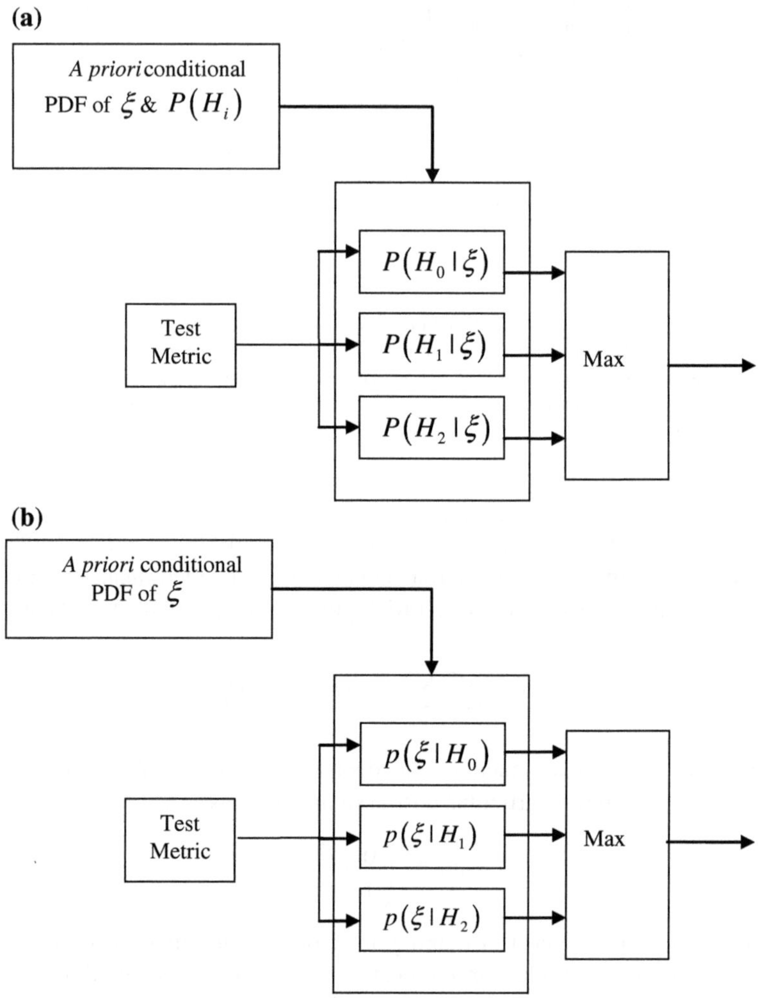

Fig. 3.16 NLOS identification using a ternary Bayesian hypothesis test. **a** MAP detector—a priori PDFs and $P(H_i)$ available. **b** ML detector—a priori PDFs are available but $P(H_i)$ assumed equal or $P(H_0) = P(H_1) = P(H_2) = 1/3$

as the M-ary maximum a posteriori (MAP) decision rule (Kay 1998). If the prior probabilities $P(H_i)$ are known to be equal then the MAP becomes the ML decision rule or decide H_k if

$$p(\xi|H_k) > p(\xi|H_i) \quad i \neq k. \tag{3.31}$$

Thus, the Bayesian approach to NLOS identification can be summarized in the system diagrams in Fig. 3.16.

The generic binary or ternary hypothesis testing is, in general, adopted by most NLOS identification techniques in literature. The difference arises in the metric used to identify the channels. The performance of NLOS identification will depend primarily on the conditional PDFs of the metric under different channel conditions. The metric with the highest conditional PDF separation will result in more robust identification. This can be seen for the binary case in Fig. 3.16. In the following we introduce some non-CIR and CIR based NLOS identification techniques.

3.2.1.1 Non-CIR Based NLOS Identification Techniques

Techniques that rely on simple measurement metrics such as the received signal strength (RSS) or the TOA estimate (variance) can be grouped into the non-CIR based techniques since they do not require estimating the CIR. They are simpler and offer low-complexity solution to the NLOS problem. They might sacrifice some accuracy in identification but can be robust in isolating the really bad channel conditions.

NLOS Identification Based on the Variance of RSS/TOA Estimate

Intuitively, one might expect that the variance of RSS (shadow fading) or TOA estimation can be different in LOS compared to NLOS. One simple NLOS identification technique was first proposed in Borras et al. (1998), where a running variance was computed using N range measurements (in time). It was assumed that the running variance is computed when the transmitter and receiver are both stationary. Given N range measurements $\hat{d}_n$, where $n \in [1, N]$, the proposed metric to detect NLOS conditions is the running variance that is given by Borras et al. (1998),

$$\sigma_{\text{rv}}^2 = \frac{\sum_{n=1}^{N} \left(\hat{d}_n - \mu_{\text{rv}}\right)^2}{N-1} \tag{3.32}$$

where μ_{rv} is the running estimate of the mean.

Thus NLOS identification can be achieved by comparing the metric against a threshold resulting in a simple hypothesis test or

$$\begin{aligned} &H_0 : \sigma_{\text{rv}}^2 < Th && \rightarrow \text{LOS} \\ &H_1 : \sigma_{\text{rv}}^2 < Th && \rightarrow \text{NLOS} \end{aligned} \tag{3.33}$$

where Th is a suitable threshold that guarantees a certain probability of detection and probability of false alarm. For TOA-based systems, a three-state hypothesis test is more suitable and it can be given by

$$\begin{aligned} &H_0 : \sigma_{\text{rv}}^2 < Th_{\text{rv}}^1 && \rightarrow \text{LOS} \\ &H_1 : Th_{\text{rv}}^1 < \sigma_{\text{rv}}^2 < Th_{\text{rv}}^2 && \rightarrow \text{NLOS} - \text{DP} \\ &H_2 : \sigma_{\text{rv}}^2 > Th_{\text{rv}}^2 && \rightarrow \text{NLOS} - \text{NDP} \end{aligned} \tag{3.34}$$

where Th_{rv}^{1} and Th_{rv}^{2} are thresholds that depend on the distributions of the variance of TOA estimation in the three different scenarios.

An even simpler NLOS identification alternative is one based on the variance of the RSS of the received signal. Since RSS is readily available in many wireless systems (WiFi, GSM, etc.), NLOS identification can be accomplished by observing the variance of RSS at a given distance. It is well known from modeling wireless propagation channels that the shadow fading is more pronounced in NLOS channels (larger variance around the mean). As a result, a simple NLOS identification algorithm can be achieved by comparing the variance of RSS against a predetermined threshold or

$$\begin{array}{ll} H_0 : \sigma_{\mathrm{RSS}}^2 < Th & \rightarrow \mathrm{LOS} \\ H_1 : \sigma_{\mathrm{RSS}}^2 > Th & \rightarrow \mathrm{NLOS} \end{array} \tag{3.35}$$

Although it is possible to devise a three state hypothesis test for RSS-based NLOS identification, the statistics of the RSS variance between LOS and NLOS-DP are very similar. As a result a two-state hypothesis test is more suitable.

NLOS Identification Based on the Statistics of the Envelope of the Received Signal

An alternative NLOS identification technique that is based on evaluating the envelope of the received signal has been proposed by Al-Jazzar and Caffery (2003). The basic idea behind this technique is that the envelope of the received signal behaves differently in LOS and NLOS environment, as is well-known from established measurement and modeling results. More specifically, the envelope of the received signal $z(t) = |r(t)|$ is typically modeled as a Rayleigh distributed random variable in NLOS because the received signal $r(t)$ is a complex Gaussian process and thus the absolute value is Rayleigh distributed, or

$$p_z(\zeta) = \frac{2\zeta}{\Omega}\exp\left(\frac{-\zeta^2}{\Omega}\right) \tag{3.36}$$

where $\Omega = E\left[\zeta^2\right]$. The distribution of the received envelope in LOS, however, is modeled as Ricean given by

$$p_z(\zeta) = \frac{2\zeta(K+1)}{\Omega}\exp\left(-K - \frac{(K+1)\zeta^2}{\Omega}\right) I_0\left(2\zeta\sqrt{\frac{K(K+1)}{\Omega}}\right) \tag{3.37}$$

where K is the Rice factor and I_0 is the modified Bessel function of the first kind. NLOS identification is achieved through comparing the statistics of the incoming observed data and inferring to which distribution it belongs. Al-Jazzar and Caffery (2003) proposes a simple Kolmogorov–Smirnov test to distinguish between a LOS and NLOS channel. More specifically, the hypothesis test can be given by

$$\begin{aligned} H_0 &: \hat{D} < D \rightarrow \text{LOS} \\ H_1 &: \hat{D} > D \rightarrow \text{NLOS} \end{aligned} \tag{3.38}$$

where $\hat{D}$ is given by

$$\hat{D} = \frac{\max|F_O - F_E|}{n} \tag{3.39}$$

and it is the test ratio used to compare the distributions. F_O is the cumulative observed absolute frequencies and F_E is the absolute frequency expected under the null hypothesis (LOS). Essentially F_E represents the statistics that is expected under the null hypothesis while F_O represents the statistics of the observation (measurements). For a level of significance 0.05, the bound D is $D = 1.358/\sqrt{n}$ (Al-Jazzar and Caffery 2003). The test ratio is basically the difference between the data and the expected distribution divided by the sample size, n.

Hybrid TOA/RSS NLOS Identification

Combining the information from TOA and RSS estimates, it is possible to have an improved NLOS identification. Intuitively, for short distances (small TOA estimates), the RSS estimate should be larger when compared to long distances (large TOA estimates). By amassing a priori knowledge of the path loss "behavior" in the different channel conditions, it is possible to implement a Bayesian approach to NLOS identification. This technique, which was introduced by Alsindi et al. (2009), models the TOA-based range estimates as

$$\hat{d}_{ij} = d_{ij} + w_{ij} + \varepsilon_{ij} = d_{ij} + w_{ij} + \begin{cases} 0, & H_0 \\ \beta_{ij}, & H_1 \\ \gamma_{ij}, & H_2 \end{cases} \tag{3.40}$$

where d_{ij} is the distance between nodes i and j, w_{ij} is the measurement noise, and ε_{ij} is the bias associated with the different channel types. Note that H_0 is LOS, H_1 is NLOS-DP, and H_2 is NLOS-NDP. The basic idea behind the hybrid TOA-RSS NLOS identification algorithm is that, given a TOA-based range estimate, $\hat{d}$, and an RSS measurement, $\hat{r}_{\text{RSS}}$, the channel is identified by computing the conditional probability $p(H_i|\hat{d}, \hat{r}_{\text{RSS}})$ for $i = 1, 2, 3$. The conditional probability can be computed using Bayes' equation

$$p(H_i|\hat{d}, \hat{r}_{\text{RSS}}) = \frac{f(\hat{r}_{\text{RSS}}|H_i, \hat{d})p(H_i|\hat{d})}{\sum_{k=0}^{2} f(\hat{r}_{\text{RSS}}|H_k, \hat{d})p(H_k|\hat{d})} \tag{3.41}$$

where $f(\hat{r}_{\text{RSS}}|H_i, \hat{d})$ is the distribution of the signal power for a given channel conditioned at an estimated distance and $p(H_i|\hat{d})$ is the probability of the channel

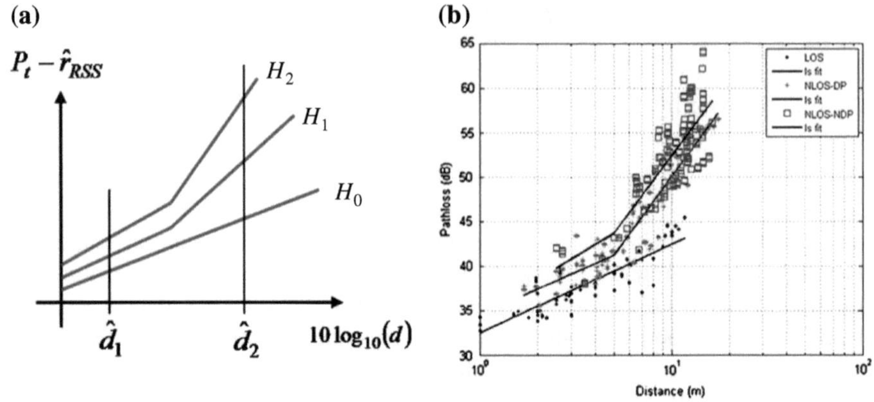

Fig. 3.17 Pathloss-distance relationship in the three different ranging scenarios. **a** Intuitive representation. **b** Results of UWB channel measurements and modeling

condition given the estimated distance. The latter can be similarly obtained using Bayes' equation

$$p\left(H_i|\hat{d}\right) = \frac{f\left(\hat{d}|H_i\right)p(H_i)}{\sum_{k=0}^{2} f\left(\hat{d}|H_k\right)p(H_k)} \tag{3.42}$$

where $f\left(\hat{d}|H_i\right)$ is the distribution of the TOA-estimated distances given the channel condition and $p(H_i)$ is the probability of the occurrence of the channel condition. This technique relies on the a *priori* information $f\left(\hat{r}_{\mathrm{RSS}}|H_i, \hat{d}\right)$ which can be obtained through channel measurements and modeling of the path loss (distance-power relationship). Figure 3.17 illustrates an example path loss-distance relationship for the three channel conditions along with results of measurement and modeling (more details can be found in Chap. 2).

When examining the power distribution for two distances $\left(\hat{d}_1, \hat{d}_2\right)$ in Fig. 3.18a, it is possible to see that the distributions will be farther separated for d_2 as illustrated in Fig. 3.18.

Thus, at shorter distances it is more difficult to distinguish between the channel conditions compared to longer distances. This, fortunately, is not an issue for two main reasons. The first is that ranging errors are typically much larger in longer distances as illustrated in Fig. 3.19.

The second is that when computing $p\left(H_i|\hat{d}, \hat{r}_{\mathrm{RSS}}\right)$, the PDFs $f\left(\hat{r}_{\mathrm{RSS}}|H_i, \hat{d}\right)$ will be weighted by $p\left(H_i|\hat{d}\right)$ which is a function of $f\left(\hat{d}|H_i\right)$ and $p(H_i)$. The pdf, $f\left(\hat{d}|H_0\right)$, is the distribution of distance for LOS channels, which is assumed to be uniform between 0 and R_c (communication range). On the other hand, $f\left(\hat{d}|H_1\right)$ and $f\left(\hat{d}|H_2\right)$ are distance dependent, with the former being a monotonic decreasing

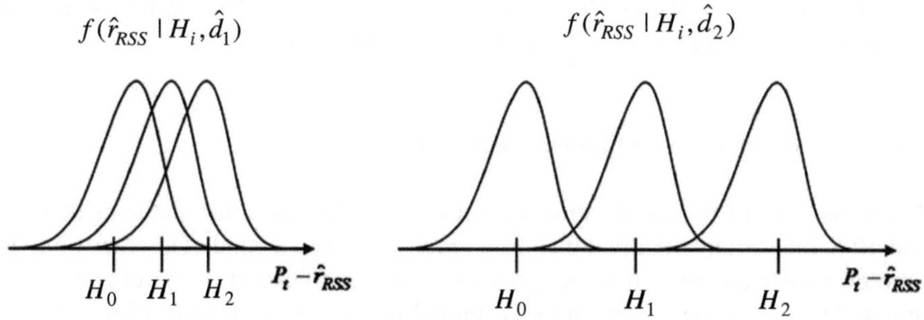

Fig. 3.18 Distribution of RSS in the three channel conditions for two distances

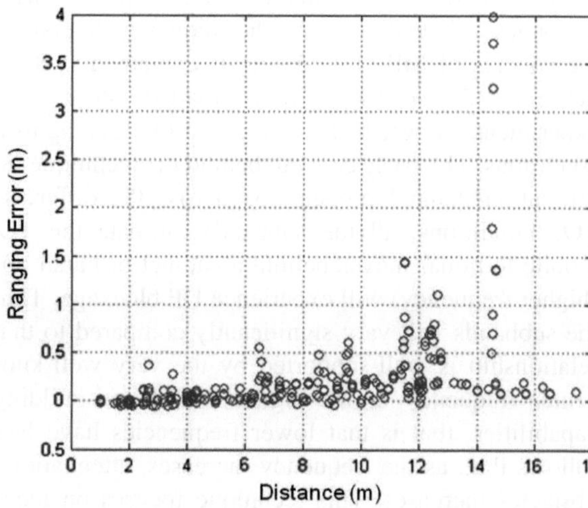

Fig. 3.19 Relationship between ranging error and distance. Results are generated from channel measurements

function of $\hat{d}$ while the latter is a monotonic increasing function of $\hat{d}$. This relationship holds because the probability of losing the DP (DP blockage) becomes more likely with increasing distance in NLOS conditions. This means that for the NLOS-DP condition the frequency of short distances is higher than longer distances. This occurs since the DP detection decreases with distances (due to obstacles attenuating the DP). As a result the distance distribution under NLOS-DP can be modeled as monotonically decreasing. For the NLOS-NDP case the DP is "lost" at a higher frequency for larger distances compared to shorter distances. Thus, it can be modeled as a monotonically increasing function.

Finally, a "hard" decision on the channel condition can be achieved by comparing the conditional probabilities for all three conditions and selecting the condition which maximizes $p(H_i|\hat{d}, \hat{r}_{\text{RSS}})$ or

$$H_k = \arg\max_k p(H_k|\hat{d}, \hat{r}_{\text{RSS}}) \tag{3.43}$$

The results of simulations have revealed that the NLOS identification algorithm has a success rate of 85 % over 40,000 simulated ranges.

Frequency Diversity-Based NLOS Identification

The previous NLOS identification techniques were based on the statistics of TOA and RSS measurements operating in a single frequency band. The variations in time or statistics across time can provide a strong indication of the channel condition. The techniques' robustness is limited due to the high probability of false alarms (missed detection), which is a caveat of their simplicity.

A different approach to NLOS identification has been proposed by Alsindi et al. (2008), where TOA estimation with frequency diversity is used to improve the robustness of identification. The frequency diversity technique, which is based on the concept of MB-OFDM signals explained in Sect. 3.1.2.2, where a bandwidth between 3.1 and 10.6 GHz is divided into a number of subbands, has been experimentally verified to successfully distinguish between different channel conditions. The basic concept behind the technique lies in the behavior of the TOA estimation/channel propagation across the different subbands. For example, in LOS conditions, all the subbands estimate the TOA fairly accurately. As the mobile terminal moves behind a cabinet or harsh obstacle, some of the subbands (higher frequency) will experience DP blockage. Thus, the TOA estimation across the subbands will vary significantly compared to that of the LOS condition. This relationship is well-supported by the very well-known relationship between the center frequency of the signal and the in-building attenuation or penetration capabilities, that is that lower frequencies have better penetration properties. It follows that, as the frequency increases, attenuation of the signals going through obstacles increases. This technique focuses on identifying between two ranging states: Presence of DP or absence of DP. Formally,

$$\xi = \begin{cases} 0, & \hat{d} = \hat{d}_{\mathrm{DP}} \\ 1, & \hat{d} = \hat{d}_{\mathrm{NDP}} \end{cases} \tag{3.44}$$

The technique therefore examines the variation of TOA estimation across subbands and integrates this information in a hypothesis testing framework to identify the channel conditions. Given N TOA estimates across N subbands $\hat{\boldsymbol{\tau}} = \left[\tau_1^1, \tau_1^2, \ldots, \tau_1^N\right]^T$, DP blockage identification can be achieved by examining the standard deviation of the TOA estimates across the subbands, or

$$s = \sigma(\hat{\boldsymbol{\tau}}) = \sqrt{\left(E[\hat{\tau}^2] - E[\hat{\tau}]\right)^2} \tag{3.45}$$

The subband TOA estimation standard deviation is directly related to the number of subbands that experience DP blockage. In the LOS case, all the subbands estimate the TOA with good accuracy and so s is fairly small. As the number

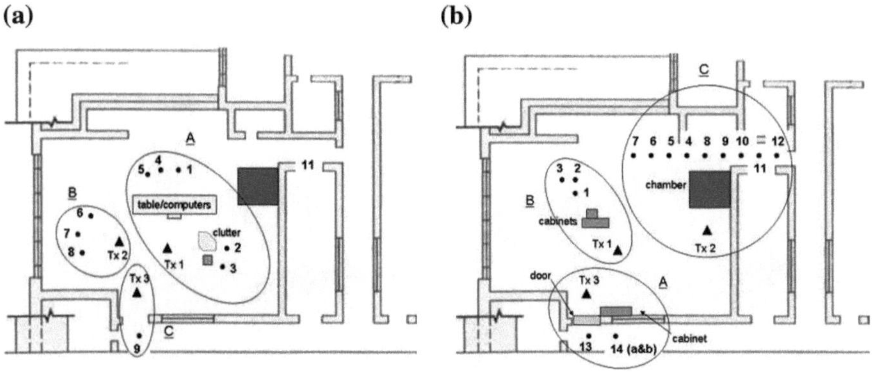

Fig. 3.20 Measurement locations: (**a**) LOS (**b**) NLOS. Triangle is the TX location and dots are the receive locations. Three different scenarios (TX-RX) have been measured for LOS and NLOS

of subbands that experience DP blockage increases, the standard deviation of TOA estimates increases. The number of subbands experiencing DP blockage can be given by

$$\upsilon = \sum_{n=1}^{N} \rho_n \tag{3.46}$$

where ρ_n is the ranging state in the nth subband. The channel condition is then identified by devising a binary hypothesis test that can be used to determine the presence or absence of DP blockage, or

$$\begin{aligned} H_0 &: \upsilon = 0 \\ H_1 &: \upsilon > 0 \end{aligned} \tag{3.47}$$

A decision can be achieved by examining the likelihood ratio

$$\frac{p_{s|H_1}(s|H_1)}{p_{s|H_0}(s|H_0)} \underset{H_0}{\overset{H_1}{\gtrless}} s_{\mathrm{th}} \tag{3.48}$$

where $p_{s|H_1}(s|H_1)$ and $p_{s|H_0}(s|H_0)$ are PDFs of the TOA standard deviation s. The NLOS identification technique has been verified through UWB channel measurements and analysis. Several LOS and NLOS measurements were conducted using the frequency domain measurement system described earlier. The measurements were conducted for a 5 GHz bandwidth that was sub-divided into 9 subbands that comply with the IEEE 802.15.3a MB-OFDM standard. TOA estimation was obtained for each subband and its standard deviation was examined across the subbands for the different LOS and NLOS scenarios. Figure 3.20 illustrates the location of the measurements and Fig. 3.21 shows the TOA estimation across the subbands for LOS and NLOS conditions.

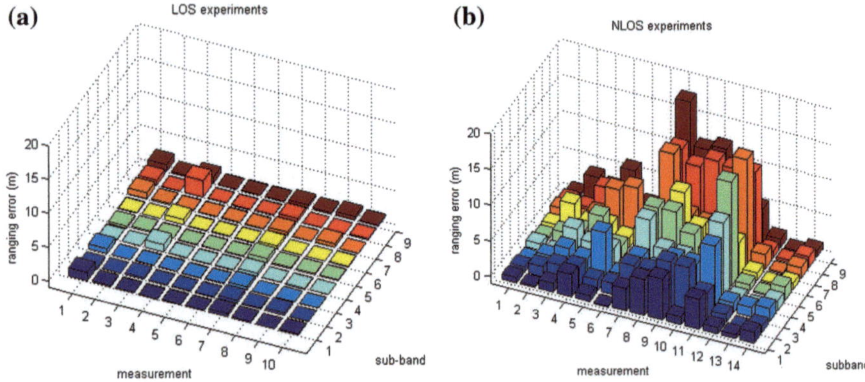

Fig. 3.21 TOA estimation (ranging error) across 9 subbands: **a** LOS, **b** NLOS

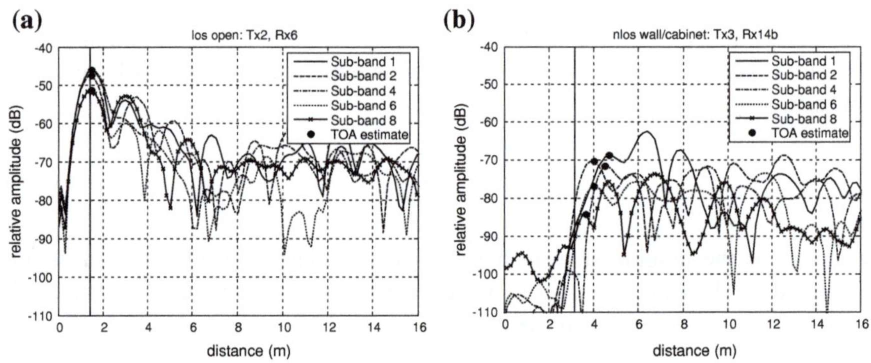

Fig. 3.22 Sample TOA/range estimation across 9 subbands: **a** LOS, **b** NLOS-cabinet/wall. Note the fluctuation of the TOA estimates across the subbands for NLOS compared to LOS

The difference in TOA behavior across the subbands can be clearly seen for LOS versus NLOS. In LOS scenarios, the variance in TOA estimation across the subbands is small. For NLOS scenarios, however, the variance in TOA estimation is directly related to the severity of the obstruction between the transmitter and receiver. The results indicate that it is possible to identify NLOS conditions through frequency diversity. Figure 3.22 illustrates two (LOS and NLOS) sample measured CIRs for different subbands.

The frequency-diversity based NLOS identification has the capability to distinguish between different NLOS scenarios. As the severity of NLOS condition increases, the number of subbands that "loses" the DP increases due to the increased penetration loss of the signal. The penetration loss will vary significantly for the subbands and this can be a very good indication of the severity of the obstruction. Figure 3.23 illustrates the relationship between the standard deviation of TOA estimation and the number of subbands that experienced blockage of the DP.

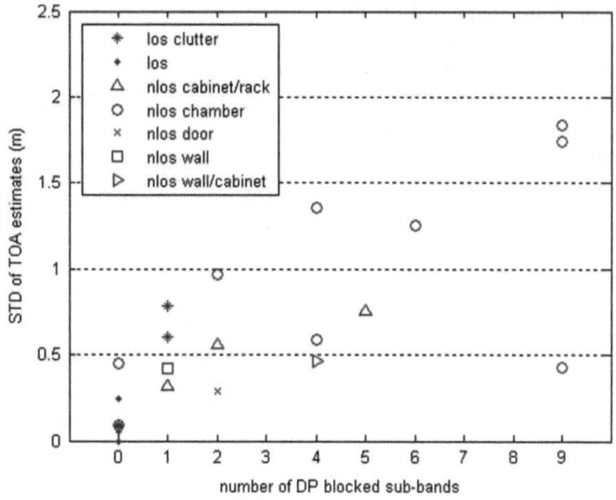

Fig. 3.23 Correlation between the numbers of DP blocked subbands and the standard deviation of the TOA estimates. In addition to identifying NLOS conditions, frequency diversity NLOS identification can provide an insight into the severity of the NLOS condition

3.2.1.2 CIR-Based NLOS Identification Techniques

TOA, RSS, and hybrid NLOS identification techniques are simple to implement and provide acceptable performance. Frequency diversity has been shown to provide enhanced identification capability to distinguish between different NLOS conditions. These techniques, however, do not exploit all the channel information. The multipath CIR contains valuable information that can be used for NLOS identification. The second class of NLOS identification is based on the CIR. The typical metrics that can be used are: the ratio of the First Path Power (FPP) to the total power, the RMS delay spread or kurtosis which is a statistical measure of the "peakedness" of the CIR. In this sub-section we will provide an overview of some of the main CIR-based NLOS identification metrics.

Ratio of First Path Power to Total power

Recall that the multipath channel can be modeled as

$$h(t) = \sum_{k=1}^{L_p} \alpha_k e^{j\phi_k} \delta(t - \tau_k) \tag{3.49}$$

The total power can be computed from the amplitude of the multipath components, or

$$P_T = \sum_{k=1}^{L_p} |\alpha_k|^2 \tag{3.50}$$

In LOS conditions, the first path is the DP and it is the strongest. As a result, the FPP is a major component of the total power. In NLOS scenarios, the relationship

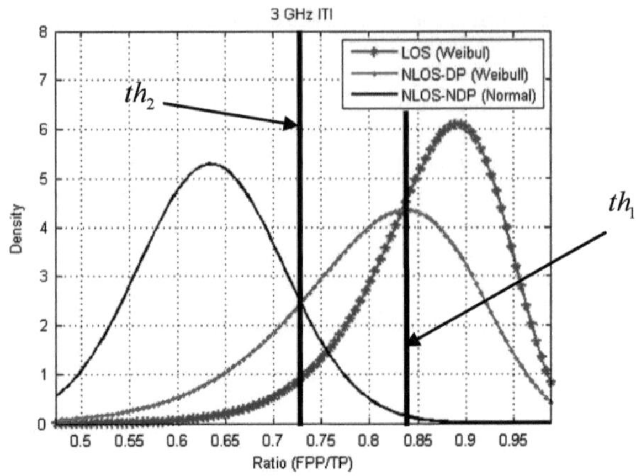

Fig. 3.24 PDFs of FPP/TP ratio under different ranging condition. The thresholds th_1 and th_2 are found by assuming that the three channel states are equiprobable

changes and the FPP is no longer the strongest. In some scenarios the FPP is not the DP. A good indication of the channel condition can be given by the ratio of the FPP and the total signal power (TP) or

$$\rho = \frac{P_{\mathrm{FP}}}{\mathrm{P_T}} = \frac{|\alpha_1|^2}{\left|\sum_{k=1}^{L_p} \alpha_k\right|^2} \tag{3.51}$$

Thus, the closer the ratio is to 1, the more likely the channel condition to be LOS. This technique has been experimentally evaluated by Alsindi et al. (2008) and a hypothesis test was devised. Based on the measurements in different indoor environments, Fig. 3.24 illustrates the PDF of the ratio FPP/TP under the three different channel conditions: LOS, NLOS-DP, and NLOS-NDP.

The figure clearly highlights the capability of distinguishing between LOS and the severe NLOS-NDP condition. The technique is simple but requires estimating the power of the first arriving path, which requires channel estimation.

RMS Delay Spread

The power ratio metric only exploits the amplitude of the first and strongest paths in the CIR. A better metric for NLOS identification is the RMS delay spread (Heidari et al. 2009; Marano et al. 2010), which is given by

$$\tau_{\mathrm{rms}} = \sqrt{\frac{\sum_{k=1}^{L_p} \alpha_k^2 \tau_k^2}{\sum_{k=1}^{L_p} \alpha_k^2} - \tau_m} \tag{3.52}$$

where $\tau_m = \sum_k \alpha_k^2 \tau_k / \sum_k \alpha_k^2$ is the mean excess delay of the channel. A hypothesis test based on the RMS delay spread can be devised similar to the tests in the previous techniques that is

$$\begin{aligned} H_0 &: \tau_{\text{rms}} < \tau_{\text{th}} \rightarrow \text{LOS} \\ H_1 &: \tau_{\text{rms}} > \tau_{\text{th}} \rightarrow \text{NLOS} \end{aligned} \tag{3.53}$$

where τ_{th} is a suitable threshold that provides a desired probability of correct detection given a certain false alarm using the Neyman-Pearson approach. When available the two-state hypothesis test can be extended to a three-state if the data is available and appropriate thresholds chosen.

Kurtosis-Based NLOS Identification

The shape of the channel impulse response in different channel conditions can be characterized by the kurtosis metric. The kurtosis of the CIR is defined as the ratio of the fourth order moment of the data to the square of the second order moment (variance) (Guvenc et al. 2007). It has been defined as "a measure of whether the data is peaked or flat relative to a normal distribution; i.e., data sets with high kurtosis tend to have a distinct peak near the mean, decline rather rapidly and have heavy tails, while data sets with low kurtosis tend to have a flat top near the mean rather than a sharp peak". Thus, the kurtosis metric can be used to identify LOS channels since they are "peaky" with respect to flatter NLOS channels. The kurtosis of the CIR is expressed mathematically as (Guvenc et al. 2007)

$$\kappa = \frac{E\left[\left(|h(t)| - \mu_{|h|}\right)^4\right]}{E\left[\left(|h(t)| - \mu_{|h|}\right)^2\right]^2} = \frac{E\left[\left(|h(t)| - \mu_{|h|}\right)^4\right]}{\sigma_{|h|}^4} \tag{3.54}$$

The kurtosis metric can then be used to devise a hypothesis test. Then a simple binary hypothesis test (LOS/NLOS) can be devised as follows

$$\frac{p(\kappa|\text{LOS})}{p(\kappa|\text{NLOS})} \underset{H_1}{\overset{H_0}{\gtrless}} 1 \tag{3.55}$$

Results presented in Guvenc et al. (2007) showed that the kurtosis outperformed the mean excess delay and RMS delay spread.

3.2.2 NLOS Mitigation Algorithms

Once the channel conditions have been identified for each range/angle measurement then this information can be used to improve the localization accuracy through NLOS mitigation. NLOS mitigation algorithms/techniques, in general, aim to reduce the impact of NLOS corrupted measurements on the location estimate. There are several major approaches to NLOS mitigation and some rely on the NLOS identification. These major techniques can be grouped into Identify and Discard (IAD), Least Squares, and Constrained Localization (Guvenc and Chong 2009). In LS techniques the information from the NLOS channel identification can be explicitly or implicitly used to improve the position estimate. Weighted Least Square (WLS) algorithms explicitly integrate the NLOS channel identification information into a weighting matrix. However the Residual Weighting Algorithm (RWA) implicitly uses the NLOS information to obtain a better estimate. Constrained Localization is generally either based on Quadratic Programming (QP) or Linear Programming (LP) and they differ by the constraint models adopted. In the following an overview of these NLOS mitigation algorithms will be presented.

3.2.2.1 Identify and Discard

This is the simplest technique in NLOS mitigation and can only be effective if there are a large number of base stations aiding the mobile device in the localization process. In this technique, the identified NLOS base stations are removed from the localization process—that is—only the base stations under LOS propagation are considered. This approach has several drawbacks. The first is that by discarding NLOS reference points, the geometrical configuration of the localization process might be affected and in turn affect the localization accuracy. In addition, if there are only a few number of base stations, then its effectiveness will be limited, and even impractical, if there are fewer than four reference points. This approach, however, is more suitable in wireless sensor networks, where there are a number of "anchors" that aid the sensor nodes in the localization; it then becomes possible to discard several NLOS anchor nodes without degrading performance. Finally IAD cannot be used when all the range measurements are in NLOS.

3.2.2.2 Weighted Least-Squares

The WLS mitigation technique improves on the LS technique in that weights are assigned in proportion to the confidence in the range measurements. Specifically, the more reliable LOS range measurements are associated with a higher weight while the NLOS range measurements are associated with a lower weight value. Therefore, the contribution of the NLOS corrupted range measurements can be dynamically incorporated into the algorithm to improve the localization performance. The WLS can be derived from the least square formulation. Recall that the least squares technique minimizes a cost function given by

$$C_{\mathrm{LS}}(\hat{\mathbf{x}}) = \left[\hat{\mathbf{d}} - \mathbf{F}(\hat{\mathbf{x}})\right]^H \left[\hat{\mathbf{d}} - \mathbf{F}(\hat{\mathbf{x}})\right] \tag{3.56}$$

where $\mathbf{F}(\hat{\mathbf{x}})$ contains the N distances given by

$$\mathbf{F}(\mathbf{x}) = \left[\sqrt{(\hat{x} - x_1)^2 + (\hat{y} - y_1)^2} \quad \cdots \quad \sqrt{(\hat{x} - x_N)^2 + (\hat{y} - y_N)^2}\right]^T.$$

The LS location estimate is then given by

$$\hat{\mathbf{x}} = \mathbf{x}_0 + \left(\mathbf{J}^H \mathbf{J}\right)^{-1} \mathbf{J}^H [\mathbf{d} - \mathbf{F}(\mathbf{x}_0)] \tag{3.57}$$

where $\mathbf{J}$ is the Jacobian of $\mathbf{F}$. When the NLOS information is available the LS formulation can be modified to incorporate the weighting information and thus the cost function becomes

$$C_{\mathrm{WLS}}(\hat{\mathbf{x}}) = [\mathbf{d} - \mathbf{F}(\hat{\mathbf{x}})]^H \mathbf{W} [\mathbf{d} - \mathbf{F}(\hat{\mathbf{x}})] \tag{3.58}$$

where $\mathbf{W} = \mathrm{diag}\{w_1, w_2, \ldots, w_N\}$ is a diagonal weighting matrix with positive elements. The elements can be chosen to distinguish the LOS or NLOS and typically they are chosen to be the inverse of the variances of the measurement noise in each condition. The WLS solution is then given by

$$\hat{\mathbf{x}} = \mathbf{x}_0 + \left(\mathbf{J}^H \mathbf{W} \mathbf{J}\right)^{-1} \mathbf{J}^H \mathbf{W} [\mathbf{d} - \mathbf{F}(\mathbf{x}_0)] \tag{3.59}$$

The weights on the range measurements need not be identical if the information is available. If the information regarding the severity of the NLOS condition is available, then the weights can reflect this information. The WLS algorithm is a practical method that relies on the NLOS information. The accuracy increases if the weights are closely related to the channel condition or the bias on the range measurements.

3.2.2.3 Residual Weighting Algorithm

An alternative technique has been introduced in Chen (1999), where the residual of the error is used as a weighting mechanism to mitigate the NLOS errors. The algorithm is based on the concept that NLOS range measurements typically produce larger residual error. Recall from (2.2) that for n_B range measurements the residual error is given by

$$R_{\mathrm{es}}(\mathbf{x}) = \sum_{i=1}^{n_B} \beta_i \left(\hat{d}_i - \|\mathbf{x} - \mathbf{x}_i\|\right)^2 \tag{3.60}$$

Thus it is expected that if more of the range measurements, $\hat{d}_i$, are corrupted by NLOS then the residual would be higher. One way to "find" the NLOS corrupted measurements is to form different combinations of range measurements and evaluate the residual. The combinations with higher residuals indicate that some of the range measurements within the combination are under NLOS. Since, ultimately we are interested in mitigating the impact of NLOS then this can be achieved by forming N_c combinations of range measurements, estimate the residual and position for each combination and then form a weighted sum of all the combinations; where the weight is chosen to be the inverse of the residual. Thus, a combination with high residual is weighted less. Formally, the algorithm forms $N_c = \sum_{i-3}^{n_B} {}_{n_B}C_i$ combinations with i BSs selected from a total of n_B BSs. For the set S_k, $k = [1, 2, \ldots, N_c]$, an "intermediate" kth position can be computed by

$$\hat{\mathbf{x}}_k = \arg\min_{\mathbf{x}} R_{\text{es}}(\mathbf{x}; S_k) \tag{3.61}$$

where $R_{\text{es}}(\mathbf{x}; S_k)$ is the intermediate residual associated with the set S_k. In here, "intermediate" means that the position or residual of a set S_k of BSs. Thus, it is not the final position estimate or residual. The location estimate is then a weighted combination of the N_c intermediate location estimates given by

$$\hat{\mathbf{x}} = \frac{\sum_{k=1}^{N_c} \hat{\mathbf{x}}_k \left[\tilde{R}_{\text{es}}(\hat{\mathbf{x}}; S_k)\right]^{-1}}{\sum_{k=1}^{N_c} \tilde{R}_{\text{es}}(\hat{\mathbf{x}}; S_k)^{-1}} \tag{3.62}$$

where $\tilde{R}_{\text{es}}(\hat{\mathbf{x}}; S_k)$ is the normalized residual or

$$\tilde{R}_{\text{es}}(\hat{\mathbf{x}}_k; S_k) = \frac{R_{\text{es}}(\hat{\mathbf{x}}_k; S_k)}{|S_k|} \tag{3.63}$$

The RWA is computationally expensive since different combination of range measurements should be used to estimate the intermediate position values. For a small number of range measurements (few BSs), then this algorithm can provide practical and robust NLOS mitigation. However, the scalability of the algorithm is its major shortcoming.

3.2.2.4 Constrained Localization: LS/Quadratic Programming

An alternative approach to solving the NLOS problem is to examine the non-linear relationship between the range measurements and the unknown locations. A QP formulation can be formed to find a WLS solution but with a set of constraints. The set of non-linear equations resulting from $n_B$measurements is given by Guvenc and Chong (2009)

$$(x - x_i)^2 + (y - y_i)^2 = \hat{d}_i^2 \tag{3.64}$$

where $i = 1, 2, \ldots, n_B$. The squared estimated distances in (3.64) at high SNR can then be given by Cheung et al. (2004)

$$\hat{d}_i^2 = (d_i + w_i)^2 \approx d_i^2 + 2d_i w_i. \tag{3.65}$$

Then this results in an error or disturbance given by

$$\varepsilon = \hat{d}_i^2 - d_i^2. \tag{3.66}$$

The definition of the following variables is necessary to have more compact expressions. Thus for the following

$$s = x^2 + y^2 \tag{3.67a}$$

$$k_i = x_i^2 + y_i^2 \tag{3.67b}$$

we represent the set of nonlinear equations in (3.64) in matrix form or

$$\mathbf{A}\boldsymbol{\theta} = \frac{1}{2}\mathbf{p} \tag{3.68}$$

where

$$\mathbf{A} = \begin{bmatrix} x_1 & y_1 & -0.5 \\ x_2 & y_2 & -0.5 \\ \vdots & \vdots & \vdots \\ x_{n_B} & y_{n_B} & -0.5 \end{bmatrix}, \boldsymbol{\theta} = \begin{bmatrix} x \\ y \\ s \end{bmatrix}, \mathbf{p} = \begin{bmatrix} k_1 - \hat{d}_1^2 \\ k_2 - \hat{d}_2^2 \\ \vdots \\ k_{n_B} - \hat{d}_{n_B}^2 \end{bmatrix}. \tag{3.69}$$

Then based on (3.68a, b) it is possible to develop a QP approach to solve the NLOS problem (Wang et al. 2003). The technique formulates a constrained LS algorithm that can be solved by using QP which is given by

$$\begin{aligned} \hat{\boldsymbol{\theta}} &= \arg\min_{\boldsymbol{\theta}}(\mathbf{A}\boldsymbol{\theta} - \mathbf{p})^T \Psi^{-1}(\mathbf{A}\boldsymbol{\theta} - \mathbf{p}) \\ s.t.\ &\mathbf{A}\boldsymbol{\theta} \leq \mathbf{p} \end{aligned} \tag{3.70}$$

where $\boldsymbol{\Psi} = [\varepsilon\varepsilon^T]$ is the covariance matrix of the disturbances in $\mathbf{p}$ (it is the *weighting* matrix that statistically characterizes the disturbances) given by Cheung et al. (2004)

$$\boldsymbol{\Psi} = \mathbf{BQB} \tag{3.71}$$

where $\mathbf{B} = \text{diag}(2d_1, 2d_2, \ldots, 2d_{n_B})$ and $\mathbf{Q} = \text{diag}(\sigma_1^2, \sigma_2^2, \ldots, \sigma_{n_B}^2)$ is the covariance of the noise vector. Thus, the QP formulation in (3.70) finds a WLSs solution to the mobile device location while the constraint $\mathbf{A}\boldsymbol{\theta} \leq \mathbf{p}$ relaxes the equality to an inequality for NLOS conditions (equality for all LOS condition) (Guvenc and Chong 2009). This approach is essentially a 2-stage ML technique with quadratic constraints. It is computationally expensive compared to the other approaches (Guvenc and Chong 2009).

3.2.2.5 Constrained Localization: Linear Programming

The basic idea behind the LP technique is that it combines the NLOS information with LS to achieve better location estimates. Essentially, the algorithm incorporates the NLOS BSs to establish a feasible region that is composed of squares (Guvenc and Chong 2009). Once the feasible region is established then the LS technique is used to find a solution within the feasible region. The approach was introduced in Venkatesh et al. (2007) where for the NLOS BSs the non-linear constraint is given by

$$\|\mathbf{x} - \mathbf{x}_i\|^2 \leq \hat{d}_i \tag{3.72}$$

The non-linear constraints can be further linearized for the *i*th BS as (Larsson 2004)

$$\begin{aligned} x - x_i \leq \hat{d}_i, \quad -x + x_i \leq \hat{d}_i \\ y - y_i \leq \hat{d}_i, \quad -y + y_i \leq \hat{d}_i \end{aligned} \tag{3.73}$$

These linearized constraints effectively relax the circular constraints into rectangular constraints which define the feasible region. The mobile device location can be estimated by minimizing the objective function with the constraint imposed by both the LOS and NLOS range measurements (Guvenc and Chong 2009). These LP techniques will be revisited in greater detail in Chap. 6, which deals with cooperative localization techniques. When compared to the QP, the LP results in a less complex constraint but a coarser solution. Thus, complexity is traded with location accuracy.

3.3 Conclusion

In this chapter, we have provided an overview of algorithms and techniques that attempt to mitigate the two major propagation problems introduced in Chap. 2: multipath fading and non-line-of-sight conditions. For the multipath problem, two main techniques can be used depending on the system implementation. If the requirement is to improve the time-resolution of existing wireless systems (WiFi, UMTS, etc.) then it is possible to integrate spectral estimation techniques such as MUSIC to achieve a higher accuracy in TOA estimation. It has been illustrated through examples that super-resolution algorithm can improve the accuracy substantially, given that the DP is strong enough. In cases where the DP is not available, TOA estimation can still be improved using super-resolution techniques, but an unavailable DP cannot be "reconstructed". On the other hand, if a new wireless system can be dedicated for the geolocation problem then a UWB system can be implemented which can provide centimeter level accuracy in LOS and improved estimation in NLOS. It has been verified both theoretically and experimentally that UWB has the potential to mitigate multipath significantly, whether Impulse Response-UWB or MultiBand-OFDM UWB approaches are used.

Multipath mitigation algorithms work very well in LOS environments, but face limitations under NLOS conditions. As so, we introduced popular NLOS identification and mitigation algorithms. NLOS identification algorithms infer the state of the channel by either analyzing the statistics of the ranging, the statistics of the received envelope or the statistics of the channel impulse response. The CIR-based NLOS identification techniques are expected to have better identification capabilities because they harness the multipath information inherent to the channel.

NLOS mitigation algorithms then integrate the NLOS identification results in different techniques to improve the position estimate. The most popular is incorporating the NLOS information in a weighted optimization approach, where LOS and NLOS are assigned different weights according to the severity of the condition. In another approach constrained optimization (linear or QP) can be used where NLOS measurements create a set of constraints that can further improve the location estimate. Results of NLOS mitigation show that the localization accuracy can be improved significantly.

Practical localization systems operating in harsh multipath environments cannot be realized with just one solution. Thus, the localization system must incorporate multipath, NLOS identification and NLOS mitigation algorithms in order to achieve acceptable accuracy. In addition, techniques and technologies such as tracking algorithms and inertial navigation systems (addressed in Chaps. 8 and 9) should be integrated in an overall geolocation solution. Thus, the strength of each technique can be harnessed to tackle the difficult challenges facing accurate localization in harsh multipath environments.

References

B. Alavi, K. Pahlavan, Studying the Effect of Bandwidth on Performance of UWB Positioning Systems. in *Proceedings of IEEE Wireless Communications and Networking Conference (WCNC)*, vol. 2, Apr 2006, pp. 884–889

S. Al-Jazzar, J. Caffery Jr., New Algorithms for NLOS Identification. in *Proceedings of IST Summit Conference,* Dresden, Germany, June 2003

N. Alsindi, *Indoor Cooperative Localization for Ultra Wideband Wireless Sensor Networks,* Ph.D. Dissertation, Worcester Polytechnic Institute, Worcester, MA, USA, Apr 2008

N. Alsindi, M. Heidari, K. Pahlavan, Blockage Identification in Indoor UWB Ranging Using Multi Band ODFM Signals. in *Proceedings of IEEE Wireless Communications and Networking Conference*, Las Vegas, NV, Apr 2008

N. Alsindi, C. Duan, J. Zhang, T. Tsuboi, NLOS Channel Identification and Mitigation in UWB TOA-Based Wireless Sensor Networks. in *Proceedings of 6th Workshop on Positioning, Navigation and Communication (WPNC)*, Hannover, Germany, Mar 2009, pp. 59–66

C. R. Berger, Z. Tian, P. Willett, S. Zhou, Precise Timing for Multiband OFDM in a UWB System. in *Proceedings of IEEE International Conference on Ultra-Wideband (ICUWB),* Waltham, MA, Sept 2006, pp. 269–274

J. Borras, P. Hatrack, N. B. Mandayam, Decision Theoretic Framework for NLOS Identification. 48th IEEE Vehicular Technology Conference, vol. 2, Ottawa, Canada, 1998, pp. 1583–1587

P.C. Chen, A Non-Line-of-Sight Error Mitigation Algorithm in Location Estimation. in *Proceedings of IEEE International Conference on Wireless Communication Networking (WCNC),* vol. 1, New Orleans, LA, Sept 1999, pp. 316–320

K.W. Cheung, H.C. So, W.K. Ma, Y.T. Chan, Least square algorithms for time-of-arrival-based mobile location. IEEE Trans. Sig. Process. **52**(4), 1121–1128 (2004)

A. G. Dabak, A. Batra, J. Balakrishnan, Ranging in Muti-Band OFDM Communications Systems. US Patent, US 2005/0050130 A1, 3 Mar 2005

D. Dardari, A. Conti, U. Ferner, A. Giorgetti, M.Z. Win, Ranging with ultrawide bandwidth signals in multipath environments. IEEE Proc. **97**(2), 404–426 (2009)

B.H. Fleury, M. Tschudin, R. Heddergou, D. Dahlhaus, K.I. Pedersen, Channel parameters estimation in mobile radio environments using SAGE algorithm. IEEE J. Select. Areas Commun. **17**(3), 434–449 (1999)

R.J. Fontana, Recent system applications of short-pulse ultra-wideband (UWB) technology. IEEE Trans. Microwave Theor. Tech. **52**(9), 2087–2104 (2004)

M. Ghavami, L.B. Michael, R. Kohno, *Ultra Wideband Signals and Systems in Communication Engineering*, 2nd edn. (Wiley, New York, 2007)

I. Guvenc, Z. Sahinoglu, P.V. Orlik, TOA estimation for IR-UWB systems with different transceiver types. IEEE Trans. Micro. Theor. Tech. **54**(4), 1876 (2006)

I. Guvenc, C.-C. Chong, F. Watanabe, NLOS Identification and Mitigation for UWB Localization Systems. in *Proceedings of IEEE Wireless Communications and Networking Conference*, Mar 2007

I. Guvenc, C.-C. Chong, A survey on TOA based wireless localization and NLOS mitigation techniques. IEEE Commun. Surv. Tutorials **11**(3), 3rd Quarter 2009

H. Hashemi, The indoor radio propagation channel. Proc. IEEE **81**(7), 943–968 (1993)

M. Heidari, N. Alsindi, K. Pahlavan, UDP identification and error mitigation in TOA-based indoor localization systems using neural network architecture. IEEE Trans. Wirel. Commun. **8**(7), 3597–3607 (2009)

M.G.M. Hussain, Ultra-wideband impulse radar—an overview of the principles. IEEE Aerosp. Electron. Syst. Mag. **13**(9), 9–14 (1998)

S.M. Kay, *Fundamentals of Signal Processing Volume II: Detection Theory* (Prentice Hall, Englewood Cliffs, 1998)

E.G. Larsson, Cramer-Rao bound analysis of distributed positioning in sensor networks. IEEE Sig. Process. Lett. **11**(3), 334–337 (2004)

Y. Lee, R.A. Scholtz, Ranging in a dense multipath environment using an UWB radio link. IEEE J. Select. Areas Commun. **20**(9), 1677–1683 (2002)

X. Li, K. Pahlavan, Super-resolution TOA estimation with diversity for indoor geolocation. IEEE Trans. Wireless Comm. **3**(1), 224–234 (2004)

D. Manolakis, V. Ingle, S. Kogon, *Statistical and Adaptive Signal Processing* (McGraw Hill Co. Inc., New York, 2000)

S. Marano, W.M. Gifford, H. Wymeersch, M.Z. Win, NLOS identification and mitigation for localization based on UWB experimental data. IEEE J. Sel. Areas Commun. **28**(7), 1026–1035 2010

W. Pam Siriwongpairat, K.J. Ray Liu, *Ultra-wideband Communications Systems—A Multiband OFDM Approach* (Wiley, New York, 2008)

E. Saberinia, A.H. Tewfik, Enhanced Localization in Wireless Personal Area Networks. in *Proceedings of the IEEE Global Telecommunications Conference (GLOBECOM'04),* vol. 4, Dallas, TX, Dec 2004, pp. 2429–2434

H. Sheng, P. Orlik, A.M. Haimovich, L.J. Cimini Jr., J. Zhang, On the spectral and power requirements for ultra-wideband transmission. in *Proceedings of IEEE International Conference Communication*, vol. 1, May 2003, pp. 738–742

L. Stoica, A. Rabbachin, I. Oppermann, A low-complexity noncoherent IR-UWB transceiver architecture with TOA estimation. IEEE Trans. Microw. Theory Tech. **54**(4), 1637–1646 (2006)

H.L. Van Trees, *Detection, Estimation, and Modulation, Part I: Detection, Estimation and Linear Modulation Theory* (Wiley, New York, 2001)

S. Venkatesh, R.M. Buehrer, NLOS mitigation using linear programming in ultrawideband location-aware networks. IEEE Trans. Veh. Technol. **56**(5), 3182–3198 (2007)

X. Wang, Z. Wang, B.O. Dea, A TOA based location algorithm reducing the error due to non-line-of-sight (NLOS) propagation. IEEE Trans. Veh. Technol. **52**(1), 112–116 (2003)

M.Z. Win, R.A. Scholtz, Impulse radio: how it works. IEEE Commun. Lett. **2**(2), 36–38 (1998)

M.Z. Win, R.A. Scholtz, Ultra-wide bandwidth time-hopping spread-spectrum impulse radio for wireless multiple-access communications. IEEE Trans. Commun. **48**, 679–691 (2000)

H. Xu, C.-C. Chong, I. Guvenc, F. Watanabe, L. Yang, High-resolution TOA estimation with multi-band OFDM UWB signals. in *Proceedings of IEEE ICC'08,* 2008, pp. 4191–4196

References

[illegible]

Chapter 4
Survey-Based Location Systems

The two main primitives in location systems—range and angle-of-arrival—were introduced in Chap. 2. By measuring either of the two between a mobile device and the base stations in a backbone network, the primitives can be respectively triangulated or angulated to an estimated location for the mobile. For indoor systems that require high accuracy and fidelity, the range primitive—when measured through time-of-arrival (TOA) techniques—is the more robust of the two to signal fading, as explained in Chap. 2. The success of time-of-arrival techniques hinges upon the predictable mapping between the TOA and the distance travelled by the signal. The range primitive can also be measured through received-signal-strength techniques, however, in harsh propagation environments, the mapping is very sensitive to fading, which is nondeterministic in nature. Hence, range-based mapping cannot be reliably exploited for RSS systems in such environments.

Yet, received signal strength has been shown to deliver decent accuracy even in indoor environments—when used in survey-based location systems. In survey-based systems, RSS is not mapped to range, but directly to location. The technique is just to assume that because signal loss occurs in the environment—not only due to path loss, but also due to penetration loss and specular effects such as reflection and diffraction—such a mapping exists. Because the mapping is so complex, there is no attempt to explicitly model the received signal strength as a function of location. Instead, the mapping is constructed by observing the "fingerprints" that the RSS "leaves" throughout the environment. From the observed fingerprints, the RSS-location mapping can be reconstructed. In practice, fingerprinting systems associate values of a physically measureable feature to discrete locations throughout a survey area. RSS is the most common feature but, as we shall see, others have become popular recently. Then, a mobile device can estimate its location based on the value it measures during a query. The feature value at a particular location is known as a fingerprint, or signature, because it can be used to identify the location.

C. Gentile et al., *Geolocation Techniques*, DOI: 10.1007/978-1-4614-1836-8_4,

One of the earlier and most simple fingerprinting systems, known as RADAR (Bahl and Padmanadhan 2000), is based on the received signal strength. The simplicity of this indoor location system stems from the fact that RSS measurements are readily available in the IEEE 802.11 standards implementation. For outdoor systems, on the other hand, the RSS is measured from cellular towers or satellites. More on cellular systems is discussed in the following chapter. Since base stations (access points) typically have overlapping coverage, the actual feature is the vector of RSS values received from all available base stations. Before the system can be operational, a radio map of the environment must be constructed in a so-called fingerprinting stage. In this stage, a discrete set of n_M candidate sites $\boldsymbol{x}_i, i = 1 \ldots n_M$, for the mobile is selected throughout the survey area a priori. At each site, the received powers from the base stations are recorded and stored in a database. Let n_B denote the maximum number of base stations from which a mobile device can receive a distinct signal. Then, the signature at location $\boldsymbol{x}_i$ is the vector of received powers denoted as $\boldsymbol{P}_i = \left[P_{i1}, P_{i2}, \ldots, P_{i,n_B}\right]^{\mathrm{T}}$. We refer to an ordered pair composed from a location and its associated signature $(\boldsymbol{x}_i, \boldsymbol{P}_i)$ as a *training pair*.

During system operation—which is known as the localization stage—a vector $\hat{\boldsymbol{P}} = \left[\hat{P}_1, \hat{P}_2, \ldots, \hat{P}_{n_B}\right]^T$ of received powers is measured at the mobile device. RADAR uses the nearest neighbor method as a mapping algorithm from the measured power to the estimated location for the mobile. Specifically, the mobile's location is determined as the location $\boldsymbol{x}_c$ of the registered site whose fingerprinted power $\boldsymbol{P}_c$ is closest to the measured power $\hat{\boldsymbol{P}}$ in terms of some similarity metric in the RSS vector space.

Location fingerprinting systems can be differentiated for the most part by the following two characteristics: (1) the feature selected to fingerprint the sites; and (2) the mapping algorithm to determine the mobile's location. In this chapter, we introduce several fingerprinting techniques. Given its prevalence, we concentrate on the RSS feature in the first part of the chapter. The same techniques, however, apply to other features as well. In the first section, an analytical model of a generic fingerprinting system is presented. The model describes how the salient parameters common to most systems affect their performance. The subsequent section showcases a number of methods to compute the similarity metric for memoryless systems—that is—systems which estimate location based on readings taken at a single time instant. Section 4.3 introduces systems with memory and shows how maintaining some historic path data can enhance location precision significantly. In the remainder of the chapter, we introduce some non-RSS features. Section 4.4 investigates the use of the channel impulse response as an alternative radio frequency signature. Conversely, Sect. 4.5 reports on non-RF features altogether—features which are available from devices such as smartphones, namely sound, motion, and color.

4.1 Analytical Models

Besides the selection of the feature (or features) in any fingerprinting system, there are a number of system factors which affect performance, most notably the number of base stations, the number of fingerprinted sites, and the spacing between the sites. Naturally, the harshness of the propagation environment also affects performance. In this section, we describe two analytical models proposed in Kaemarungsi and Krishnamurthy (2004) to investigate these factors. Again, although the proposed models are specific to RSS-feature systems, the principles apply to all types of fingerprinting systems. As in the RADAR system, it is assumed that the vector of received signal strengths from the base stations is used to fingerprint the sites.

4.1.1 A Stochastic Model for the Similarity Metric

A popular similarity metric between the measured power, $\hat{\boldsymbol{P}}$, and fingerprinted power at a particular site indexed as i, $\boldsymbol{P}_i$, is the square Euclidean norm in the n_B-dimensional space (Liu et al. 2007):

$$\begin{aligned} \rho_i &= \left\|\hat{\boldsymbol{P}} - \boldsymbol{P}_i\right\| \\ &= \sum_{j=1}^{n_B} \left(\hat{P}_j - \hat{P}_{ij}\right)^2. \end{aligned} \tag{4.1}$$

The units for the power are in dBm. In practice, a mobile device may not receive a signal from all n_B base stations. This is because when a mobile device is far away from a site—especially in large deployment areas—the set of base stations from which it receives may differ from the set registered at the site. In this case, the similarity metric can compare only the signal strengths from the common base stations. So, a penalty term ρ_0 is added to (4.1) instead for each base station which is not common to both sets, where ρ_0 is a system-specific tuned constant. Figure 4.1 illustrates a simple case for which the penalty term is functional. Site a is registered to both stations whereas Site b, since it lies beyond the coverage area of Base 2, is only registered to Base 1. Since the mobile is at Site b, it also cannot receive from Base 2. As such, the similarity metric is computed only from Base 1. Since both sites are equidistant from Base 1, without the penalty term they would have equal similarities; on the other hand, by penalizing Site a because there is no reception from Base 2, the mobile's location can be successfully resolved to Site b.

In the fingerprinting stage, the sites are selected on a square grid throughout the deployment area. The fingerprinted power at a site is the expected value of the received power at the location, i.e., neglecting the stochastic effects of shadowing. The measured power at the mobile device during a location query can be modeled as the sum of three terms. The sum is expressed as

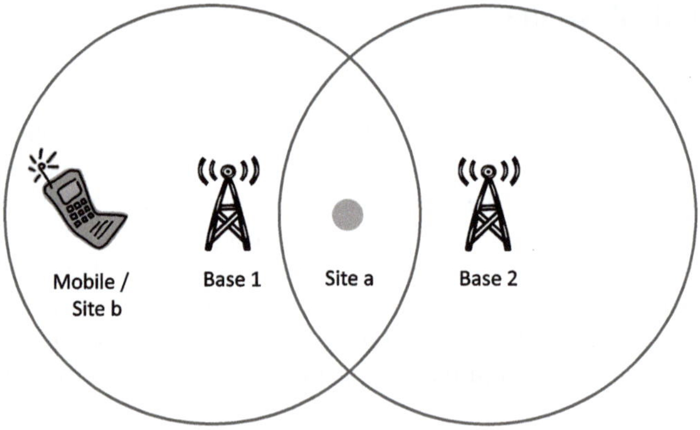

Fig. 4.1 Site a is within the coverage areas of both base stations while Site b is only within the coverage area of Site b. Although both sites are equidistant from Base 1, because the mobile only receives from Base 1, the mobile's location can be resolved to Site b

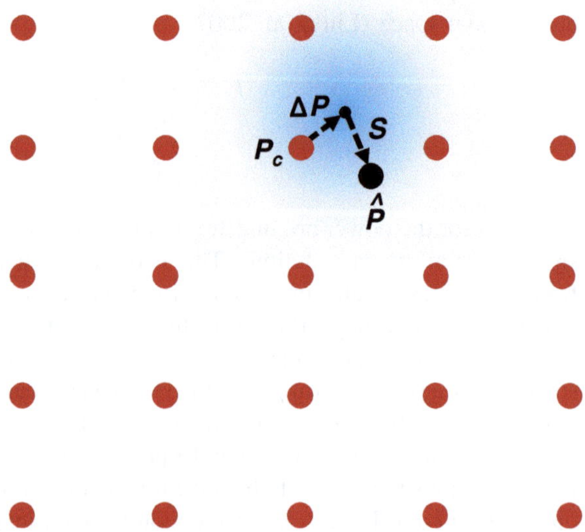

Fig. 4.2 A 5 × 5 grid of fingerprinted sites (*red*). The actual location of the mobile device is shown in *black*. The expected location of the mobile is in the middle of the radial pattern, however, due to shading, the mobile may be found anywhere. The probability of finding the mobile decreases as the radial pattern fades away

$$\hat{\boldsymbol{P}} = \boldsymbol{P}_c + \Delta\boldsymbol{P} + \boldsymbol{S}. \tag{4.2}$$

The first term is the signature power of the site which is most similar to the measured power. This site is indexed as site *c* because it indicates the correct estimate for the mobile's location. The second term is the offset between the mobile's expected power and the fingerprinted power at site *c*. And the third term is the fluctuation of the signal due to shadowing. Figure 4.2 illustrates these three components. Shown is a 5 × 5 grid of fingerprinted sites (red). The expected location of the mobile is at the center of the radial pattern. Due to shadow fading,

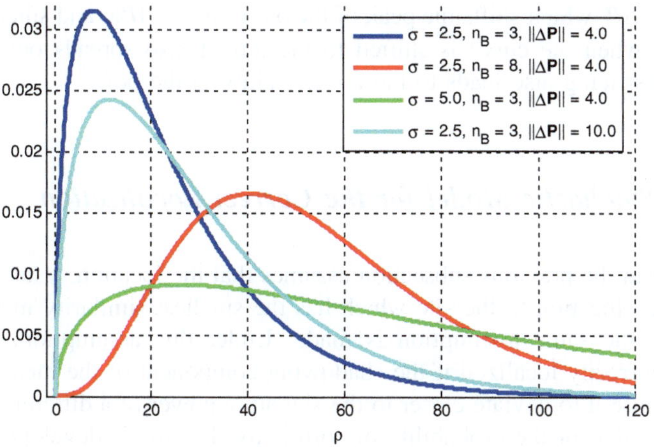

Fig. 4.3 The non-central Chi-squared distribution represents the pdf of the similarity metric ρ_c: σ is the standard deviation of the shadow fading parameter, n_B is the number of base stations in the system, and the parameter $\| \Delta P \|$ is proportional to the grid spacing between the fingerprinted sites in the area

however, the mobile (black) may be found anywhere. The probability of finding the mobile decreases as the radial pattern fades away.

Since the power is measured on a logarithmic scale, the n_B individual components of $\boldsymbol{S}$ are distributed normally as $S_j \sim N(0, \sigma)$ [see Chap. 3]. By substituting Eq. (4.2) into (4.1), the latter then reduces to $\rho_c = \|\boldsymbol{\Delta P}\| + \|\boldsymbol{S}\|$. The resultant distribution for the similarity metric of the correct location is the non-central Chi-square probability density function (pdf) with n_B degrees of freedom:

$$f_{\rho_c}(\rho) = \mathrm{e}^{-\frac{(\|\boldsymbol{\Delta P}\|+\rho)}{2\sigma^2}} \frac{1}{2\sigma^2} \left(\frac{\rho}{\|\boldsymbol{\Delta P}\|}\right)^{\frac{(n_B-2)}{4}} \mathrm{J}_{\frac{n_B-2}{2}}\left(\frac{\sqrt{\|\boldsymbol{\Delta P}\|\rho}}{\sigma^2}\right), \quad \rho \geq 0, \tag{4.3}$$

where $J_\vartheta(\cdot)$ is the ϑ-th order Bessel function of the first kind.

The effects of σ, n_B, and $\boldsymbol{\Delta P}$ on the similarity metric are illustrated in Fig. 4.3. Naturally, the pdf spreads out as the amount of shadowing, represented by σ increases. Adding base stations to the system magnifies this effect since there will be more shadowing components in $\boldsymbol{S}$. The latter phenomenon is captured in Eq. (4.3) through the associated parameter n_B, which spreads the curve out yet further. Although with additional stations the similarity metric is more susceptible to shadowing, the enhanced identifiability that the stations bring to the sites delivers better performance overall. This is highlighted in the following subsection.

The maximum achievable offset power occurs when the mobile device lies as far as possible from any one of the fingerprinted sites, i.e. at the midpoint of the square formed by the four sites closest to the mobile. By increasing the grid spacing, this maximum displacement will also increase. Hence, $\|\boldsymbol{\Delta P}\|$ is proportional to the grid spacing. The non-centrality of the distribution is attributed to the offset term, $\Delta \boldsymbol{P}$, which shifts the peak of the pdf to $\rho = \|\Delta \boldsymbol{P}\|$; and since $f_{\rho_c}(0) = 0$ invariably, when the curve is shifted to the left, it also spreads out. Therefore, larger grid spacing also leads to more uncertainty in the pdf.

4.1.2 A Stochastic Model for the Correct Localization

The model in the previous subsection assumes that the location system associates the mobile's location to the site which has the smallest similarity metric. In this subsection, the same assumption is made. Under this assumption, the mobile device is correctly localized if the shadowing component of the measured power does not cause it to deviate closer to the signature power of a different site. In the sequel, a model for the probability of correct localization is developed.

4.1.2.1 Model Description

Formally stated, the system correctly localizes the mobile device if the measured power, $\hat{\boldsymbol{P}}$, is more similar to the fingerprinted power of site c, $\boldsymbol{P}_c$, than to the fingerprinted power of any another site i. The marginal probability of correct localization when considering a single site i can be expressed as

$$p(\rho_c \le \rho_i) = p\left(\sum_{j=1}^{n_B} \left(\hat{P}_j - P_{cj}\right)^2 \le \sum_{j=1}^{n_B} \left(\hat{P}_j - P_{ij}\right)^2\right). \tag{4.4}$$

By expanding and collecting terms, the expression can be reduced to

$$p(C_i \le 0), \tag{4.5}$$

where $C_i = \sum_{j=1}^{n_B} 2\hat{P}_j \beta_{ij} + \Delta P_{ij}$ is a newly defined random variable with associated constants $\Delta P_{ij} = \left(P_{ij} - P_{cj}\right)$ and $\beta_{ij} = \left(P_{cj}^2 - P_{ij}^2\right)$. Note that vector $\Delta \boldsymbol{P}_i$ is the offset power between the fingerprints of sites i and c. The vector $\boldsymbol{\beta}_i$ is a second-order offset. It follows that since $\hat{P}_j$ is normally distributed due to shadowing, C_i is also normally distributed, however with mean and variance

$$\begin{aligned} \mu_{C_i} &= \sum_{j=1}^{n_B} 2P_{ij}\beta_{ij} + \Delta P_{ij} \\ \sigma_{C_i}^2 &= \sum_{j=1}^{n_B} \left(2\beta_{ij}\sigma\right)^2. \end{aligned} \tag{4.6}$$

Now, the total probability of correctly localizing the mobile device to site c—total here implies when considering all of the other $n_M - 1$ sites, not just site

i—can be computed. This probability, $p(C)$, can be expressed as the joint probability of all the other sites having a greater similarity metric than site c:

$$p(C) = p(C_1 \leq 0, C_2 \leq 0, \ldots, C_{c-1} \leq 0, C_{c+1} \leq 0, \ldots, C_{n_M} \leq 0). \tag{4.7}$$

Note, however, that the $n_M - 1$ events above are interdependent. This can be seen by considering a simple example with only one base station in the deployment area for which $P_{c1} < \hat{P}_1 < P_{i1} < P_{i+1,1}$. It follows that $p(C_i \leq 0)$ implies $p(C_{i+1} \leq 0)$. Unfortunately, computing the joint probability in Eq. (4.7) results in a complicated expression. Rather, as an approximation, the events are considered to be independent. As such,

$$p(C) \approx \prod_{\substack{i=1 \\ i \neq c}}^{n_M} p(C_i \leq 0). \tag{4.8}$$

The validity of this approximation is examined in the paper. It shows that for $n_B > 2$, which is the case in most practical implementations, the approximation holds very well. This demonstrates that adding base stations to the system decorrelates the events. The events were further decorrelated because the experiments were conducted in non-line-of-sight conditions—conditions for which the size of the random component (shadow fading) is yet larger. The details of the experiments are included next.

4.1.2.2 Performance Evaluation

The probability of correctly localizing the mobile device—the performance metric of the system—was analyzed by considering an example deployment with 25 sites arranged as a 5 × 5 grid (see Fig. 4.2). The grid spacing was 1 m. The n_B base stations were positioned randomly at grid points around the outermost square and the mobile device was fixed at the grid center. The simple path loss model from (2.24) was employed. In it, the path loss is modeled as a deterministic function of the distance, d_{ij}, between site i and base station j:

$$L(d_{ij}) = L^0 + 10\alpha \log_{10}\left(\frac{d_{ij}}{d^0}\right). \tag{4.9}$$

The reference path loss and the reference distance were set respectively to $L^0 = 37.7$ dB and $d^0 = 1$ m and the base stations all transmitted at $P^{TX} = 15$ dBm. Neglecting the transmitter and receiver antenna gains, the received power is computed as the transmit power minus the path loss, plus a shadowing component, or:

$$P_{ij} = P^{TX} - L(d_{ij}) + S. \tag{4.10}$$

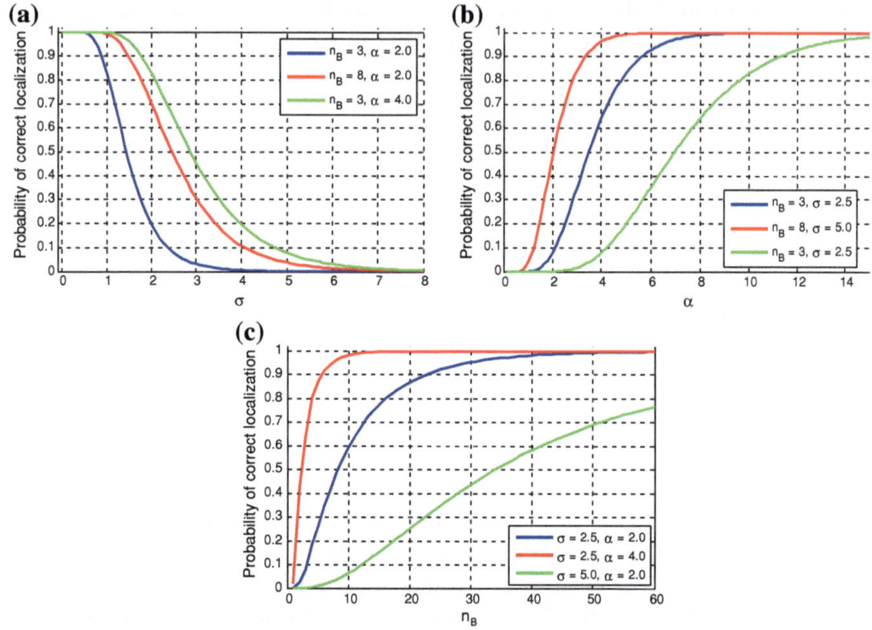

Fig. 4.4 The probability of correct localization. **a** Probability of correct localization as a function of the standard deviation of shadow fading for different numbers of base stations and path loss exponents. **b** Probability of correct localization as a function of path loss exponent for different numbers of base stations and shadow fading standard deviations. **c** Probability of correct localization as a function of the number of base stations for different shadow fading standard deviations and path loss exponents

The shadowing was assumed to be distributed as a zero-mean Gaussian with standard deviation σ. In the fingerprinting stage, the feature vector of a site was calculated as the deterministic received power—i.e., with zero shadow fading—from each base station. Only $n_M = 9$ of the sites were fingerprinted: the center grid point and its eight adjacent points.

Figure 4.4a investigates the effect of the shadowing parameter σ on the probability of correct localization. The set of parameters associated with the green curve could represent the outdoor propagation environment ($\alpha = 4$ is a typical value for the path loss exponent) with three base stations. For these parameters, the performance of the system is shown to degrade rapidly—from a probability above 0.8 to a value below 0.2—as the standard deviation increases from 2 dB beyond 4 dB. But typical values of σ outdoors can be as high as 10 dB in urban environments; even indoors, the experiments in Gentile et al. (2008) report values in the range 2.8–5.4 dB. This demonstrates that a localization resolution of 1 m is practically unattainable for these parameters—which correspond to the best-case scenario of the three shown—even with such fine grid spacing, which in practice would require a laborious fingerprinting stage. In fact, most physical implementations of memoryless fingerprinting systems using RSS report errors above 2 m (Bahl and

Padmanadhan 2000; Brunato and Battiti 2005). As we shall see in later sections, implementing memory systems or using different features, namely the channel impulse response to better characterize the sites, can greatly improve results.

Figure 4.4b investigates the effect of the path loss exponent on the system. The performance is shown to improve as α increases, meaning that the system localizes better in a harsher propagation environment. The explanation for this is that the fingerprinted power between sites dropped off more rapidly, enhancing the discriminatory properties of the system, thereby decreasing the chances of misclassification. Hence, the fingerprinting technique exploits the very weaknesses of the RSS ranging technique—which is intended for quasi-line-of-sight conditions only—described in Chap. 2. So, in fact, the two techniques are complementary.

The probability of correct localization—that is selecting the correct nearest neighbor grid point—improves as the grid spacing increases. The effect seen through the model is equivalent to increasing the path loss exponent. This is because, since the sites are farther apart, there is more variability in the signal strength between them. In fact, when plotting the performance metric as a function of grid spacing, the curves look very similar to the ones in Fig. 4.4b. Of course, the disadvantage of larger grid spacing is that it lowers the maximum attainable resolution.

Finally, the number of base stations in the system was varied. As seen in Fig. 4.4c, the performance of the system for the parameters corresponding to the red curve, which assumes harsh propagation environment ($\alpha = 4$) and low shadow fading standard deviation ($\sigma = 2.5$), stabilizes at $n_B = 5$. Beyond this value it continues to increase, but at a diminishing rate. In a more favorable propagation environment, there is a bit more benefit from adding base stations (blue curve), and with higher shadow fading (green curve) there is more consistent benefit. As mentioned earlier, the effect of shadowing weighs more heavily on the system as the number of base stations increases; this is seen by the shallower slope of the green curve with respect to the blue. Yet, this effect is offset by the benefit of greater identifiability; hence the performance continues to improve monotonically.

4.2 Memoryless Systems

The analysis in the previous section assumes the nearest neighbor method—the most simplistic of mapping algorithms—is employed to determine the mobile's location from a measured feature. However, more sophisticated methods, such as the *k*-nearest neighbor method (*k*NN), probabilistic methods, neural networks, support vector machines, and the smallest M-vertex polygon method in Liu et al. (2007) can enhance localization accuracy. For example, Agiwal et al. (2004) introduced the LOCATOR algorithm, which is an RSS-based fingerprinting technique incorporating a number of different approaches. Specifically, in the fingerprinting stage, the radio map is subdivided into clusters to reduce the computational cost in the localization stage. The authors further use RSS distribution functions at the sites and interpolations to improve performance. In Moustafa and Ashok (2005), the Horus RSS-based system models the RSS distribution received from base stations through parametric and non-parametric distributions, exploiting this information to reduce temporal variations in the radio map. Also, Fang et al. (2008) demonstrated further improvements by using an RSS averaging technique on a logarithmic scale to mitigate noise resulting from multipath.

The purpose of this section is to provide an overview of some of the aforementioned methods. Specifically, we present the comparison which was published in Brunato and Battiti (2005) between the weighted k-nearest neighbor method, support vector machines, Bayesian inference, and neural networks. As mentioned earlier, although these methods implement the received-signal-strength feature, they can be readily extended to features such as the channel impulse response or the frequency channel coherence function.

4.2.1 The Weighted k-Nearest Neighbors Method

The first of the mapping algorithms we investigate is the *k-Nearest Neighbor* method, which is just an extension of the nearest neighbor method providing enhanced robustness to shadowing. Precisely, rather than map the mobile's location to the single nearest neighbor site, the k nearest neighbor sites are employed, where k is a fixed constant. In practice, the mobile's location is estimated as the centroid of the k site locations—together these sites form subset $\boldsymbol{K}$—which have the smallest similarity metrics among all the sites. A refinement of the method is the weighted kNN method proposed in Brunato and Battiti (2005), which scales the contribution of each by the reciprocal of the similarity metric. Specifically, the mobile's location is estimated as a linear combination from the subset:

$$\tilde{\boldsymbol{x}} = \frac{\sum\limits_{i \in K} \frac{\boldsymbol{x}_i}{\rho_i + \rho_0}}{\sum\limits_{i \in K} \frac{1}{\rho_i + \rho_0}}. \tag{4.11}$$

As such, the location will fall within the convex hull of the site locations. By associating to location $\boldsymbol{x}_i$ a weight inversely proportional to the similarity metric ρ_i, greater importance is given to sites whose signature power is closer to the measured power. The constant ρ_0 is a small quantity added to ensure numerical stability when the similarity metric is close to zero, and the denominator of (4.11) serves to normalize the weights such that their sum is equal to one.

4.2.2 Support Vector Machines

Support vector machines (SVM) were developed in the area of supervised machine learning in order to solve nonlinear regression and statistical classification problems. RSS-based fingerprinting methods based on support vector machines have been reported in Wu et al. (2004), Li Wu et al. (2007). In Brunato and Battiti (2005), they provide a direct mapping from the measured power at the mobile device to its estimated location through nonlinear regression[1]—nonlinear regression on the training pairs $(\boldsymbol{x}_i, \boldsymbol{P}_i), i = 1 \ldots n_M$. Two mappings from the measured power vector, $\hat{\boldsymbol{P}}$, to the estimated (x, y)-coordinates of the mobile location, $\boldsymbol{x}$—denoted as $x(\hat{\boldsymbol{P}})$ and $y(\hat{\boldsymbol{P}})$—are generated separately. Henceforth, we concentrate on the x-mapping, as the method applies equivalently to the y-mapping.

The x mapping can be expressed as a weighted sum of M prescribed nonlinear functions, $g_m(\cdot), m = 1 \ldots M$, or

$$x(\hat{\boldsymbol{P}}) = \sum_{m=1}^{M} w_m g_m(\hat{\boldsymbol{P}} - \overline{\boldsymbol{P}}) + \overline{x}, \tag{4.12}$$

where $\overline{x} = \sum_{i=1}^{n_M} x_i$ is the x-centroid and $\overline{\boldsymbol{P}} = \sum_{i=1}^{n_M} \boldsymbol{P}_i$ is the mean power vector. The solution to the regression yields values for the weights w_m. For instance, if $g_m(\cdot) = (\cdot)^{m-1}$ is selected, $x(\hat{\boldsymbol{P}})$ is represented by an $(M-1)$th-degree polynomial, where the weights form the associated set of coefficients.

The regression is obtained by solving a convex quadratic program with the following objective function:

$$C \sum_{i=1}^{n_M} \xi_i + \frac{1}{2} \sum_{m=1}^{M} w_m^2. \tag{4.13}$$

The main objective of the program, which is embodied by the first term, is to find the mapping which yields the best fit to the training pairs; this is achieved by minimizing the sum of residuals $\xi_i = |x_i - x(\boldsymbol{P}_i)|$ over all the pairs. The secondary objective, embodied by the second term, is to reduce the complexity of the mapping such that it can be represented in the lowest dimensional space, where M is the maximum dimension; this is achieved by minimizing the norm of the weights. The constant, C, balances the importance of the two objectives. Then the quadratic program can be stated completely as:

$$\begin{aligned} \min \quad & C \sum_{i=1}^{n_M} \xi_i^+ + \xi_i^- + \tfrac{1}{2} \sum_{m=1}^{M} w_m^2 \\ \text{subject to} \quad & \begin{cases} x_i - x(\boldsymbol{P}_i) \le \xi_i^+ \\ x(\boldsymbol{P}_i) - x_i \le \xi_i^- \\ \xi_i^+, \xi_i^- \ge 0 \end{cases} \end{aligned} \tag{4.14}$$

By decomposing the residuals into positive and negative components, as $\xi_i = \xi_i^+ - \xi_i^-$, the absolute values on the residuals are removed such that the

[1] Their application to statistical classification is similar.

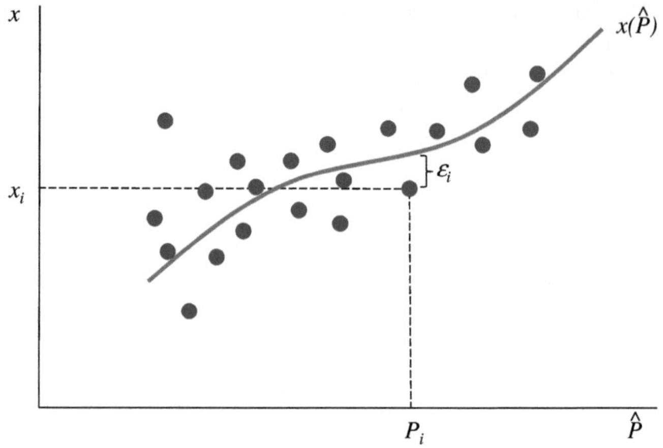

Fig. 4.5 The Support Vector Machine (*SVM*) mapping between the measured received signal power space, $\hat{P}$, and the mobile location space, x

problem can be written in standard form. Figure 4.5 illustrates an example regression for the function $g(P) = P^3$ in the one-dimensional power vector space.

As is often the case in convex programing, here it is more practical to solve the dual quadratic program in (4.15) instead by introducing Lagrange multipliers, $(\lambda_i^+, \lambda_i^-), i = 1 \ldots n_M$ (Smola and Schoelkoepf 2004):

$$\begin{aligned} \text{max} \quad & -\frac{1}{2}\sum_{i=1}^{n_M}\sum_{k=1}^{n_M}(\lambda_i^+ - \lambda_i^-)K(\boldsymbol{P}_i, \boldsymbol{P}_k)(\lambda_k^+ - \lambda_k^-) + \sum_{i=1}^{n_M} x_i(\lambda_i^+ - \lambda_i^-) \\ \text{subject to} \quad & \sum_{i=1}^{n_M}(\lambda_i^+ - \lambda_i^-) = 0, \\ & 0 \le \lambda_i^+, \lambda_i^- \le C \end{aligned} \tag{4.15}$$

where $K(\boldsymbol{P}_i, \boldsymbol{P}_k) = \sum_{m=1}^{M} g_m(\boldsymbol{P}_i - \boldsymbol{P})g_m(\boldsymbol{P}_k - \boldsymbol{P})$ is known as the kernel function. The solution to the dual problem yields the values for $(\lambda_i^+, \lambda_i^-)$. From them, the components of the weight vector in (4.12) can be found as

$$w_m = \sum_{i=1}^{n_M}(\lambda_i^+ - \lambda_i^-)g_m(\boldsymbol{P}_i - \boldsymbol{P}). \tag{4.16}$$

4.2.3 Neural Networks

In contrast to the well-defined mathematical formulation provided by Support Vector Machines, a "black box" approach for generating the mapping between the measured power vector space and the estimated mobile location space is through the application of neural networks. Neural networks are powerful tools for solving ill-posed problems, i.e., problems for which the causes of certain observations are either not understood or too complex to define mathematically, and for which there is often no unique solution. This applies to fingerprinting systems in that an observed radio frequency feature value—depending on the number of base stations in the system—may correspond to multiple locations in the survey area; moreover, the value depends on the structure of the environment (building blueprint, wall materials, furniture characteristics, etc.). For such systems, there is no attempt to explicitly model the complex propagation environment which causes these observations. Rather, it is simply acknowledged that such a nonlinear relationship exists. By observing the RF signatures at specific locations, the neural network learns the relationship. Neural network methods for RSS-based location fingerprinting have been reported in Battiti et al. (2002), Edgar et al. (2004), Brunato and Battiti (2005).

A neural network is a network composed from entities, known as neurons, which have multiple input ports and a single output port. In Brunato and Battiti (2005), the multilayer perceptron neural network is implemented. The multilayer perceptron, in particular, is a feedforward network partitioned into distinct layers. Feedforward means that the input of a neuron in one layer is connected only from the outputs of a neuron in the immediate lower layer. Each connection has an associated weight which serves to scale the output value between the two layers. In the RSS-fingerprinting application, the inputs to the lowest layer of the network are the n_B elements of the measured power vector—there is one neuron for each base station. Likewise, the outputs of the highest layer are the two coordinates of the location vector—there is one neuron for each coordinate dimension. Figure 4.6 shows a diagram of the network.

The role of the individual neuron in the network is simply to compute its output value from the collection of its inputs. This is executed by summing over the input values and then mapping the sum to the output through an activation function. By choosing a nonlinear function for the neuron, the network is capable of representing any nonlinear function for the network as a whole. In fact, the Cybenko theorem, also known as the universal approximation theorem, states that a feedforward network with a single hidden layer and a finite number of neurons can approximate any continuous function (assuming a "well-behaved" activation function) (Cybenko 1989). A commonly used activation function for the perceptron is the sigmoid function:

$$h(z) = \frac{1}{1 + e^{-\lambda z}}, \tag{4.17}$$

where λ controls the linearity of the function. Figure 4.7 illustrates the curve for several values of the parameter. As in the SVM framework, neural networks can be implemented for both regression problems and classification problems. For small values of λ, the function is linear around $z = 0$ and then saturates at the extrema. This range of λ is applicable to regression problems, such as ours, so that the

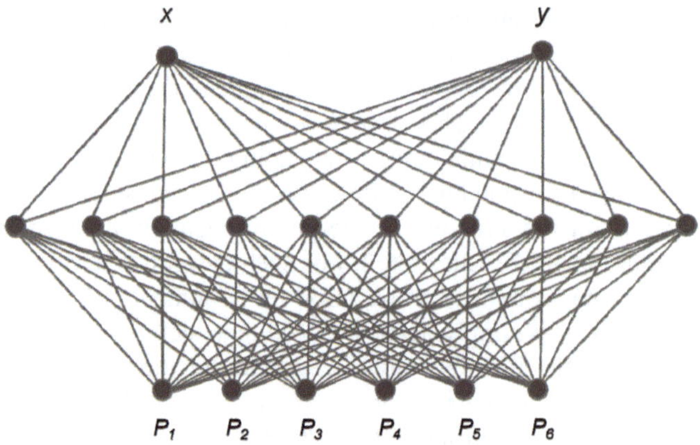

Fig. 4.6 A three-layer feed forward neural network. The inputs to the first layer are the measured received signal strengths from six base stations. The outputs are the two-dimensional coordinates of the estimated location of the mobile device. The network has one hidden layer

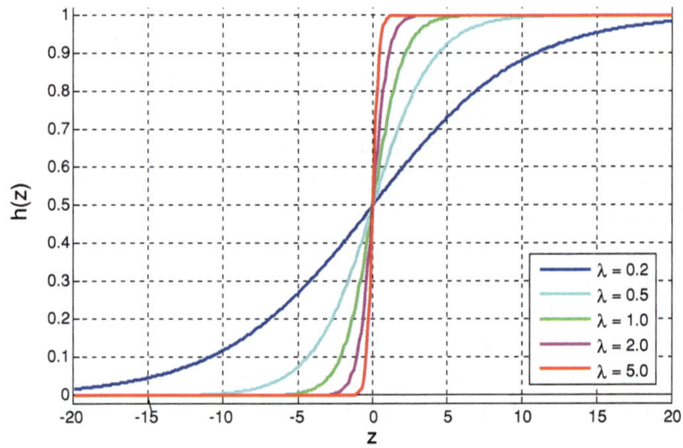

Fig. 4.7 The sigmoid function

output can assume continuous values. As λ approaches infinity, $h(z)$ becomes a step function. This range is applicable to classification problems for which the output is either a one or a zero, meaning that the input either belongs to a certain class or belongs respectively to a different class.

The network learns the mapping through an iterative algorithm, such as the well-known Backpropagation Algorithm, which tunes the connection weights according to the input/output excitations. During each iteration, by clamping the inputs and outputs of the network with the values of the training pair $(\boldsymbol{x}_i, \boldsymbol{P}_i)$, the weights are adjusted such that the network yields $\boldsymbol{x}_i$ as an output given $\boldsymbol{P}_i$ as an input. The details of network design and training can be found in Bose (1995).

4.2.4 Bayesian Inference

Probabilistic approaches for estimating the location of a mobile device in fingerprinting systems have been reported in Roos et al. (2002), Youssef et al. (2003), Fox et al. (2003), Madigan et al. (2005), Kushki et al. (2007). As in the *k*NN, SVM, and neural network frameworks, the estimated location is not constrained to any one of the discrete fingerprinted sites, meaning that it can assume any position throughout the deployment area. The problem is posed in the framework of Bayesian inference: given the measured power vector, $\hat{\boldsymbol{P}}$, the posterior probability (or simply the posterior), $p(\boldsymbol{x}|\hat{\boldsymbol{P}})$, that the mobile is located at position $\boldsymbol{x}$ is calculated for all candidate positions in the area. Then, from this probability, the mobile's location is estimated either through Maximum Likelihood as

$$\tilde{\boldsymbol{x}} = \max_{\boldsymbol{x}} \arg p\left(\boldsymbol{x}|\hat{\boldsymbol{P}}\right) \tag{4.18}$$

or as an expected value over the area:

$$\tilde{\boldsymbol{x}} = \int_{\boldsymbol{x}} \boldsymbol{x} \cdot p(\boldsymbol{x}|\hat{\boldsymbol{P}}) d\boldsymbol{x}. \tag{4.19}$$

The posterior probability, $p(\boldsymbol{x}|\hat{\boldsymbol{P}})$, can be viewed as a mapping from the power vector space to the mobile location space. In the SVM and neural network frameworks, such mappings are computed through some sort of nonlinear regression on the training pairs. In the Bayesian framework, however, a mapping is first computed in the opposite direction, i.e., from $\boldsymbol{x}$ to $\hat{\boldsymbol{P}}$. This inverse mapping, denoted as $p(\hat{\boldsymbol{P}}|\boldsymbol{x})$, is known as the likelihood function (or simply the likelihood) and effectively serves as the RSS signature for the site. It is the probability function that the power vector $\hat{\boldsymbol{P}}$ will be measured if the mobile device is at $\boldsymbol{x}$. The benefit of this approach is that likelihood can be computed directly from the training pairs. For example, in (Roos), (Kuschki), (Fox), (Madigan) the likelihood function is constructed from the histogram of RSS values registered at each site. (More details about how to generate the histogram are provided in Sect. 4.3). Once the likelihood is computed, it is related back to the posterior probability through Bayes' Rule, as we shall see in the sequel.

As an alternative to constructing histograms at the discrete sites, (Brunato and Battiti 2005) invoke a path loss model to calculate the likelihood function. The path loss model enables generating RSS signature values at continuous points throughout the survey area—rather than at discrete points only—in the hope of improving localization resolution. The path loss model employed is similar to the traditional model in (4.9), however, it also accounts for the attenuation of the walls between a base and mobile pair. This more comprehensive path loss model can be expressed as:

$$L_j(\boldsymbol{x}) = L_j^0 + 10\alpha_j \log_{10}\left(\frac{d_j(\boldsymbol{x})}{d^0}\right) + L_j^w \cdot n_j^w(\boldsymbol{x}) \tag{4.20}$$

Note that, in order to represent the radio environment more precisely, each base station j has it own path loss, $L_j(\boldsymbol{x})$. The first term is the associated reference path loss, L_j^0, and the second term is the propagation loss, where α_j is the loss exponent and $d_j(\boldsymbol{x})$ is the distance between the base station and the mobile device. The last term is the penetration loss due to walls, with $n_j^w(\boldsymbol{x})$ denoting the number of walls between the base and the mobile and L_j^w denoting the penetration loss per wall.

The unknown parameters of the path loss model can be extracted through the data points given by the training pairs $(\boldsymbol{x}_i, \boldsymbol{P}_i), i = 1 \ldots n_M$. To this end, recall from Eq. (4.10) that the deterministic received power at the mobile is given from the loss as

$$P_j(\boldsymbol{x}) = P^{TX} - L_j(\boldsymbol{x}), \tag{4.21}$$

where P^{TX} is the known transmit power. Then for each base station j, the training pair $(\boldsymbol{x}_i, \boldsymbol{P}_i)$ furnishes exactly one linear equation with three unknowns from (4.21). The system of n_M equations, which is overdetermined for $n_M > 3$, can be solved for the values of (L_j^0, α_j, L_j^W) through Least Squares Regression. Note that it is also possible to assume the same loss model for all base stations by removing the index j in (4.20), however the authors report that this causes degradation in performance.

With the parameters of the path loss model in hand, the likelihood function can now be obtained. Recall that in the shadow fading model, the measured power, $\hat{P}_j$, from base station j at location $\boldsymbol{x}$ deviates from the deterministic power, $P_j(\boldsymbol{x})$, by the random variable S. In other words,

$$\hat{P}_j = P_j(\boldsymbol{x}) + S \tag{4.22}$$

Since S is a zero-mean normally distributed random variable, the likelihood that $\hat{P}_j$ was measured at location $\boldsymbol{x}$ is given through the Gaussian kernel[2]:

$$p(\hat{P}_j|\boldsymbol{x}) = \frac{1}{\sqrt{2\pi}\sigma} e^{-\frac{\left(\hat{P}_j - P_j(x)\right)^2}{2\sigma^2}}. \tag{4.23}$$

For simplicity of computation—although not always true in practice—the signals from the n_B base stations are assumed to experience independent and identically distributed shadowing such that the received powers from each base station are also statistically independent. Note that this is directly related to the independence assumption in (4.8), which was experimentally shown to be a valid approximation for $n_B > 2$. This assumption is also made in (Roos), (Kuschki), (Fox), (Madigan). As a result, the likelihood of the measured power vector can be calculated as the product of the measured powers from each of the base stations:

[2] On a similar note, in (Roos), (Kuschki), (Fox), (Madigan) mathematical expressions which are close to the Nadaraya-Watson Kernel regression are developed.

$$p(\hat{\boldsymbol{P}}|\boldsymbol{x}) = \prod_{j=1}^{n_B} p(\hat{P}_j|\boldsymbol{x}) \tag{4.24}$$

Finally, the likelihood, $p(\hat{\boldsymbol{P}}|\boldsymbol{x})$, is related back to the posterior probability, $p(\boldsymbol{x}|\hat{\boldsymbol{P}})$, through Bayes' Rule:

$$p(\boldsymbol{x}|\hat{\boldsymbol{P}}) = \frac{p(\hat{\boldsymbol{P}}|\boldsymbol{x}) \cdot p(\boldsymbol{x})}{p(\hat{\boldsymbol{P}})} \tag{4.25}$$

If all the locations throughout the survey area are visited with equal frequency, the prior probability or simply the prior, $p(\boldsymbol{x})$, is uniformly distributed. Otherwise, if certain locations have higher or lower frequencies, the prior will be distributed proportionately; as a simple example, in most households more time is spent in the living room than in the attic. The value of $p(\hat{\boldsymbol{P}})$ is computed through the law of total probability:

$$p(\hat{\boldsymbol{P}}) = \int_x p(\hat{\boldsymbol{P}}|\boldsymbol{x}) \cdot p(\boldsymbol{x})d\boldsymbol{x} \tag{4.26}$$

4.2.5 Comparison of Methods

The specific parameters implemented for comparing the four methods described in this section are provided in Brunato and Battiti (2005). The test experiments were conducted in a deployment area of roughly 750 m^2. The area was partitioned into five rooms and in each room a separate Wi–Fi base station was deployed. While for the most part LOS conditions existed within the individual rooms, the walls throughout the area between the base stations and the mobile device created NLOS conditions. The fingerprinted sites were spaced at about 3.5 m apart, for a total of 257 sites in the area. For the parameter settings in the paper, the weighted *k*-nearest neighbor and the support vector machine methods delivered the best performance, both averaging a location error of about 3 m. While the computational complexity for training the *k*NN is lower than that of the SVM, the latter boasts a much lower complexity in the localization stage. The average location error for the neural network method was about 3.2 m, but the time required for tuning the 60 weights was the highest among all methods; once tuned, however, the neural network localized the quickest. The Bayesian interference was both computationally inefficient and also sustained the worst average error of 3.35 m. The authors attributed the poor performance to the adopted path loss model in (4.20) with a total of only 20 tunable parameters—four for each of the five base stations. The Bayesian method required only a few training points for parameter fitting but, once fit, providing more training points did not improve the results further. On the other hand, the neural network, with a total of 60 tunable weights, offered a better degree of fitting.

Another performance evaluation of different RSS-based fingerprinting methods is presented in Lin and Lin (2005). The authors compare the *k*NN, a probabilistic method, and neural networks. The results of the analysis and experiments reveal that the *k*NN reports the best overall performance for indoor positioning. The performance of histogram, nearest neighbor, parametric and kernel location fingerprinting methods were evaluated in Honkavirta October (2008), Honkavirta et al. March (2009). Again, the results revealed that the *k*-nearest neighbor method fared the same or better than the other methods depending on the environment.

In practice, it is difficult to rank the performance of the four methods described here because they are sensitive—each to one extent or another—to the choice of implementation parameters. As a matter of fact, the methods are more similar to each other than they are different—all essentially just fit a curve to the training pairs. The parameters selected for each determine to what degree the fitting can be achieved. In support vector machines, the degree of freedom increases with the order of the nonlinear functions $g_m(\cdot), m = 1 \ldots M$ in (4.12) and with the number M of functions itself. In addition— and more explicitly—by increasing the parameter C in Eq. (4.13), greater importance is given to the minimizing the fitting error; in contrast, by decreasing this parameter, the system order is minimized. In neural networks, the relationship to the system order is even more explicit: increasing the number of neurons in the hidden layer increases the degree of freedom. In Bayesian inference, as just mentioned, the degree of freedom is dictated by the number of unknowns in the path loss equation in Eq.(4.20). Finally, in the *k*-nearest neighbor method the estimated location is determined as a curve interpolated between the *k*-nearest neighbors. By increasing the value of *k*, although more robust to measurement error, the estimated location is more constrained.

For illustrative purposes only, Fig. 4.8 shows three curves fit to a set of training pairs (red). The orange curve represents a function which is overfit; in order to reduce the fitting error to zero, a high-order curve is allowed. While the error is zero for the set of training pairs, the curve does not interpolate well between the training pairs. The large oscillations indicate that a small change in the measured RSS vector maps to a completely different location, making for an unstable system. On the other hand, the green curve represents an underfit function; because the function has only a few degrees of freedom, it is very robust to fluctuations in signal strength. At the same time, the poor fitting to the training pairs can also lead to large location errors. Lastly, the blue curve presents a good balance between location accuracy and robustness.

4.3 Memory Systems

Thus so far we have considered only memoryless systems, which estimate the location of a mobile device based solely on the received signal strength observed at a single instant in time. While these systems may deliver acceptable performance

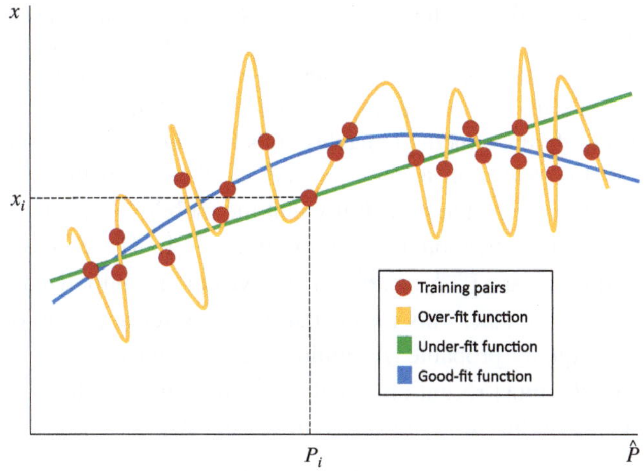

Fig. 4.8 The four methods presented in this section—each through a different algorithm—generate some mapping between the signal strength space and the location space. The parameters of each determine the degree of fitting to the training pairs. Shown here are three fits for illustrative purposes

for some applications, by integrating observations available from previous time instants as well, both precision and stability can be enhanced. In this section, we describe techniques first developed to solve the *wake-up robot problem* (Burgard et al. 1996) which have been adapted to fingerprinting. The scope of wake-up robot problem is for a robot, which is placed in an arbitrary environment, to discern its position by gathering and processing sensory data with no prior knowledge. Chapter 9 is completely dedicated to these techniques—often referred to as Simultaneous Localization and Mapping (SLAM)—with specific application to inertial based localization. In the following, we first investigate a technique which is an adaptation of the Bayesian interference method introduced in Sect. 4.2.4 to memory systems. We then present an evolution of this technique, known as grid-based Markov localization, which delivers enhanced stability.

4.3.1 Bayesian Inference in Memory Systems

In this section, we consider an application for which the orientation of the mobile device, in additional to its location, is estimated. This is achieved by augmenting the fingerprinted information gathered at site i—previously only the location coordinates, $\boldsymbol{x}_i$, were fingerprinted—with an orientation identifier denoted as θ_i. The orientation identifier can assume one of two values: $\theta_i = 1$ signifies that the user is facing a designated direction at the site while $\theta_i = -1$ signifies that the user is facing the opposite direction. Of course more than just two orientations can be incorporated, if desired. We now define a *state* variable $s_k = \{\boldsymbol{x}_k, \theta_k\}$ for the mobile, which indicates both its location and its orientation. The mobile can lie in any of n_s possible states indexed through $k = 1\ldots n_s$. Note that for a total n_M fingerprinted sites with two orientations per site, $n_s = 2n_M$.

The Bayesian inference method enables constructing a time-varying posteriori probability for the state of a mobile device. This probability, denoted as $p\left(s_k|\hat{\boldsymbol{P}}^t,\ldots,\hat{\boldsymbol{P}}^0\right)$, represents the probability that the mobile lies in state k given the observations from initialization $(t = 0)$ to time $t-1$. These observations are indexed accordingly as $\hat{\boldsymbol{P}}^{t-1},\ldots,\hat{\boldsymbol{P}}^0$. The system is initialized by setting $p\left(s_k|\hat{\boldsymbol{P}}^0\right) = 1/n_s, k = 1\ldots n_s$. This means that in the absence of any observations, all locations are equally probable. Assuming the posterior at time $t-1$ has been computed, when the most recent observation, $\hat{\boldsymbol{P}}^t$, becomes available, Bayes' Rule is applied to compute the posterior at the next time step:

$$p\left(s_k|\hat{\boldsymbol{P}}^t,\hat{\boldsymbol{P}}^{t-1},\ldots,\hat{\boldsymbol{P}}^0\right) = \frac{p\left(\hat{\boldsymbol{P}}^t|s_k,\hat{\boldsymbol{P}}^{t-1},\ldots,\hat{\boldsymbol{P}}^0\right)\cdot p\left(s_k|\hat{\boldsymbol{P}}^{t-1},\ldots,\hat{\boldsymbol{P}}^0\right)}{p\left(\hat{\boldsymbol{P}}^t|\hat{\boldsymbol{P}}^{t-1},\ldots,\hat{\boldsymbol{P}}^0\right)} \tag{4.27}$$

The denominator in the equation above follows from the law of total probability as $p\left(\hat{\boldsymbol{P}}^t|\hat{\boldsymbol{P}}^{t-1},\ldots,\hat{\boldsymbol{P}}^0\right) = \sum_{k=1}^{n_s} p\left(\hat{\boldsymbol{P}}^t|s_k,\hat{\boldsymbol{P}}^{t-1},\ldots,\hat{\boldsymbol{P}}^0\right)\cdot p\left(s_k|\hat{\boldsymbol{P}}^{t-1},\ldots,\hat{\boldsymbol{P}}^0\right)$, i.e. the sum of the k-indexed numerator over all the n_s states. The denominator effectively serves as a normalizing factor such that $\sum_{k=1}^{n_s} p\left(s_k|\hat{\boldsymbol{P}}^t,\hat{\boldsymbol{P}}^{t-1},\ldots,\hat{\boldsymbol{P}}^0\right) = 1$, meaning that the mobile device will lie necessarily in one of the n_s states at time t. The likelihood, $p(\hat{\boldsymbol{P}}^t|s_k,\hat{\boldsymbol{P}}^{t-1},\ldots,\hat{\boldsymbol{P}}^0)$, in the numerator is the probability that the signal strength vector $\hat{\boldsymbol{P}}^t$ will be observed when the mobile lies in state s_k. Since this probability is assumed to be stationary—meaning that the observed power when the mobile user is at a particular location and in a particular orientation is static over time—the readings from previous time instants have no bearing on it. This assumption can be stated mathematically as $p\left(\hat{\boldsymbol{P}}^t|s_k,\hat{\boldsymbol{P}}^{t-1},\ldots,\hat{\boldsymbol{P}}^0\right) = p\left(\hat{\boldsymbol{P}}^t|s_k\right)$.

The probability $p\left(\hat{\boldsymbol{P}}|s_k\right)$ represents the distribution of the received signal strength vector, $\hat{\boldsymbol{P}}$, when the mobile is in state s_k. In this application, the distribution acts as the RSS signature for the corresponding location and orientation of the mobile. Indeed it has been shown in Gentile and Klein-Berndt (2004), Ladd et al. (2005) that the signature varies not only by location but also by orientation. While the distribution of the received power from a single base station is often assumed to be normal, in fact it is typically more complex and even multimodel. Rather, for a more accurate characterization of the RSS signature, the same authors propose generating a histogram of signal strength values empirically from a training set, $\boldsymbol{P}_k^l, l = 1\ldots L$ of L readings gathered at the mobile over a fixed window of time during the fingerprinting stage. Let $h_{kj}(\zeta)$ stand for the histogram of signal strength values collected from base station j when the mobile is in s_k. The histogram can be expressed mathematically as

$$h_{kj}(\zeta) = \frac{1}{L}\sum_{l=1}^{L}\delta(P_{kj}^{l} - \zeta), \tag{4.28}$$

where δ is the Kronecker delta function and ζ is an indicator variable which spans the range of all possible signal strength values. The range depends on the specifications of the equipment used.

When the most recent observation becomes available, the likelihood probability is computed as a product of the measured power mapped by the histogram, or:

$$p(\hat{\boldsymbol{P}}^{t}|s_k) = \prod_{j=1}^{n_B} h_{kj}(\hat{P}_j^{t}). \tag{4.29}$$

To improve the accuracy of the system, the implementation in Gentile et al. (2004) actually fingerprints the signal strengths of packets both to and from the base stations as two separate readings. Each site will then have two histograms per base station rather than one, doubling the factors in Eq. (4.29). The expression is based on the same assumption, as in Eq. (4.24), of independent RSS value between the n_B base stations. In reality, the histograms of different states will be correlated to some degree, however the independence assumption yields good results regardless.

As suggested in Ladd et al. (2005), the stability of the system can be enhanced through a simple post-processing step, where a modified posterior probability is generated at each update as

$$\tilde{p}(s_k|\hat{\boldsymbol{P}}^{t}, \hat{\boldsymbol{P}}^{t-1}, \ldots, (\hat{\boldsymbol{P}}^{0}) = \left(p\left(s_k|\hat{\boldsymbol{P}}^{t}, \ldots, \hat{\boldsymbol{P}}^{0}\right) + u_1\right) \cdot \left(p\left(s_k|\hat{\boldsymbol{P}}^{t-1}, \ldots, \hat{\boldsymbol{P}}^{0}\right) + u_2\right). \tag{4.30}$$

The modified posterior filters any spurious values which may appear as spikes in the system at a single time instant due to glitches or erroneous observations. The values of (u_1, u_2) are small constants which keep the modified probability from collapsing to zero. Then from the modified posterior, the estimated state of the mobile at time t is given through the Maximum Likelihood Estimation as

$$\tilde{s}_k = \arg\max_k \tilde{p}\left(s_k|\hat{\boldsymbol{P}}^{t}, \ldots, \hat{\boldsymbol{P}}^{0}\right). \tag{4.31}$$

4.3.2 Grid-Based Markov Localization

By integrating observations over a period of time, Bayesian inference can deliver enhanced stability over static localization. However, the method is still susceptible to large fluctuations in received signal strength due to fading, even while the mobile remains in the same state. So that these fluctuations are not converted into random motion, a sort of temporal averaging mentioned above is incorporated by post-processing the estimated output; in addition, estimated locations that do not support human motion, such as hops between mutually distant sites in the deployment area are filtered out. While filtering [e.g. Kalman filtering (Kalman 1960)] can improve location tracking in simple experiments—for instance, a mobile user walking up and down a hallway—it fails with more complex trajectories such as turning from a corridor into a room.

As an alternative to post-processing, in this subsection, the problem is cast in the framework of a Markov random process through which the dynamics of human movement can be encoded intrinsically as transition probabilities. In this framework, the system can be tuned for fluid motion and direction while still providing for abrupt changes where appropriate.

4.3.2.1 Motion Dynamics

In order to capture the system dynamics, the definition of a state s_k, which indicates a unique location and orientation of the user, is extended to a *sequence* $\mathbf{s}_k = \{s_k^1, \ldots, s_k^n\}$. A sequence is defined as a set of states ordered in time representing the last n states traversed by the mobile device, from time $t-n+1$ to time t. Accordingly, n_s now denotes the total number of possible sequences. Integrating more than a single state captures not only the location and orientation of the mobile at consecutive instants in time, but also the dynamics of the motion between the states. How these sequences are composed is discussed later in the subsection.

At time step t, the localization algorithm calculates the posterior probabilities of the sequences, $p(\mathbf{s}_k|\hat{\boldsymbol{P}}^0, \ldots, \hat{\boldsymbol{P}}^t)$, given the observations since initialization. A first-order Markov process governs the transition of the sequences from step $t-1$ to the next (Fox et al. 1999):

$$p\left(\mathbf{s}_k|\hat{\boldsymbol{P}}^0, \ldots, \hat{\boldsymbol{P}}^t\right) = \eta^t \cdot p\left(\hat{\boldsymbol{P}}^t|\mathbf{s}_k\right) \sum_{\tilde{k}=1}^{n_s} p\left(\mathbf{s}_k|\mathbf{s}_{\tilde{k}}\right) \cdot p\left(\mathbf{s}_{\tilde{k}}|\hat{\boldsymbol{P}}^0, \ldots, \hat{\boldsymbol{P}}^{t-1}\right) \tag{4.32}$$

Note the similarity of the expression to Eq. (4.27). The only difference is the incorporation of the sequence transition probabilities, $p(\mathbf{s}_k|\mathbf{s}_{\tilde{k}})$: in the Bayesian framework, the posterior of $\mathbf{s}_k$ at t is computed only from the posterior of $\mathbf{s}_k$ at $(t-1)$. In the Markov framework, rather, it is computed from the posteriors of all n_s sequences at $t-1$ through the sequence transition probabilities. The algorithm reports the output state of the system at each step t as $\tilde{s}^n, \tilde{\mathbf{s}} = \arg\max_k p(\mathbf{s}_k|\hat{\boldsymbol{P}}^0, \ldots, \hat{\boldsymbol{P}}^t)$. Again, the normalization factor $\eta^t = 1/\sum_{k=1}^{n_s} p(s_k^t|\hat{\boldsymbol{P}}^0, \ldots, \hat{\boldsymbol{P}}^t)$ enforces the law of total probability and, since an observation at time t affects only the state of a sequence corresponding to the same time instant s_k^n, the likelihood can be simplified to $p(\hat{\boldsymbol{P}}^t|\mathbf{s}_k) = p(\hat{\boldsymbol{P}}^t|s_k^n)$. As such, the value of $p(\hat{\boldsymbol{P}}^t|s_k^n)$ is given from Eq. (4.29).

We now turn our attention to computing the sequence transition probabilities as in Gentile et al. (2004). First of all, in order to ensure spatiotemporal consistency between back-to-back sequences, when the mobile is in sequence $\mathbf{s}_{\tilde{k}}$ the sequences s_k to which it can transition at the next time step are restricted. This is implemented by setting $p(\mathbf{s}_k|\mathbf{s}_{\tilde{k}}) = 0$ if $\mathbf{s}_k$ does not meet the condition $s_k^{l-1} = s_{\tilde{k}}^{l}, l = 2, \ldots, n$; in other words, s_k must be a left-shift of $s_{\tilde{k}}$ with replacement of only the nth state, s_k^n, with a new state. Then, the other sequences, the so-called *allowed* sequences, are assigned a nonzero transition probability. The probability is assigned in order to promote fluid motion—that is motion which follows a predictable trajectory, such as the mobile moving down a corridor at a fixed velocity or slowing to a stop. If the ordered states of a sequence reflect fluid motion, the sequence is assigned a high probability and vice versa. The fluidness is characterized through an $(n - 1)$-tap filter. The filter is employed to predict the most likely n^{th} location in the sequence from the trajectory of the first $n - 1$ locations:

$$\hat{\boldsymbol{x}} = \sum_{l=2}^{n} \alpha^l \cdot \boldsymbol{x}_{\tilde{k}}^l = \sum_{l=1}^{n-1} \alpha^l \cdot \boldsymbol{x}_k^l, \tag{4.33}$$

where α^l are the filter coefficients. Other non-finite impulse response filters, in particular the popular Kalman filter, may be applied alternatively. A Gaussian kernel maps the difference—between the actual location of the nth state, $\boldsymbol{x}_k^n$, and its predicted location, $\hat{\boldsymbol{x}}$—to the sequence transition probability:

$$p(\mathbf{s}_k|\mathbf{s}_{\tilde{k}}) = \frac{1}{\gamma\sqrt{2\pi}} \mathrm{e}^{-\frac{1}{2\gamma^2}\left\|\boldsymbol{x}_k^n - \hat{\boldsymbol{x}}\right\|^2}. \tag{4.34}$$

A small difference (high probability) indicates that the sequence conforms well to the motion dynamics represented by the filter and a large difference (low probability) the opposite. The parameter γ controls the degree of Gaussian rolloff. Reducing the value of γ makes the sequence filtering more selective.

Even by restricting the sequences which are allowed, the number n_s may still grow exponentially large with n. Hence grid-based Markov localization can suffer from computational overhead and/or overcommitment of the memory requirements for the sequence space. Both, indeed, can present significant issues for location devices which are often very compact in size. The CONDENSATION algorithm, which falls into the general class of particle filters, offers a solution. Essentially, rather than maintaining the posterior probability for each discrete sequence in the model, the algorithm maintains only an abridged set of the most likely $n_c \ll n_s$ sequences, i.e. the ones with the relatively largest associated values of $p\left(\mathbf{s}_k|\hat{\boldsymbol{P}}^{t-1}, \ldots, \hat{\boldsymbol{P}}^0\right)$. At the next step, these posteriors are updated to time t, as normal. And again, only the n_c sequences which have the relatively largest updated values are retained. The CONDENSATION algorithm has proven to be a powerful tool in recent years in the context of Bayesian estimation and computer vision. The details of the algorithm can be found in Isard and Blake (1998).

4.3.2.2 Motion Constraints

Certain applications, such as in emergency response, require precision location discrimination—knowing whether a firefighter lies in a particular room or in the one adjacent to it can make all the difference in saving a life. While the received signal strengths from multiple base stations alone may not suffice to make this distinction—depending on the material and thickness of a wall, the RF signature may differ little on its opposite sides—tracking the path of the mobile user as he or she enters the room may. As we shall see, in the framework of Markov localization, mobile user tracking can be realized by restricting the allowed sequences further by applying motion constraints.

The percentage of walking space in a typical office environment, which is furnished with desks, bookshelves, cubicles, and other furniture and equipment, ranges between 25 and 40 % of the total deployment area. The same is true in many other environments, namely in residential environments and in public environments such as libraries, supermarkets, etc. The presence of these obstacles severely constrains the paths along which humans can move about. Figure 4.9 illustrates a typical office environment. Six paths, displayed in different shadings of gray, connect any two fingerprinted sites in the environment. The pair of numbered arrows represents the two states corresponding to the opposite orientations at each site, splitting each path into two *tracks*. By fingerprinting each site with the antenna orientation aligned with the heading of the person, a mobile user walking forward on a path follows the states on either one track or on its complementary track. Under the assumption that a human walks only forward and that the antenna orientation remains constant with respect to the person's heading, motion constraints can be imposed such that the mobile can be localized as moving only along the tracks.

Motion constraints are applied to the Markov model such that a state can transition only to a spatially adjacent state from one time instant to the next. Consequently, the mobile must traverse a sequence of adjacent states or *neighbors* in order to reach any one state in the model from another. This is implemented by assigning the appropriate sequence transition probabilities a zero value. Recall that the same was explained earlier in application to restricted sequences. This mechanism, which allows only those sequences in the model which conform to the motion constraints (and restricts those which do not), turns out to be a highly effective manner to reconstruct a path from a series of observations during the localization stage. Classical Kalman filtering may predict the trajectory of a human advancing through a wall because it considers only the locations on the trajectory; motion constraints, rather, provide a blueprint of the area encoded through the sequence transition probabilities. The desired effect is that the system realizes that humans must go through doors in order to reach locations on opposite sides of a wall.

We now turn to the description of how neighbors and tracks are encoded in the Markov model. Most states have three neighbors: (1) itself—to allow stationary motion in time; (2) the next state on the same track—to allow motion in the same

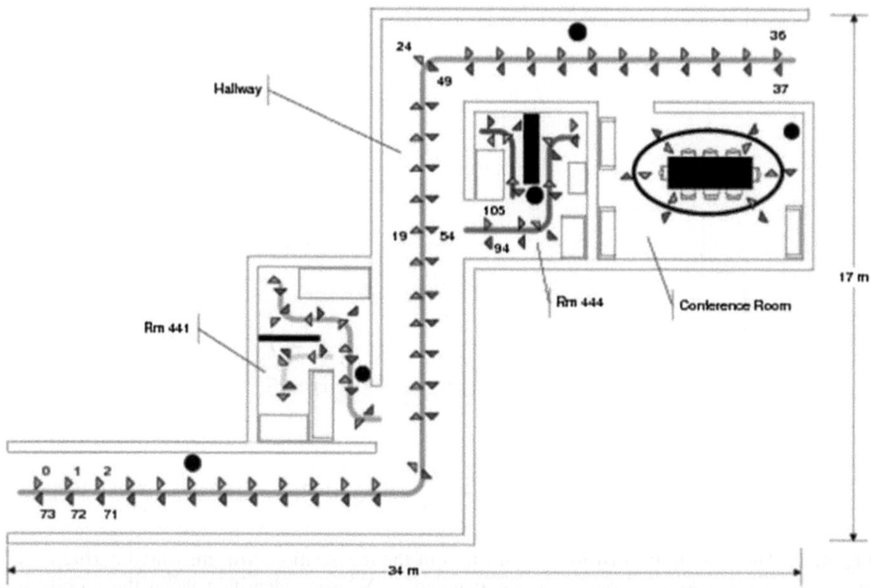

Fig. 4.9 A typical office environment. Shown here are 124 numbered states divided into six paths—each path is shaded differently. The five base stations in the survey area are labeled as solid circles. Each path is split into tracks which allow motion in opposite directions, as indicated by the arrows

direction; (3) the state at the same location on the opposite track—to allow a change in direction. Exceptions occur for states at the end of tracks with no next state; they have only two neighbors. Another exception is for states falling at T-junctions or crossroads between two paths; additional neighbors enable the mobile to switch paths. In order to promote motion along tracks, sequences which contain more than one track transition are restricted. Moving backwards on a track is also not permitted in this particular implementation; such motion, however, is actually common in some applications; for example, firefighters walk backwards pulling hoses and crawl backwards downstairs. Of course the system can be tuned accordingly. Also, in large, open areas, a grid of states, rather than tracks, can be created and the appropriate motion constraints applied.

As an experiment, in Gentile et al. (2004) a system with sequence length $n = 5$ was tested against a benchmark system with length $n = 1$ in the office environment depicted in Fig. 4.9. For each trial, the localization error was recorded either as (1), the distance between the estimated location and the ground-truth location; or as (2), a logical error X when the mobile was localized in a wrong room or on the wrong side of a partition, bookcase, or table within the same room. Figure 4.10 shows the cumulative distribution function of the localization error for both systems in the *Conference Room*—the area in the office environment where the greatest disparity in performance between the two systems was observed. Because there was only free space between sites on opposite sides of the table, the RF signatures there were too

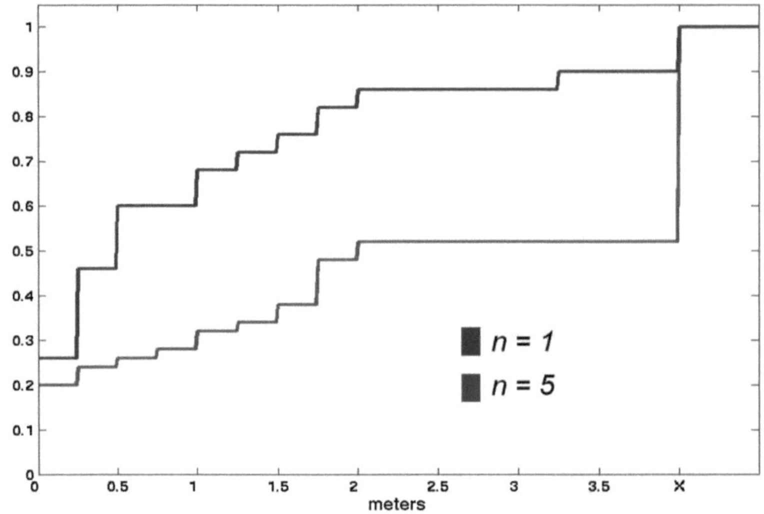

Fig. 4.10 The cumulative distribution function of the localization error measured during testing in the *Conference Room*. Logical errors, denoted as X, were identified when the mobile was localized in a wrong room or on the wrong side of a partition, bookcase, or table within the same room

similar for robust discrimination between them. In fact, in the case where a memoryless system model is used ($n = 1$), almost 50 % of the time it reported a logical error whereas in the case where a five-state sequence was used, the error was reported only 10 % of the time. Clearly, information about a single state alone does not suffice to correctly identify the trajectory; rather, information provided from multiple states taken collectively must be considered. This experiment underscores the strength of the Markov localization technique using sequences.

4.4 Channel Impulse Response Fingerprinting

Thus far we have considered only the received signal strength feature for fingerprinting. RSS is typically used for localization in WLAN and 3G cellular networks (cellular localization systems are presented in Chap. 5). In these networks, signal strength is measured as the carrier power sensed over a period of time. While it varies by technology, the period is normally the duration of a packet. The power sensed is that of the transmitted signal arriving along the direct path from the base station. Also sensed are copies of the transmitted signal which arrive along other propagation paths. As explained in detail in Chap. 2, the collection of copies is referred to as multipath. The multipath copies arrive delayed with respect to the direct path due to the characteristics of the propagation environment. When indexed according to delay, the power is referred to as the power delay profile or

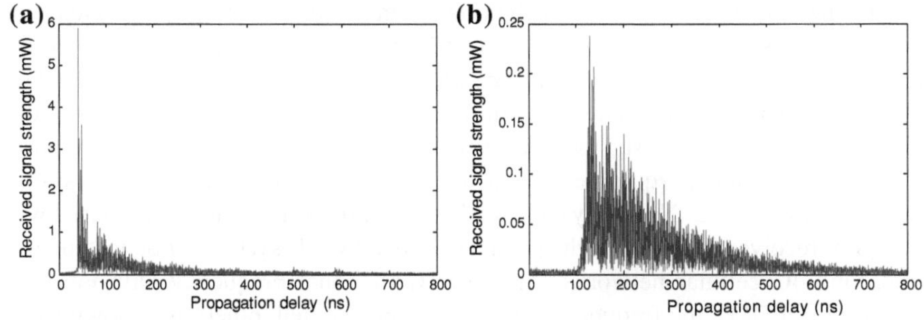

Fig. 4.11 **a** Channel impulse responses for a 6 GHz signal a line-of-sight (*LOS*) conditions. **b** non-line-of-sight (*NLOS*) conditions

the channel impulse response (CIR). Figure 4.11a illustrates the CIR for a 6 GHz signal and Fig. 4.11b illustrates another channel CIR from the same base station, however with the mobile displaced to a different location (more examples can be found in Chaps. 2 and 3). Notice the distinct properties between the two profiles. The profiles provide unique characterization of the mobile sites whereas the RSS signatures at the two sites—which is essentially just the integration of the power across the profile—may be very similar. As such, the channel impulse response can serve as an alternative fingerprint with enhanced identifiability. The CIR-fingerprinting technique was first introduced by US Wireless Corporation of San Ramon, California (Koshima and Hoshen 2000). Fingerprinting using the channel impulse response was also proposed for cellular UMTS localization (Ahonen and Eskelinen 2003a, b).

Channel impulse responses first appeared in localization in time-of-arrival based systems. Ideally, the first multipath arrival in the power delay profile will correspond to the direct propagation path between the base and the mobile. This then begs the question: if the channel impulse response is available, why not just use it to extract time-of-arrival? While TOA systems are capable of delivering accuracy on the order of several centimeters in line-of-sight conditions, the accuracy can degrade significantly in non-line-of-sight depending on the number, size, and material type of the obstacles between the radios. For example, in industrial environments rich in metal scatterers, deflection of the direct path off the straight line between the TX to the RX can cause a significant delay in the first arrival; or in subterranean mines, being that the walls are impenetrable by the direct path, the first arrival detected must necessarily correspond to some other path with a longer delay. In these cases, CIR fingerprinting offers a viable solution. In fact, radio frequency fingerprinting systems thrive on environments rich in scattering, such as the industrial environment. The scattering helps create distinctive signatures even between trained sites in close proximity. Since the multipath signature is unique and varies from one location to the next, it is even possible to implement a fingerprinting system with a single base station.

One benefit of survey-based techniques is that they can be implemented with existing infrastructure, necessitating no proprietary location and tracking equipment. For example, RSS fingerprinting systems exploit Wi–Fi base stations, which are both cheap and evermore ubiquitous worldwide, reducing deployment costs significantly. In the past, measuring the channel impulse response required expensive laboratory equipment such as the Vector Network Analyzer system described in Chap. 2. Nowadays, channel impulse responses can be measured with complex receivers used in high-speed, wide-bandwidth systems. For instance, in wideband 4G cellular networks which use Orthogonal Frequency Division Multiplexing (OFDM), the frequency response of the channel, otherwise known as the Channel Transfer Function (CTF), is measured from the preamble of a packet in order to enable channel estimation and equalization. The channel impulse response can then be recovered from the CTF by converting it to the delay domain through the inverse Fourier Transform. The wider the bandwidth of the telecommunications system, the better its capacity to resolve multipath (Gentile and Kik 2007). In fact, because narrowband systems have poor resolution, the different arrivals appear as if they were grouped all as one. As a result, the power from the different paths cannot be discriminated and it is detected, rather, as a single quantity over the period, i.e. as the RSS value.

Two CIR-based systems are considered in this section. In Sect. 4.4.1, a system which was implemented inside a mine tunnel is described. The system only processes the magnitude information of the multipath delay components. An improvement to the implementation, in which a nonparametric regression technique also exploits the phase information, is described in Sect. 4.4.2.

4.4.1 Mapping Using a Neural Network

Since mine shafts are typically void of objects, they tend to have poor scattering properties. Then for the reasons explained earlier, received signal strength fingerprinting may deliver unacceptable resolution. As demonstrated in Nerguizian et al. (2006), channel impulse response fingerprinting in the mine environment, instead, can achieve good performance. In this subsection, we provide an overview of this paper. In the fingerprinting stage, the CIR was recorded at a number of sites throughout the mine using a Vector Network Analyzer, which acted as the sole base station[3]. In reference to Eq. (3.38), the channel impulse response can be represented mathematically as a train of uniformly sampled complex amplitudes, $h(\tau_k)$, indexed according to delay $\tau_k, 1 \leq k \leq L_p$. Each fingerprinted site was characterized by seven representative features extracted from the channel impulse response. The main features, which have already been introduced in Chap. 3, are the received signal power

[3] Details of the VNA are described in Chap. 2.

$$P = \sum_{k=1}^{Lp} |h(\tau_k)|^2, \tag{4.35}$$

the mean excess delay

$$\overline{\tau} = \frac{1}{P} \sum_{k=1}^{Lp} \tau_k \cdot |h(\tau_k)|^2, \tag{4.36}$$

and the root mean square delay spread

$$\tau_{\text{RMS}} = \sqrt{\frac{1}{P} \sum_{k=1}^{Lp} (\tau_k - \overline{\tau})^2 \cdot |h(\tau_k)|^2}. \tag{4.37}$$

The other features are the number of components L_{SNR} whose power is above a designated signal-to-noise ratio threshold T_{SNR}, i.e. $L_{\text{SNR}} = \underset{|h(\tau_k)|^2 \geq T_{\text{SNR}}}{\text{sum}} k$, the time-of-arrival $\tau_1 = \underset{|h(\tau_k)|^2 \geq T_{\text{SNR}}}{\min} \tau_k$ the power of the first arrival $P_1 = |h(\tau_1)|^2$, and the maximum arrival time $\tau_{\text{MAX}} = \underset{|h(\tau_k)|^2 \geq T_{\text{SNR}}}{\max} \tau_k$.

For the experiment in Nerguizian et al. (2006), close to 400 sites were fingerprinted throughout a surveyed mine. The multilayer perceptron, which was described in Sect. 4.2.3, was utilized to perform the mapping from the CIR-feature space to the location space. The seven features extracted for each of the sites, coupled with the two-dimensional location coordinates of each site, were used to train the perceptron. Accordingly, the neural network had seven inputs and two outputs. In this application, only one hidden layer with ten neurons was sufficient for training. The results for this method are presented in a side-to-side comparison in the next subsection.

4.4.2 Mapping Using a Gaussian Kernel

In the work presented above, the seven features of the channel impulse response described were deemed sufficient to discriminate the sites throughout the survey area. Aside from the benefits of a compact representation, the authors in Jin et al. (2010) argue that, by extracting these features only, useful information available for location identification is discarded. In their paper, the authors show that by exploiting the unreduced CIR, results can be improved significantly. To this end, let the channel impulse response at site i from base station j, denoted as $\boldsymbol{h}_{ij}$, be a vector of L received power values sampled at uniform delay intervals. The signature at site i is then given through the collection of the CIR vectors from each of the n_B base stations; together they form the concatenated *supervector* $\boldsymbol{H}_i = \left[\boldsymbol{h}_{i1}\boldsymbol{h}_{i2} \ldots \boldsymbol{h}_{i,n_B}\right]^T$ of length $n_B \, x L$.

Once the sites have been fingerprinted, the mobile location is estimated as a linear combination of the locations of the n_M trained sites:

$$\tilde{\boldsymbol{x}} = \frac{1}{n_M} \sum_{i=1}^{n_M} \rho_i \boldsymbol{x}_i \tag{4.38}$$

The weight associated with each site is the similarity metric—between the CIR supervector, $\hat{\boldsymbol{H}}$, measured during localization and the fingerprinted CIR supervector, $\boldsymbol{H}_i$. In the paper, the similarity metric is selected as the Gaussian kernel function

$$\rho_i = \frac{1}{\sqrt{(2\pi)^{n_M} |\sum|}} \mathrm{e}^{-\frac{1}{2}(\hat{\boldsymbol{H}}-\boldsymbol{H}_i)^T \sum^{-1} (\hat{\boldsymbol{H}}-\boldsymbol{H}_i)}, \tag{4.39}$$

where $\sum$ is the sample covariance matrix of the n_M fingerprinted supervectors. The sample covariance matrix is defined as

$$\sum = \frac{1}{n_M} \sum_{i=1}^{n_M} \boldsymbol{H}_i \cdot \boldsymbol{H}_i^T \tag{4.40}$$

In practice, fingerprinted sites throughout a survey area will be statistically correlated. For instance, adjacent sites collocated in a hallway will receive correlated multipaths—both in strength and in delay—from the same base station, especially those multipaths reflected into the hallway from the same direction. The key strength of the Gaussian kernel function as a similarity metric is that through the covariance matrix, the statistical correlation between the CIR supervectors of the fingerprinted sites is captured. This is a departure from the independence assumptions of Eq. (4.8) in Sect. 4.1.2.1 and of Eq. (4.24) in Sect. 4.2.4. Note that in Nerguizian et al. (2006), this statistical correlation is also captured, however, more implicitly through a neural network. Whereas in the latter the CIRs are processed on a linear scale, in Jin et al. (2010) the CIRs are processed on a logarithmic scale for the following reason. Since arrivals with larger delay are significantly attenuated with respect to the direct path, the logarithmic scale serves to leverage the contribution of each arrival; otherwise the contribution of the later arrivals would be dwarfed by the earlier arrivals. Figure 4.12 shows the same channel impulse response on a linear scale in (a) and on a logarithmic scale in (b).

We now present the results from the performance comparison in Jin et al. (2010). The paper compares the method described in this subsection, referred to as *LOG-ACIR-NKR*, to the method described in the previous subsection, referred to as *ACIR-GRNN*. Also compared was the *RSS-Kernel* method which draws on the same Gaussian kernel function in (4.39), however by replacing the CIR-supervector features, $\hat{\boldsymbol{H}}$ and $\boldsymbol{H}_i$, with the RSS-vector features, $\hat{\boldsymbol{P}}$ and $\boldsymbol{P}_i$ (the RSS value, P, was computed from the generated CIR as in (4.35)). The three methods were implemented using raytracing software: given the three-dimensional CAD model of a building together with the thickness and the dielectric properties of the walls,

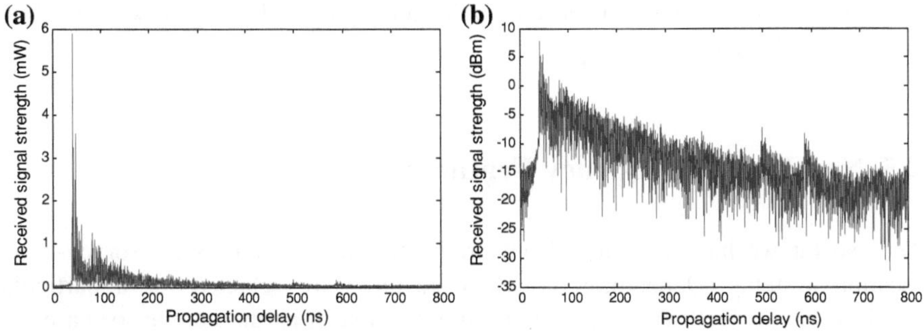

Fig. 4.12 The channel impulse responses for a 6 GHz signal in line-of-sight conditions. **a** The signal power on a linear scale. **b** The signal power on a lograithmic scale

the software can generate high-resolution CIRs based on the configured positions of the base stations and the mobile device throughout the environment. The bandwidth of the system was set to 60 MHz and there were two base stations and 173 fingerprinted sites spaced at 1.5 m in the deployment area. Based on the cumulative distribution function of the location error for all three methods, the *LOG-ACIR-NKR* method achieves an average location error as low as 1 m, followed by the *ACIR-GRNN* method with an average error of 1.7 m, and then by the *RSS-Kernel* method with an error of 3.2 m. The *LOG-ACIR-NKR* method outperforms the *ACIR-GRNN* method mainly because it operates on a logarithmic scale as opposed to on a linear scale, making full use of distinctive properties of even the weakest arrivals in the CIRs. As expected, the *LOG-ACIR-NKR* method easily outperforms the *RSS-Kernel* method because it exploits the supplemental information provided by the CIR, with the arrivals sorted by delay rather than grouped as a single value.

4.4.3 Variations of CIR Fingerprinting

Analogous to the time domain channel impulse response, the channel transfer function can be used alternatively for fingerprinting. The CTF contains the same multipath channel information, however in the form of complex samples in the frequency domain. Similarly, the CTF correlation function, known as the Frequency Channel Coherence Function (FCF), is also proposed as a signature in (Malik and Allen Nov. 2006). The paper shows that the FCF is more stable and has superior performance to the CTF. A patent application proposes a similar technique that integrates FCF-based fingerprinting in existing OFDM-based systems (such as WLANs) (Bevan et al. 2010). Finally, the multipath characteristics alone can be further enhanced by incorporating an antenna array at the receiver. The antenna array enables the spatial characteristics, not just the temporal, of multipath—indexed according to both arrival angle and delay—to be captured. This results in a richer signature defined as the power spatial delay profile (PSDP) (Triki et al. 2006; Gentile and Braga 2008).

4.5 Non-Radio Frequency Features

Thus so far we have considered only radio frequency features for survey-based location systems. Depending on which features are selected for a particular application—as well as other design parameters such as the number of base stations and the number of fingerprinted sites in a survey area—systems typically deliver localization accuracy between 1 and 3 m. For many applications, the expenses associated with the equipment and infrastructure necessary to deliver this level of accuracy are too high. For such applications, precise physical location within a certain environment is not required; rather, determining whether a mobile user lies within a confined space with a high degree of reliability takes priority. This type of *logical* localization is important in environments such as stores, museums, libraries, gas stations, etc. For instance, in a museum the appropriate automated tour can be offered as the visitor approaches the entry of an exhibition. In a grocery store, location-based services can notify a shopper of available coupons for items while walking along the aisles where they are stocked; in this environment, even if a device can deliver accuracy up to 2 m, this accuracy may not suffice to determine whether the mobile is in one aisle or the one adjacent to it.

Other features can be used to supplement, or even substitute, radio frequency fingerprinting. In this section, we investigate the application of the features described in Azizyan et al. (2009), namely the non-RF features of sound, motion, and color and the RF-feature of connectivity. For example, in a Laundromat the authors observed that sound is characterized by moving mechanical parts while in a library, on that other hand, it is very quiet. Analogously, typical motion in a supermarket involves walking up and down aisles with periodic pauses to select items; this contrasts static motion in a restaurant where the customer remains mostly seated for the duration of the stay. The chromatic features take advantage of the fact that many stores have trademark colors which are accentuated throughout the environment, such as red and white in *Target* © or pink and orange at *Dunkin' Donuts* ©. Finally, regarding connectivity, a mobile device can form a radio link only with base stations within the vicinity of the environment; hence, connectivity alone—as opposed to the degree of connectivity expressed by received signal strength—can be exploited as a signature for the environment. While the individual features may be similar from environment-to-environment, combining all four of the features together can prove to be highly discriminatory. The authors show that it is possible to achieve logical-localization accuracy of up to 87 % in 51 different environments using only these features which are accessible on most smartphones. In the remainder of this section, we described these four features in greater detail.

4.5.1 Sound Features

Sound in an environment is characterized through the temporal distribution of its volume intensity. This distribution is represented by a histogram of the intensity values recorded over a 1 min segment. The histogram is divided into 100 bins of equal size ranging from the minimum to maximum volume of the mobile device. In the Laundromat environment, for example, the histogram is very sharp at the center, indicating a constant buzz of medium intensity; this can be attributed to the rotation of the internal parts of the washers and dryers. Conversely, the distribution in a coffee shop is wider by virtue of the traits of human conversation, composed from a greater range of volumes—from the baristas preparing the items and calling out orders to conversational chatter in the background. The histogram vector serves as the signature for an environment. When compared against a vector measured during the localization stage, the inverse of the Euclidean distance—the distance is computed in the 100-dimensional space of the histogram vector—is used as the similarity metric. Notice that, as opposed to the previous sections in this chapter, the authors' convention in Azizyan et al. (2009) is that a larger similarity metric is more favorable.

4.5.2 Motion Features

Most smartphones now offer location services. When GPS is available, mainly outdoors and in some indoor environments—especially indoor environments with many windows through which the signal can penetrate—GPS can furnish the location of the mobile device. In GPS-denied areas, however, accelerometers can be employed to interpolate between the GPS readings (details of inertial-based systems are provided in Chap. 8). It turns out that in application to fingerprinting, accelerometers can be exploited to a second end in order to classify the types of motion which are common in an environment. This is accomplished by extracting signatures from the accelerometer readings. In Azizyan et al. (2009), each reading is the output of one of three-dimensional accelerometer axes. When sampled over time, two sequences are generated from each one of the three readings: one is the moving averages of the readings and one is their moving variances. The sequences are then fed to a Support Vector Machine (see Sect. 4.2.2) which classifies the sequences into one of two categories: either moving or stationary. The SVM is trained from the samples of the two sequences.

The actual signature is then computed as the quantity r, which is the ratio of time the mobile device is moving to the time it is stationary—over some observation window. The signature is then categorized into three classes: sitting for $0 \leq r \leq 0.2$, browsing for $0.2 < r \leq 2.0$, and walking for $r > 2.0$. By associating one or more classes to each of the fingerprinted environments, the signature can be used for the purpose of discriminating between the environments in which the different classes occur. Then, the similarity metric between the signature measured during the localization stage and the signature of a fingerprinted environment is a value between 0 and 3. The value indicates the number of classes which the two signatures have in common.

4.5.3 Color Features

In order to capture the color of an environment, the floor is chosen as the target area. The reason is twofold: first of all the view of the floor (tiles, carpeting, wood, etc.) is relatively static over time, changing mainly due to obstructions from pedestrian walking. The ceiling would make for an even better candidate since it lacks such obstructions, however smartphones usually have their camera installed on the back of the device and so the camera is seldom pointed toward the ceiling. This makes for the second reason why the floor is chosen. Based on how the smart phone is held, the phone can discern one of its six possible orientations. From its orientation, the phone can determine at which times it is pointed towards the floor.

The camera's charge-couple device digitizes the captured image into RGB (Red–Green–Blue) pixels. Because the RGB chromatic space was found to be too sensitive to shadows and reflections, the pixels were transformed into the more robust HSL (Hue-Saturation-Lightness) space. The fingerprinting procedure consisted of the following steps. First the HSL pixels from all the images taken in the same environment were captured and grouped into clusters in the three-dimensional space via the *k-Nearest Neighbor* method (see Sect. 4.2.1). The number of clusters in any environment ranged from 3 to 7. Each cluster was represented by its HSL centroid and the signature of the environment was designated as the centroids of the clusters identified. Finally the similarity metric was computed as the distance between the centroids of the image(s) captured during localization and the centroids of environment i. Specifically, the similarity metric is expressed as:

$$\rho_i = \sum_k \sum_l \frac{1}{d_i(k,l)} \left(\frac{\hat{N}^k}{\hat{N}}\right) \left(\frac{N_i^l}{N_i}\right) \tag{4.41}$$

where $d_i(k,l)$ is the Euclidean distance in the HSL space between centroid k of the captured image(s) and centroid l of environment i. The value $\hat{N}^k$ indicates the number of pixels in cluster k and the value N_i^l indicates the number in cluster l. Analogously, the value $\hat{N}$ indicates the total number of pixels over all the clusters of the captured image(s) while N_i indicates the total number over all the clusters of environment i. Note that each term in Eq. 4.41 corresponds to a cluster pair in the measured and fingerprinted spaces. Since the similarity metric is inversely proportional to the distance between the pair, the distance between the closest pair will have the greatest influence on the metric; likewise, the cluster pairs farthest apart will have the least influence. Each term is also weighted by the relative number of pixels in the respective clusters of the pair. This downplays the contribution of smaller clusters which may just be outliers representing background noise.

4.5.4 Connectivity Features

As the mobile device moves from one environment to another, its connectivity will change. Rather than measure the degree of connectivity between a base station and a mobile device through a similarity metric between their received signal strengths, a hard limit can also be used—that is—whether a radio link between the two can be established or not. Since the mobile device pings the surrounding base stations periodically in order to register their MAC addresses, the frequency of acknowledgment can be used as a similarity metric of connectivity instead. The authors quantify the frequency of the connectivity, f, as the fraction of acknowledgments the mobile receives from a particular base station to the total number it receives from all the n_B base stations during a fixed time period. The vector of connectivity values $\boldsymbol{f}_i = [f_{ij}], j = 1 \ldots n_B$, that a mobile registers within environment i is designated as the signature of the environment. Then, the similarity metric between signature $\hat{\boldsymbol{f}} = [\hat{f}_j], j = 1 \ldots n_B$ measured during localization and signature i is:

$$\rho_i = \sum_{j=1}^{n_B} (\hat{f}_j + f_{ij}) \cdot \frac{\min(\hat{f}_j, f_{ij})}{\max(f_j, f_{ij})} \tag{4.42}$$

Term j is large when the measured and fingerprinted connectivities to station j have a comparably high frequency; if they are not comparable then the min over max factor will be very small and will attenuate the weight of the term in the sum.

There are many ways in which the sound, motion, color, and connectivity features described in this section can be leveraged in order to estimate the mobile's location. For example, their individual similarity metrics can be weighted in a linear combination to formulate a comprehensive similarity metric[4]. Alternatively, the authors in Azizyan et al. (2009) chose to apply the features in the following manner. First of all, the sound, motion, and connectivity features were used sequentially to filter out unlikely environments. For each, an appropriate threshold value was set and any candidate environment with a corresponding similarity metric below that value was discarded. Once the initial filtering was performed, the color feature was selected to ultimately determine the location of the mobile as the one with the largest similarity metric among the remaining environments.

---

[4] This requires normalization of the individual similarity metrics such that each of their minimum and maximum values falls between 0 and 1, respectively.

4.6 Remarks

Some of the most practical fingerprinting techniques and system applications have been described in this chapter. While survey-based techniques may benefit from exploiting existing wireless infrastructure or from the deployment of low-cost nonproprietary equipment, they still suffer from the drawbacks of a required fingerprinting stage. One drawback is that the system cannot be deployed ad hoc, rather necessitating hours, weeks, or even months of training for large-scale networks, e.g., cellular. (Apropos, the following chapter investigates geolocation systems in cellular networks). Another drawback is that the radio frequency characteristics of the sites vary with any environmental changes. Such changes may arise from the movement of furniture, partitions, and any other objects, altering the path loss exponent and shadow fading in the environment; also, the addition or removal of base stations modifies the structure of the database. It is worth mentioning that some recent effort has been dedicated to research in automatic fingerprint training (Kim et al. 2010; Eleryan et al. 2011).

The drawbacks of fingerprinting preclude mission-critical applications for which localization services are vital, such as in emergency response and for firefighting in particular. Even if, in theory, a fingerprinting system could be trained in advance and its database updated automatically, the signature characteristics of the environment would change drastically during the rapid progression of a fire. Walls and floors may collapse and the water from the fire extinguishing hoses (which has the property of high RF reflectivity) and the ambient smoke are suspected to change the propagation characteristics of the environment. Other factors may be the flame itself as well as any dust produced from the deteriorating structure. Studies of these factors are currently underway at the National Institute of Standards and Technology.

References

A. Agiwal, P. Khandpai, H. Saran, LOCATOR: Location Estimation Systems for Wireless LANs, International Workshop on Wireless Mobile Applications and Services on WLAN Hotspots, ACM, pp. 102–109, 2004

S. Ahonen, P. Ekelinen, Performance Estimations of Mobile Terminal Location with Database Correlation in UMTS Networks. International Conference on 3G Mobile Communications Technologies, pp. 25–27, (2003a).

S. Ahonen, P. Eskelinen, Mobile terminal location for UMTS. Aerosp. Electron. Sys. Mag. **18**(2), 23–27 (2003b).

M. Azizyan, I. Constandache, R.R.Choudhury, Surroundsense: Mobile Phone Localization via Ambience Fingerprinting. International Symposium on Mobile Ad hoc Computing and Networking, ACM, pp. 261–272, 2009

P. Bahl, V. Padmanadhan, RADAR: an in-building rf-based user location and tracking system. IEEE INFOCOM **2**, 775–784 (2000)

R. Battiti, T.L. Nihat, A. Villani, Location-Aware Computing: A Neural Network Model for Determining Location in Wireless LANs. Deptartment of Information and Communications Technology, Technical Report DIT-020083, University of Trento, Italy, 2002

D.D. Bevan, L. Averin, D. Lysyakov, RF Fingerprinting for Location Estimation 0311436, United States, 2010

N.K. Bose, *Neural Network Fundamentals with Graphs, Algorithms, and Applications* (McGraw-Hill, New York, 1995)

M. Brunato, R. Battiti, Statistical learning theory for location fingerprinting in wireless LANs. Comp. Netw. ISDN Syst. **47**(6), 825–845 (2005). Elsevier

W. Burgard, et al., Estimating the Absolute Position of a Mobile Robot Using Position Probability Grids Conference on Artificial Intelligence, AAAI, pp. 896–901, 1996

G. Cybenko, Approximations by superpositions of sigmoid funcitons. Math. Control, Signals, Sys. **3**(4), 303–314 (1989). Springer

A.M. Edgar, C. Raul, F. Jesus, Estimatiing user location in a WLAN using back propagation neural networks. Lect. Notes Comp. Sci. **3315**, 737–746 (2004)

A. Eleryan, M. Elsabagh, M. Youssef, AROMA: Automatic Generation of Radio Maps for Localization Systems. International Conference on Mobile Computing and Networking, ACM, pp. 93–94, 2011

S-H. Fang, T-N. Lin, K-C. Lee, A novel algorithm for multipath fingerprinting in indoor WLAN environments , Trans. Wireless Commun.IEEE, 7(9) 2008

D. Fox, W. Burgard, S. Thrun, Markov localization for mobile robots in dynamic environments. J. Artif. Intell. **11**, 391–427 (1999)

D. Fox et al., Bayesian filtering for location estimation. Pervasive Comput. IEEE. **2**(3), 24–33 (2003)

C. Gentile, A.J. Braga, A comprehensive evaluation of joint range and angle estimation in indoor ultrawideband location systems. EURASIP J. Wireless Commun. Networking **2008**, 248509 (2008). Hindawi

C. Gentile, A. Kik, A comprehensive evaluation of indoor ranging using ultra-wideband technology. EURASIP J. Wireless Commun. Networking **2007**, 86031 (2007). Hindawi

C. Gentile, L. Klein-Berndt, Robust Location Using System Dynamics And Motion Constraints. International Conference on Communications, IEEE. Paris, France, pp. 1360–1364, 2004

C. Gentile, S.M. Lopez, A.A. Kik, Comprehensive spatial-temporal channel propagation model for the ultra-wideband spectrum 2–8 GHz. IEEE Trans. Antennas Propag. **58**(6), 2069–2077 (2008)

V. Honkavirta et al., A comparative survey of wlan location fingerprinting methods. Workshop on Positioning, Navigation and Communication. pp. 243–251, March 2009

V. Honkavirta, Location fingerprinting methods in wireless local area networks. Master of Science Thesis, Tampere University of Technology, Finland, Oct 2008

M. Isard, A. Blake, condensation: conditional density propagation for visual tracking. Int. J. Comput. Vision **1**, 5–28 (1998)

Y. Jin, W.-S. Soh, W.-C. Wong, Indoor localization with channel impulse response based fingerprint and nonparametric regression. IEEE Trans. Wireless Commun. **9**(3), 1120–1127 (2010)

K. Kaemarungsi, P. Krishnamurthy, Modeling of Indoor Positioning Systems based on Location Fingerprinting. INFOCOM. IEEE, pp. 1012–1022, 2004

R.E. Kalman, A new approach to linear filtering and prediction problems. Trans. AMSE: J. Basic Eng. **82**, 35–45 (1960)

Y. Kim, Y. Chon, H. Cha, Smartphone-based collaborative and autonomous radio fingerprinting. IEEE Trans. Syst. Man, Cybern. Part C: Appl. Rev. IEEE, **2**(1), 112–122 (2010)

H. Koshima, J. Hoshen, Personal locator services energy. Spectrum. **37**(2), 41–48 (2000)

A. Kushki, K.N. Plataniotis, A.N. Venetsanopoulos, Kernel-based positioning in wireless local area networks. Mobile Computing. **6**(6), 689–705 (2007)

A.M. Ladd et al., Robotics-based location sensing using wireless ethernet. Wirel. Networks **11**, 189–204 (2005). (Springer Science + Business Media, Inc.)

Z. Li Wu et al., Location estimation via support vector regression. Trans. Mobile Comput. IEEE **6**(3), 311–321 (2007)

T.-N. Lin, P.-C. Lin, Performance comparison of indoor positioning techniques based on location fingerprinting in wireless networks. Wireless Netw., Commun.Mobile Comput. **2**, 1569–1574 (2005)

H. Liu et al., Survey of wireless indoor positioning techniques and systems. IEEE Trans. Syst. Man Cybern. Part C: Appl. Rev. **37**(6), 1067–1080 (2007)

D. Madigan et al., Bayesian Indoor Positioning Systems, vol. 2. INFOCOM. IEEE, pp. 1217–1227, 2005

W.Q. Malik, B. Allen, Wireless Sensor Positioning with UWB Fingerprinting. European Conference on Antennas and Propagation, pp. 1–5, Nov 2006

Y. Moustafa, A. Ashok, The Horus WLAN Location Determination System. International Conference on Mobile Systems, Applications and Services, pp. 205–218, 2005

C. Nerguizian, C. Despins, S. Affes, Geolocation in mines with an impulse response fingerprinting technique and neural networks. IEEE Trans. Wireless Commun. **5**, 603–611 (2006)

T. Roos et al., A probabilistic approach to WLAN user location estimation. Int. J. Wireless Inf. Networks **9**, 155–164 (2002)

A.J. Smola, B. Schoelkoepf, A tutorial in support vector regression. Stat. Comput. **14**, 199–222 (2004). (Kluwer Academic Publishing)

M. Triki et al., Mobile Terminal Position via Power Delay Profile Fingerprinting: Reproducible Validation Simulations. Vehicular Technology Conference, IEEE, pp. 1–5, Sept 2006

C.L. Wu, L.C. Fu, F.L. Lian, WLAN Location Determination in e-Home via Support Vector Classification. International Conference on Networking, Sensing and Control, IEEE, pp. 1026–1031, 2004

M.A. Youssef, A. Agrawala, A.U. Shankar, WLAN Location Determination via Clustering and Probability Distribution. International Conference in Pervasive Computing and Communications, IEEE, pp. 143–150, 2003

Chapter 5
Cellular Localization

Mobile network operators offer location based services for their customers as well as for third party customers that support such applications. The high density of mobile users in urban and indoor environments drive the need for such location based services, especially in areas that are GPS-denied. Providing location services for mobile devices, such as position tracking, is an important objective for cellular network operators. Example applications for which location information is critical include:

- Providing ubiquitous coverage across their service areas, network operators track locations of mobile stations to compile maps of proven coverage.
- Government agencies, such as the Federal Communications Commission in the United States, require that network operators provide the locations of wireless callers to emergency services. As wireless callers will often not be at their registered address, network operators need to determine their current location at the time a call is made.
- An increasing number of commercial applications for cell phones and other mobile devices require location services, such as navigation applications. Navigation applications inform the user about his whereabouts and combine this with detailed information about his surroundings (arrival time of the next train, local advertisement, local weather, etc.).

This chapter outlines the motivation for cellular network operators and vendors to apply different positioning techniques. Section 5.2 presents the structure of a cellular network and provides some of the positioning techniques that are used in cellular networks. In Sect. 5.3, we focus on standardized cellular network systems, such as GSM, WCDMA, and LTE and provide details about which technologies are applied in the different standards.

C. Gentile et al., *Geolocation Techniques*, DOI: 10.1007/978-1-4614-1836-8_5,

Table 5.1 E-911 location requirements (FCC 2010)

Wireless E-911 location accuracy requirements		
Terminal-based or terminal-assisted	50 m	67 % of outgoing calls
	150 m	95 % of outgoing calls
Network-based	100 m	67 % of outgoing calls
	300 m	95 % of outgoing calls

5.1 Motivation

Historically, there have been three main driving forces for the development of location-based services in cellular networks. The first, an application of paramount importance, is the localization of emergency callers in distress. In the United States, the Federal Communications Commission (FCC) is the governing body that sets the requirements for localizing wireless phones for emergency calls—known as E-911—for cellular network operators. A wireless caller who dials 911 will get his call routed to a public-safety answering point (PSAP), a call-center that collects all emergency calls. The FCC mandated that the location of the caller be identified with specific location accuracy and within a certain response time. A wireless caller, unlike a wired caller, may not be present at the registered address of the caller's ID. Therefore, there is a need to localize mobile terminals within a dedicated area that emergency services can support.

In the late 1990s (Zhao 2002; Drane et al. 1998), the FCC mandated that cellular operators fulfill the requirement to localize mobile terminals with a dedicated accuracy of at least 50 m (67 %)–150 m (95 %). Table 5.1 shows the different accuracy requirements for terminal-based or network-based positioning for the case in which a mobile terminal makes an emergency call for counties or PSAP areas. The testing guidelines are outlined in an early document issued from the FCC (FCC, OET BULLETIN No. 71: Guidelines for testing and verifying the accuracy of wireless E-911 location systems 2000).

Besides the FCC requirements for E-911, another driving force for location services is the demand for commercial applications. Skyhook, a wireless company offering location services, analyzed how many applications which have been developed for devices from Apple (iPhones series with the operating system iOS) and Google (Android based smartphones) use location information (Skyhook Inc. 2012). Applications range from commercial applications such as friend-finders (a common name for applications that allow sharing your own geo-location information with your friends or other groups of people) to location aware advertisement. The number of applications had increased exponentially in time since the mid of June 2008, the release date of the second generation of the Apple iPhone (which had a GPS receiver onboard), till mid of 2009, The study from Skyhook (Skyhook Inc. 2012) shows that in the first year till July 2009 3,000 of the 50,000 available smartphone applications for the iPhone use and request geolocation information. Outdoors, in rural areas, the GPS receiver onboard provides positioning information. However, most calls (50–70 %) are initiated indoors, see

(Chandrasekhar et al. 2008, and references inside), where GPS does not work well, or even fails. Therefore, cellular localization needs to complement GPS particularly indoors. Several mobile applications use location information; users share this information for their own benefit and for the benefit of the application and content developers. Content developers understand and learn where users are regularly requesting their applications.

Examples of some application types using location information are:

- Friend-finder applications are adding additional features such as local information around mobile terminals about (e.g., touristic point of interest) objects are available [e.g., the foursquare.com app (Foursquare Labs Inc. 2012)]. This could be enhanced by using visually augmented reality applications that also consider pose in conjunction with location information to inform the user about details of what is in front of him [e.g., the wikitude app (Wikitude 2012)].
- Location aware advertising allows offering users the right coupon at the right time and at the right place. An article in the New York Times (Stross 2010) pointed out that inside a mall GPS is no longer reliable so that the application relies on other positioning technologies to guide the user to an appropriate shop. In malls the fixed infrastructure is controlled by the mall owner and therefore, survey based non-RF location techniques that are explained in detail in Sect. 4.5, could support the cellular technologies inside the smartphone.

The third driving force for location services is the network operator itself. Ongoing research for future radio networks focuses on self-organizing network functionalities, such as adapting to changes in network load or partial breakdowns of the network. Location information can support cellular operators in understanding where a network breaks down so that they can react, e.g., by increasing the coverage area of neighboring cells to compensate for the coverage area of the disabled cell.

5.2 Cellular Networks

In this section we introduce the commonalities of selected cellular network standards. First, we present a general network structure for cellular networks and then show positioning techniques that are proposed for different cellular network standards.

Since the first generation of cellular networks standardization, groups define the basics of the standard that are either mandatory for every participant or optional. The standards define parameters and general procedures of the network, but avoid defining detailed algorithms. The first generation of cellular networks was an analog cellular network and did not offer any location services. In this chapter, we focus on the next three generations of cellular mobile radio systems that are all actively deployed worldwide:

1. Global System for Mobile Communications (GSM) standard (1st standard released in 1990) was the first digital cellular network system and is also called 2G (G = generation) cellular network. The first 2G network started operating in 1991.
2. Universal Mobile Telecommunications System (UMTS)—(1st standard released in 1999) is also called 3G cellular network. There are three different standards that are covered by the UMTS standard. There is namely UMTS-FDD (Frequency Division Duplex), which is also called Wideband-CDMA (W-CDMA) and has the most dominant penetration worldwide as it is used in the USA, Japan, and Europe. The other two standards are entitled UMTS-TDD and TD-SCDMA and not addressed in this chapter. The WCDMA network started to operate in 2001.
3. 3GPP Long Term Evolution (LTE)—(1st standard released (entitled 3GPP Release 8) in 2008) is the successor of the 3G standard, but sometimes called 3.9G as it does not fulfill all requirements that the International Telecommunication Union (ITU) described for the fourth generation. The first public network started at the end of 2009. However, companies have started to advertise it as the fourth generation (4G) of cellular networks as the air interface is incompatible to the predecessor the 3G cellular network. The upcoming successor of LTE is LTE-advanced and is officially called a 4G cellular network by the ITU.

5.2.1 Cellular Network Structure

Figure 5.1 shows a diagram of a generalized cellular network. Any such network consists of two key components: the mobile terminal (MT) and the network entities. The MT is a mobile unit that shares wireless links between itself and at least one serving base station (BS). The MT may have a GPS receiver onboard to position the MT when the mobile terminal receives the GPS signals well (at least outdoors). The BSs are part of the network and are connected to a core network unit that forwards the calls or data connections to the mobile switching unit. The mobile switching unit collects the different connections from different central network units and forward them to the public network. In a cellular network, two different links are distinguished: the downlink is defined as the link from the BS to the MT and the uplink from the MT to the BS. The uplink is normally served by a single BS. However, e.g., to initiate and accomplish a handover from one BS to another BS, multiple uplinks could be active in parallel. In the downlink the MT listens to multiple BSs to be aware of the different BSs.

To position an MT in a two-dimensional plane, at least three receivable base stations are required. A typical assumption is that these BSs are synchronized in time. If this is not implemented, an additional location measurement unit (LMU) can measure the time differences between the BSs itself and the MT. LMUs are part of the fixed infrastructure and know the coordinates of themselves and of the BSs.

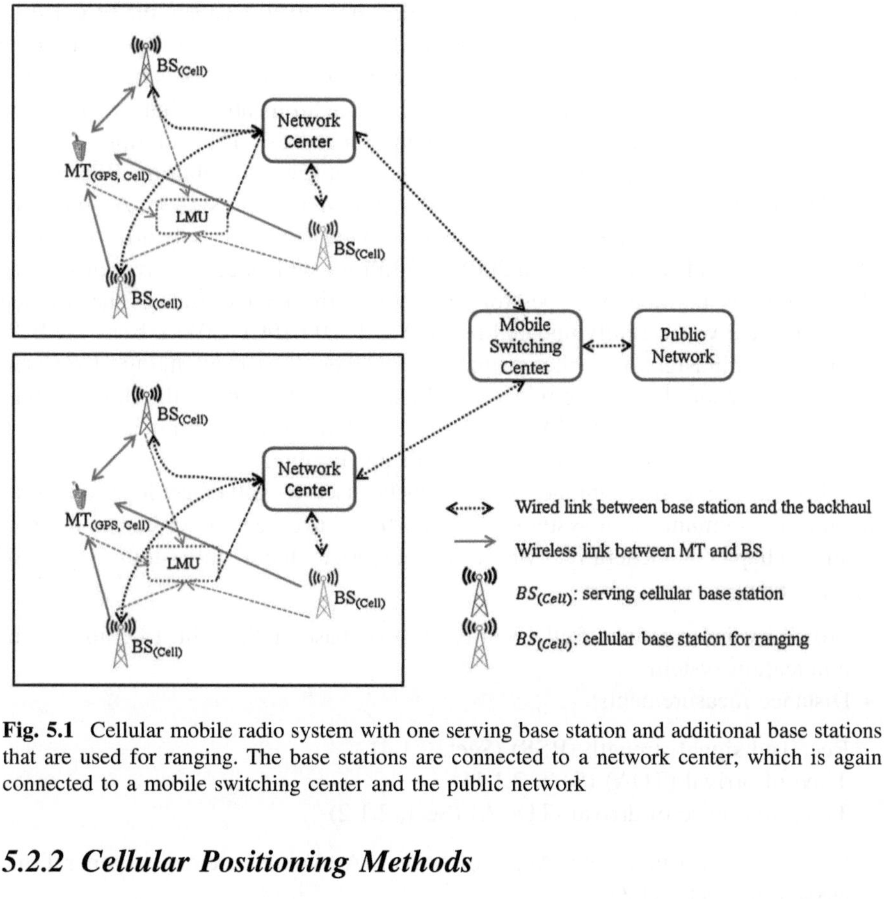

Fig. 5.1 Cellular mobile radio system with one serving base station and additional base stations that are used for ranging. The base stations are connected to a network center, which is again connected to a mobile switching center and the public network

5.2.2 Cellular Positioning Methods

The evolution of cellular positioning systems reflects the increasing interest of network operators on location services. Wireless transmission systems had been traditionally designed and standardized for either communications or positioning. This trend has been changing throughout the years as users and network operators have focused on integrating both services in a single device. One reason is due to the lack of spectrum to accommodate the complementary systems, keeping in mind that more and more smartphones now incorporate satellite based positioning systems, such as GPS, into their bandplan. When location services first appeared in the earlier Global System for Mobile Communications (GSM) radio system, the interest on localizing mobile terminals was low. The GSM standard was released in 1990 and it was far too early to consider the request from the FCC for locating emergency callers in 1996. The GSM standard considered only basic cellular ID. There were no additional positioning signals defined to improve positioning information during the process of refining the GSM standard that started in 2000.

The third generation of cellular networks (3G), the W-CDMA standard, or Universal Mobile Telecommunications System (UMTS)-FDD, was released in

1999 and the first systems were actively deployed in 2001. By then, the FCC E-911 mandated requirements had been published. However, network operators soon realized that the time-based methods, upon which the standard was based, did not perform well due to the interference of the synchronization signals that were simultaneously broadcasted between neighboring BSs. The solution was the insertion of idle periods to avoid interference between neighboring BSs, but to the detriment of the cellular capacity of the communication system. As such, the solution was standardized, but not made mandatory for network operators. Repeated demands by the FCC in 2005 to fulfill the dedicated requirements raised the interest by the network operators, at least in the United States. The original 1996 mandate was recently updated in 2010 and 2011 (FCC 2011a, b) as the FCC realized in hindsight that the original goals were too challenging. However, changing a standardized system is a complicated task. To address the shortcomings of the GSM (2G) and WCDMA (3G) cellular systems, the recent releases of 3GPP LTE added dedicated reference symbols for positioning.

There are several techniques which can be used to gather geolocation information in communication systems—all of which have been described in the preceding chapters. Generally, we can categorize them into four types of measurements:

- Proximity information: Cell identity (ID) of base stations in a cellular communication system
- Distance measurements:

- Received signal strength (RSS) (Sect. 2.1.4)
- Time of arrival (TOA) (Sect. 2.1.1)
- Time difference of arrival (TDOA) (Sect. 2.1.2)

- Directional measurements: Angle of arrival (AoA) measurements (general discussion in Sect. 2.1.3)
- Survey information (fingerprinting) based on past signal strength measurements (see Chap. 4)

These techniques (except the Cell ID) are all optional in the cellular network standards for the mobile phone industry players. In the following, we present relevant techniques that were, and currently are, used in cellular networks to position the MT. Finally, we describe a combination of different types of measurement data.

5.2.2.1 Cellular ID

Cell identity (CID) is a simple and fast method to position the mobile terminal. It uses the location of the base station that serves the mobile terminal. This is the simplest positioning method in cellular communication systems. The response time is very fast as it maps the coverage of cell to an area. It is effective because of the instantaneous response together with the cellular coverage. Figure 5.5 shows an MT that is located at the cell edges of BS1 and BS2. Depending on which of these cells serves the MT, the MT will decide its own position. Figure 5.5 shows the general scheme where IDs of two different base stations are detected. Obviously, the cell size matters significantly, where cell size could also be determined by the transmit power. If the MT in Fig. 5.5 connects to both BSs, the position estimated could be improved (e.g., by interpolating) using the location of both base stations. The CID method is available in all cellular standards. The different cell sizes are shown in Table 5.2 for the different standards. The most recent standard, LTE, offers the smallest cell sizes that are femtocells which have a diameter of several tens of meters to cover e.g., apartments.

5.2.2.2 Synchronization Methods

In mobile radio communication systems the receiver needs to synchronize with the transmitted data stream. There are different reasons for synchronizing, such as avoiding interference between different data streams of different users, or detecting the start of a data stream to decode it successfully. In case the receiver is synchronized with the transmitter, it has (approximately) the same time base and can calculate the distance between both. In the following, two methods are presented that are used to synchronize the communication streams, but in addition are used to estimate the distance between the MT and the BS.

Timing Advance

GSM is a frequency-division (FDMA) and time-division multiple access (TDMA) system. The TDMA component is used to synchronize up- and downlinks for the MTs. The BS considers the propagation delay to synchronize different users and to minimize the interference between them. Figure 5.2 shows the timing advance procedure for the TDMA part. The base station transmits eight slots within one transmit frame. The signal propagation delay is estimated from the synchronization frame, resulting in a resolution of 64 bits. Each bit corresponds to a propagation delay of 3.69 μs, which translates to a distance of 553.5 m given the speed of light. The timing advance procedure relies on an established connection of the MT with the BS—that means it requires handovers to multiple BSs to estimate the range to them. Therefore, communication resources such as spectrum and power have to be available twice in both cells at the same time.

Synchronization with Multiple Base Stations

An alternative approach to timing advance procedure is to use synchronization sequences from different BSs without performing a hard handover of the MT from

Table 5.2 Overview of basic cellular network parameters

Cellular mobile radio system	Typical cell sizes (diameter)	Synchronization	Bandwidth	Typical carrier frequency (varies for different countries)	Antenna setup	Downlink/Uplink access scheme	Handover scheme
GSM	Up to 35 km, Macro > 2 km, Micro 200 m–2 km	Timing advance	0.2 MHz	900 MHz, 1,800 MHz	MT: Single BS (rarely): Sectored	Single carrier (TDMA and FDMA)	Hard
WCDMA	Like GSM, additionally Pico: < 200 m	Synchronization symbols	5 MHz	900 MHz, 1900–2,200 MHz	MT: Single BS: Diversity and sectored	Wideband-CDMA	Soft
3GPP LTE	Like WCDMA, additionally Femto (homes): < 100 m	Several different: Primary/secondary synchronization sequence, reference signals for specific channels, positioning reference signals	1.4 MHz, 3 MHz, 5 MHz, 10 MHz, 15 MHz, 20 MHz	700–800 MHz (rural), 2,600 MHz (urban)	MT and BS: Multiple antennas on transmitter and receiver for diversity and spatial multiplexing	OFDMA/SC-FDMA (Single carrier)	Soft

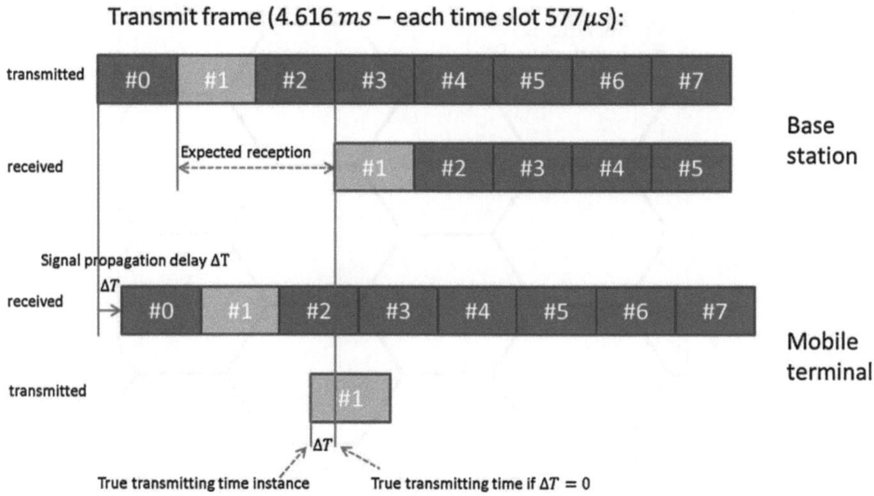

Fig. 5.2 Timing advance in the GSM frame structure of the TDMA system

one BS to another. The MT tries to synchronize with each BS by using the known sequences. In a cellular network, the network operator intends to reallocate the same spectrum for different links (MT-BS) at the same time but at different locations. The closer the locations of the BSs to each other, the better the spectral efficiency of the network.

Figure 5.3 shows a cellular system where each cell has a hexagonal shape and each BS serves three cells in parallel by using sectored antennas. The gray shadowed cell is covered by BS1 (which is called the serving BS) and the light gray dots represent different positions of MTs inside the gray cell. In general, if the MT attempts to synchronize with a BS, we call it the dedicated BS. The synchronization sequences of the neighboring cells could be used by the MT in the gray cell as well. However, these sequences will be interfered by the sequences broadcasted from the serving BS as these sequences are broadcasted at the same time and at the same frequency. The synchronization sequences are different for the neighboring cells to differentiate them by the sequence itself. The closer the MT is to the dedicated BS, the better the MT can synchronize as the signal strength of the dedicated BS increases and interfering power of the synchronization sequences of the other BSs decreases. Vice versa, the further away the MT is from the dedicated BS, the worse the performance of the synchronization will be because of the higher interference by the synchronization sequences of the neighboring BSs and because of the lower signal strength of the dedicated BS. To reduce the impact caused by the interference, an additional scheduled idle period was introduced for each BS. With this the spectral efficiency of the network is reduced, but the MT gains the ability to also listen and synchronize successfully to the neighboring BSs. If the MT synchronizes successfully with the BS, it can estimate the range.

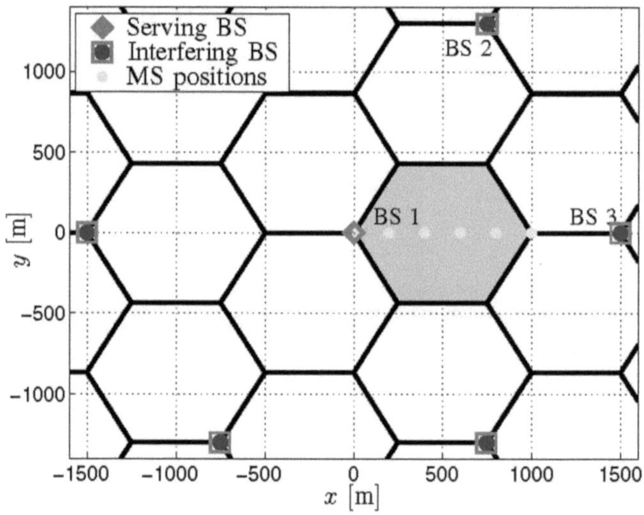

Fig. 5.3 Cellular system with a single mobile terminal (*light gray dots inside the shaded hexagonal cell*) at different distances from BS 1 (*diamond*) as serving BS and the neighboring BSs (BS 2 and BS 3 in *quadrats with an inner circle*) as non-serving BS. The MT attempts to synchronize with all three BSs to estimate the range (Mensing et al. 2010)

In Mensing et al.(2010), the authors investigated the impact of interference on synchronization sequences from different BSs. The interference is reported as negligible, therefore it causes no noticeable effect on the communication service. However, for positioning, the MT needs at least three BSs to which it is connected. The authors proposed an interference cancellation scheme to improve the connectivity to the dedicated BS. Figure 5.3 shows a cellular network with an MT that is represented at various locations by the light gray dots inside the gray shaded hexagonal cell and served by BS 1—the MT is always in the gray shadowed cell. The MT attempts to synchronize to all BSs (BS 1, BS 2 and BS 3). Figure 5.4 shows the corresponding positioning performance at the different locations of the MT. The timing estimates of BS 2 and BS 3 are only of good quality if the MT is far away from its serving BS 1 (around 1,000 m). The reason is that the interference of BS 1 on the synchronization sequences of the neighboring BS is weak. Close to the BS 1, the quality of the timing estimates for BS 2 or BS 3 is insufficient for precise positioning due to interference from the serving BS (BS 1). The authors proposed an interference cancellation scheme, which improved the timing estimation accuracy drastically. Especially close to the serving BS (BS 1), the performance gains were very high. As in these situations, the serving BS 1 can be detected with high quality, resulting in high positioning accuracy. In WCDMA, the problem was recognized and the proposed solution was much simpler. WCDMA introduced idle periods for the different BSs to avoid interference of the synchronization sequences between neighboring cells. Using idle periods shows similar performance to cancelling the known synchronization sequences, but comes at the expense of communication capacity.

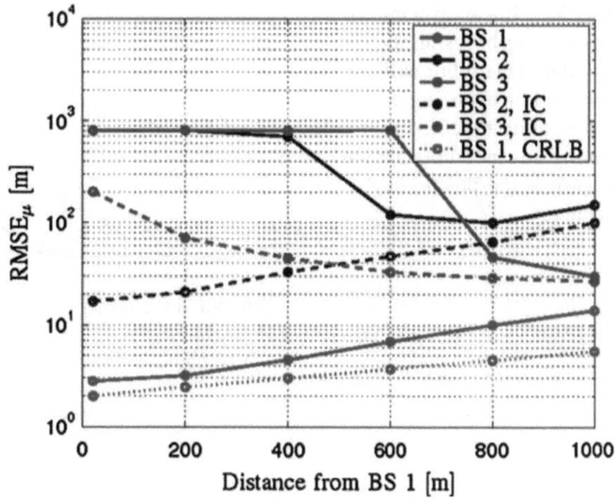

Fig. 5.4 The corresponding positioning performance using the RMSE comparing with and without interference cancellation. The *blue dotted* represents the lowest possible performance with the given signals using the Cramer Rao Lower Bound (CRLB) and the dashed lines the performance with an interference canceller of the corresponding BS (Mensing et al. 2009)

5.2.2.3 Combination of Positioning Techniques

In the following, we show how multiple techniques, such as Cell ID, synchronization techniques, and angle of arrival can complement each other to improve positioning performance. The MT in Fig. 5.5 receives, in addition to the Cell ID information about the ranging time, by using, e.g., the timing advance procedure. The timing advance procedure is a function of the distance between the BS and the MT and is therefore used to estimate the distance between the BS and the MT. The timing advance depends on the technology that is used. GSM uses the synchronization sequence and the granularity of each bit is 3.69 μs. In UMTS, timing advance was not part of the standard. In LTE, the timing advance value is about 0.52 μs. The uncertainty of the timing advance (TA) procedure is indicated by the two circles around BS_1. The timing procedure requires an existing or established communication link between the MT and the BS. The BS broadcasts its cellular ID by a regular schedule that is synchronized with neighboring cells, such as BS_2 to avoid interference between BS_1 and BS_2 broadcasts. The broadcast of BS_2 can be detected by the MT that is in range. However, BS_2 provides only its cellular ID to support determining the position of the MT. In addition, BS_1 also employs directional or sectored antennas to enhance the positioning accuracy even further. Instead of the sketched map of Fig. 5.5, Fig. 5.6 shows complementary information about the environment. A building blocks the directional link between BS_1 and the MT and a reflector causes a non-line-of-sight path between BS_1 and the MT. Both cause the position estimator, having no additional support of BS_2, to calculate the position with a significant error. The estimated angle of arrival (AoA) information from BS_1 is wrong and BS_2 cannot determine the AoA as it has an omnidirectional antenna pattern. Today, many base stations have multiple antenna systems and serve only a sectored area around the base station. Therefore, the sector information limits the coverage area and, hence, could be easily used to position the mobile terminal quickly and more accurately. A more advanced

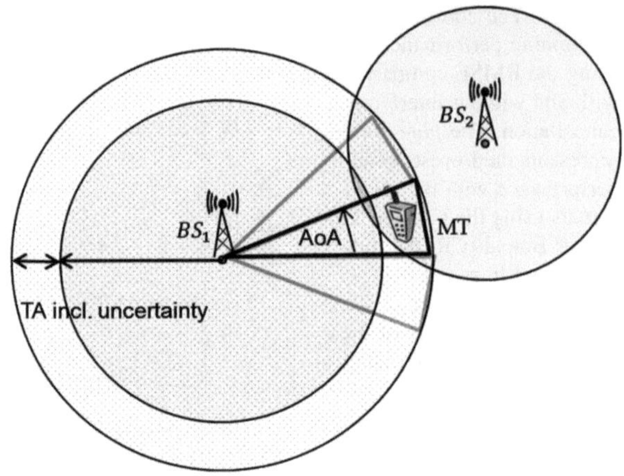

Fig. 5.5 Basic principle of positioning by using the cellular ID. Additionally, sectored or even angle information together with timing advance information improves the positioning accuracy

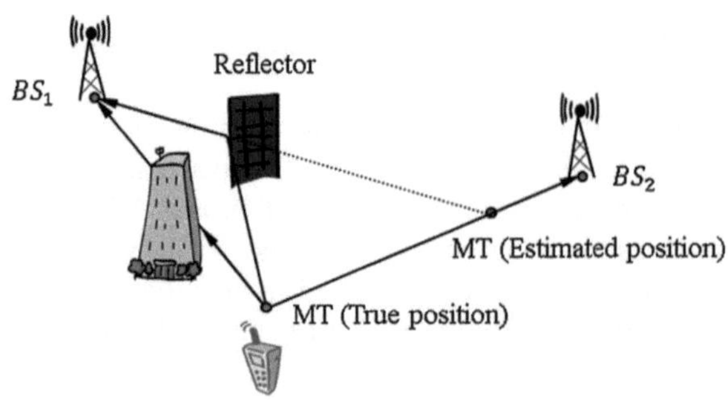

Fig. 5.6 Estimated position compared to true position based on E-CID positioning

method would be to calculate the angle of arrival (AoA) at the base station (see more details in Sect. 2.1.3). This requires a precise estimation of the spatial channel impulse response. The combination of additional information coming from synchronization methods such as the timing advance procedure or from other sources such as the angle of arrival together with the cell-ID is called Enhanced Cellular-ID (E-CID). E-CID requires more information and therefore the response time is slightly larger than for the simple cell-ID method. The additional information is requested by the mobile terminal. More about the combination of time-of-arrival and angle-of-arrival measurements is described in the next chapter which deals with cooperative sensor networks.

5.3 GSM, WCDMA and LTE Cellular Networks

In this section, we outline the details about the cellular network standards with a focus on the parameters that impact the positioning performance. Table 5.2 presents parameters for the different cellular standards that are relevant for positioning.

The change in these parameters from one cellular network generation to another is driven by the potential improvements of the communication networks.

In LTE networks, the cell size could be significantly decreased compared to WCDMA and GSM cellular networks, e.g., by using femtocells or small cells to improve the throughput (the MT is close to the serving BS). This means installing multiple cells instead of a single macro (large) cell. The benefit for positioning is that any error e.g., based on the cell ID is also significantly lower compared to the positioning performance using e.g., the macro cell. Synchronization has improved for the benefit of communications to allocate the resources for more users and to avoid interference. The improvement in synchronization also supports the ranging performance and therefore the positioning performance as well. Furthermore, explicitly dedicated signals for positioning are integrated in the most recent Release 9 of the LTE standard. The bandwidth is important in communication networks to achieve high throughput and high data rates. In comparing GSM and LTE, the available bandwidth increased a hundredfold from the former to the latter.

The carrier frequency is relevant for the penetration of transmitted signal: the lower the carrier frequency, the better the penetration of the signal. Therefore, lower carrier frequencies are used in rural areas and in urban areas (for better indoor coverage) to exploit the higher penetration of the signal. However, in urban areas cells get smaller and smaller. Therefore, the penetration could also be obstructive as it causes interference in neighboring cells. Recently, macro cells with rather large cell sizes use lower frequencies and for smaller cells, such as femtocells higher carrier frequencies are used. Multiple antennas allow estimating the angle of arrival and controlling the angle of departure. Both techniques support the communication needs to reduce interference, but also support the positioning performance by e.g., reducing ambiguity.

The access scheme and the handover procedures of the recent standard, such as LTE compared to GSM, offer a higher flexibility in a cellular networks. The additional flexibility e.g., allows to support multiple MTs at the same time with individual requests for their positioning needs. This improves the response time for each MT to achieve a position estimate compared e.g., to the GSM standard, where a single MT could allocate all resources of a cell. GSM uses a time division multiple access (TDMA) scheme to serve up to eight users in a single frequency band of 0.2 MHz. WCDMA uses direct sequence code division multiple access scheme (DS-CDMA) to spread the data of multiple users over the same frequency channel of 5 MHz. A key to the success of CDMA systems is power control of the individual data streams of the different MTs. Power control reduces the transmitted power of a signal from a close-by MT and increases the transmitted signal of a distant MT. With this the network tries to obtain a similar power level of the different MTs at the BS. The consequence of varying transmit power is that a MT may not be heard by other BSs that are outside of the cell. LTE uses orthogonal frequency division multiplexing (OFDM) as access scheme in the downlink. In the uplink, single carrier frequency division multiple access (SC-FDMA) is used. It is similar to OFDMA with an additional operation such that the transmitter can operate power more more efficiently. Both up- and downlink schemes offer flexibility to use different bandwidths for different users. The varying bandwidth also impacts the positioning performance because the positioning reference signals are part of the resource block.

5.3.1 GSM Cellular Networks

GSM networks have been operating since 1991. They use a rather low bandwidth of 200 kHz for each link compared to LTE networks that can use up to 20 MHz for each link. Since GSM is a voice oriented standard, this low bandwidth is sufficient. However, in Chap. 2 (Sect. 2.2.1 and Eq. 2.44) it was shown that the bandwidth has a significant impact on the variance and the error (Eq. 2.45) of the time-based ranging measurements.

At first, the GSM network used a carrier frequency around 900 MHz to offer good penetration of the transmitted signal. Later, new carrier frequencies around 1,800 MHz were additionally used, each with less penetration and lower transmit signal power in response to the smaller cell sizes. The cell sizes served from a BS are large and could range up to 35 km in rural areas with low customer density. The goal of the GSM system at the beginning was high coverage of mainly vehicular users with velocities up to 50 m/s. Mobile phones were bulky at the beginning of the 1990s and, as a result, mobile users operated mainly from the inside of cars. Large cell sizes are an advantage for communications as the handover procedure in GSM was a hard handover procedure. A hard handover works in the following manner: the established connection from the first BS will be dropped and, only after the drop, the connection to the second BS will be established. There was a significant risk of losing the connection in case of a voice call in between.

The low bandwidth and the large cell sizes offered rather limited accuracy performance for the positioning techniques that were built on the physical constraints of the communication system. An early paper (Reed et al. 1998) in 1998 draws the conclusion that the performance requirements of the FCC would be hard to fulfill. Reasons were the limited coverage (or hearability) of multiple base stations at the same time. The second reason was that even if multiple base stations were available, their geometrical constellations may not have been favorable (see derivation of Eq. 2.17). The final reason was that the base stations were not synchronized for time-based positioning methods such as time-of-arrival TOA (see Sect. 2.1).

The cell-ID principle, explained in Sect. 5.2.2, was applied in GSM without any further enhancements. GSM uses the timing advance procedure to synchronize the different time slots for multiple users. Timing advance was first introduced (and used for positioning) in GSM. GSM is a time-division multiple access system where multiple users share the same frequency. Therefore, the network assigns different time slots to multiple users. The network or base station informs the mobile terminal about the timing advance to align the time slots of the different users in the uplink to avoid interference. However, the mobile terminal is forced to perform multiple hard handovers in GSM to learn the different timing advances of the different base stations. As the density of base stations was low and the cell size was large, it was not always possible to find enough BSs to position the MT without ambiguity.

The technique of observed time difference positioning is derived from a procedure for handover. The mobile terminal observes the time differences between two base stations. As such, the mobile terminal can estimate the propagation delay to the serving base station. The success of this procedure was foreseen as limited as the mobile terminal had to meet stringent timing requirements (Reed et al. 1998). However, the original idea was picked up again for the UMTS/WCDMA cellular networks.

5.3.2 WCDMA Cellular Networks

WCDMA networks have been operating since 2001. The WCDMA standard uses a bandwidth of 5 MHz for the downlink and the uplink. The access scheme is a direct sequence (DS)-CDMA scheme that spreads the data of multiple users over the whole jointly used bandwidth. The carrier frequency that is used in WCDMA networks is usually between 1,900 and 2,200 MHz. In WCDMA networks, various localization methods are applied, namely cell-ID and enhanced cell-ID—enhanced with the timing advance procedure of the GSM standard. An overview paper from 2002 (Zhao 2002) summarizes the different technologies for 3G systems.

A new positioning specific hardware component introduced in WCDMA networks was the LMU. An LMU is a fixed unit to make radio measurements and offers two functionalities to the network. The first functionality is that an LMU is used to estimate the time differences between two base stations, which is important in mixed cellular networks with GSM and UMTS. The second functionality is that the LMU supports each MT individually with measurement results obtained at the LMU from the MT. LMUs support time-based positioning methods such as Observed-TDOA (the basics of time difference of arrival (TDOA) are explained in Sect. 2.1.2). The TDOA technique is necessary as only the base stations in WCDMA with the support of the LMUs and GPS receiver are synchronized with each other.

5.3.2.1 (Downlink) Observed Time Difference of Arrival (OTDOA)

The goal of the TDOA method (see Sect. 2.1.2) is to observe and examine time differences of measurements at reference points to estimate the range differences between the MT and the BS (Medbo et al. 2009). Only the BSs are synchronized with each other, e.g., using GPS receiver, but the MT is unsynchronized with the BSs. As such, the mobile terminal observes time differences of signals from different base stations. Typical signals used for TDOA are reference signals of control channels that are, for example, needed to make the MT aware of neighboring BSs or

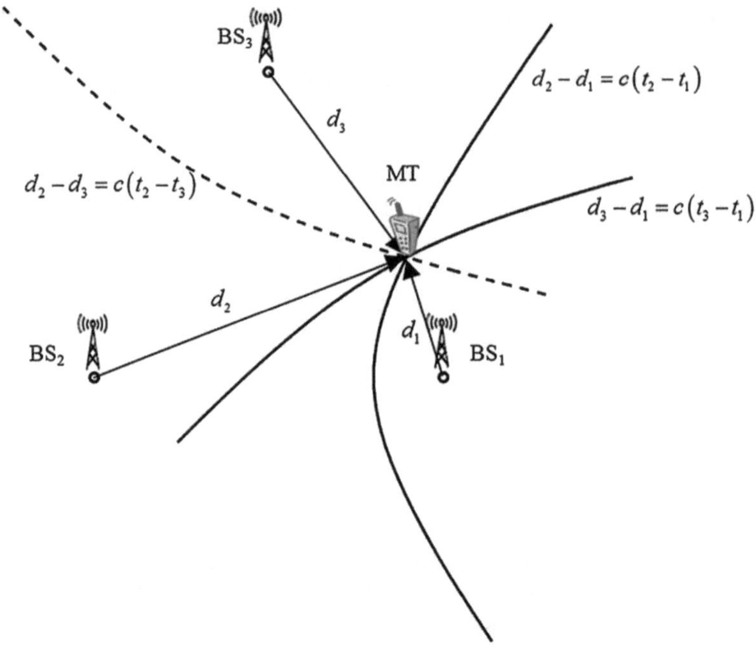

Fig. 5.7 Observed time-difference of arrival (OTDOA) method

signals that are needed to prepare a handover. Compared to GSM in WCDMA, the reference signals are used for positioning without performing a handover. For TDOA, it is required that the time differences between the BSs are known or zero (perfectly synchronized between BSs). The differences of the clocks could be measured and solved by additional LMUs at known locations. LMUs are placed at a fixed position and perform additional measurements to track the time offset of the individual BS. This was the preferred method in UMTS/WCDMA. The synchronization has to be fairly accurate; for example, a 10 ns error corresponds to an uncertainty of 3 m for the position estimate.

Figure 5.7 shows how the different links are considered by calculating the relative distances depending on the measurements t_x, where t_x describes the time measurement between the mobile terminal and the x_{th} BS. The positions of constant time differences are hyperbolas (hyperboloids) with their foci at the location of the corresponding BSs. Their intersection provides the position of the mobile terminal. The MT receives multiple signals from different BSs in parallel to perform measurements. In WCDMA, the MT may suffer from the interference of these signals with each other. Especially if the MT lies close to its serving BS, the signals of the neighboring (Kim et al. 2005) BSs are weak but needed to have multiple time measurements available. To detect the different signals, WCDMA introduced an idle period for the downlink (IPDL). The idle period improved the hearability of the neighboring base stations as it avoided any interference at this

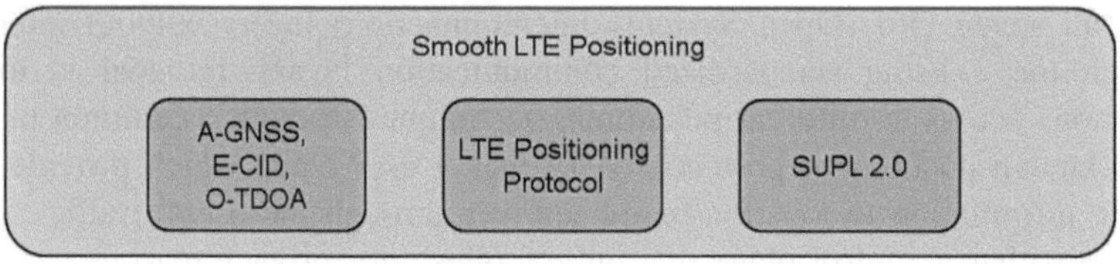

Fig. 5.8 Overview of the different technologies that are covered by the LTE standard

time. The length of the idle period influences the positioning accuracy, as the mobile terminal could integrate longer, but it also reduces the throughput of the communication system. Originally the slots were randomly allocated to perform measurements. Ludden and Lopes (2000) introduced a scheduled procedure that was time aligned to have a common idle period with less interference and mobile terminals perform measurements at the same time slot. The additional cost comes from the synchronization effort between the base stations to schedule when they need to be idle and when each BS can perform measurements.

5.3.3 3GPP LTE Cellular Networks

While Release 9 (end of 2009) of LTE introduced new positioning features, the primary goal was to offer a smooth transition from the existing cellular standards (2G (GSM) and 3G (WCDMA)) to LTE. Figure 5.8 depicts the different technologies that are used to address positioning in LTE networks. Depending on which network is additionally active (LTE base stations also work together with WCDMA and GSM networks), different protocols are applied. The left block outlines three techniques (A-GNSS, ECID, O-TDOA) that are used today in cellular systems, such as GSM and WCDMA directly. A-GNSS stands for Assisted Global Navigation Satellite System and describes the assistance of the cellular network to support GNSSs. Well-known examples of a GNSS are NAVSTAR GPS, or the European Galileo system, or the Russian Glonass system. A-GNSS is an evolution of assisted-(A-)GPS.

In addition, the LTE positioning protocol uses methods that are uniquely developed for LTE, e.g., based on the positioning reference signals (PRS 3GPP TS 25.305, 2011). SUPL 2.0 is a third protocol that establishes an overhead protocol to interface different techniques that would not work with each other directly. (Open mobile alliance 2012a, b). With this, it ensures that several air interface technologies can exploit their solutions without the need of a defined standardized interface (it includes the named three cellular standards and also supports others such as WiMAX, WiFi 802.11, etc.). Open Mobile Alliance also provides enablers of interoperable services working across the world with different operators and different mobile terminals. The relevant standard for positioning is called the Secure User Plane Location (SUPL). A plane allows establishing a connection using all seven OSI (Open Systems Interconnection) layers without interfering through the existing standardized communication layers. Its goal is to offer ubiquitous access to different positioning techniques through a common interface and a known positioning protocol. SUPL has a user plane which provides positioning information to location-based services through the users' traffic channel. SUPL also has a control plane that only applies—for privacy reasons—to a very limited number of use cases addressed by law enforcement and network monitoring tools and is used by the operator traffic channel.

A-GPS has an additional data stream (useable in some networks even without a specific data contract) of the cellular network that broadcasts recent information about the GPS system that is relevant inside the local cell. Receiving the recent information via GPS requires up to 12.5 min as the data rate is 50 bps. Therefore, the alternative solution is to use A-GPS which improves the time to first fix. Today, A-GPS is the only active system that is used in cellular networks, but the LTE protocol includes support to assist navigation systems, such as Galileo or GLONASS when they exist. There are two fundamental modes that are supported for A-GNSS:

- MT-assisted method: The MT receives information that includes visible satellite list, reference time, and other assistance data. A significant part of the assistance data is usually valid for only a few minutes. The MT receives satellite signals and transmits the measured data back to the network. With this, the network or the location server calculates the position of the MT and shares it with the MT.
- MT-based method: A GNSS receiver is part of the mobile terminal. During the start-up phase, satellite orbital elements, i.e., ephemeris, reference time, and other data are provided to the MT. The MT uses these assistance data to calculate its own position.

The concept of A-GNSS has the following advantages. The acquisition time is significantly reduced because of coarse knowledge about the position of the mobile terminal through the cell. In particular, GSM has a maximum cell size of approximately 35 km, whereas in LTE cells such as femtocells are much smaller (less than 100 m). In Monnerat (2008), the author pointed out several additional advantages of A GNSS, viz calculating the satellite Doppler frequency, improving pre-synchronization with the synchronized time between the GPS time and the network time resulting in a reduced complexity of the synchronization algorithm. The time-to-first-fix is significantly reduced from about 60 s to less than 20 s. Together with a mobility model, the accuracy of the tracking algorithm can also be improved.

The three individual techniques mentioned above will have different limitations in different operating environments. In order to provide ubiquitous positioning, the multiple techniques are leveraged. An adaptive solution has the flexibility to choose a selection of positioning technologies depending on the requested quality of service.

A recent development combined the idea of RF fingerprinting localization (see Chap. 4) with angle-of-arrival and cell-ID. Wigren (2007) outlined a multi-step approach, called adaptive enhanced cell-ID (AECID) fingerprinting. The method is an iterative method that improves by requesting more information if needed. The mobile terminal requests in which cell it is located over multiple steps and then it collects round-trip time measurements between the serving base stations. Next to the ECID it builds on fingerprints that are collected a priori. Furthermore, in AECID a continuous collection of updates allows the terminal to build and refine a database of dedicated high-precision measurements and computes confidence information for that data. The proposed method initiates auxiliary measurements if needed and refines by this the positioning estimate. Furthermore, the authors proposed a quantized measurement to improve the consistence of the fingerprints between different mobile terminals. This is especially helpful for mobile terminals that are affected by shadowing from the user—e.g., the RSS values easily vary by 5–10 dB. The response time is slightly higher than for E-CID in case auxiliary measurements are needed.

3GPP LTE uses different synchronization symbols that are mainly used for communication purposes. The primary synchronization signal (PSS) is based on a Zadoff-Chu sequence (TS 36.211, 2011) and is used to achieve a coarse synchronization. The Zadoff-Chu sequences result in a zero-cross-correlation if the sequences are shifted. The secondary synchronization sequence (SSS) is of two length-31 binary sequence that is used to acquire the communication signal more precisely. LTE aims for a reuse factor of one, which means neighboring cells shall allocate the same spectrum to different users. As such, to obtain reasonable ranging measurements with the neighboring cells, Release 9 of the standard proposed positioning reference signals (PRS). Figure 5.9 depicts the scattered signals over a part of an LTE frame in the downlink. The subcarrier index represents the frequency direction. Each subcarrier is generally 15 kHz broad. Each OFDM symbol is 66.7 μs long. The signals are dedicated to positioning and intercellular interference is avoided by carefully scheduling the symbols between the different base stations. Additionally, when interference is unavoidable, neighboring base stations can also stay mute to further reduce intercellular interference and to improve hearability.

LTE uses different bandwidths depending on the downlink data-rate requested by the MT. As shown in Chap. 2, generally the performance of ranging depends on the bandwidth of the signal. The different bandwidths range as 1.4, 3, 5, 10, 15, and 20 MHz. The synchronization signals (PSS and SSS) for communication are all allocated in the narrow band (1.4 MHz). However, LTE offers a more precise positioning as it also adds more PRS depending on the actual bandwidth used. The PRS are part of the resource frame.

In addition to O-TDOA discussed previously, LTE also considers uplink TDOA (U-TDOA). In theory, time-of-arrival (TOA) measurements could be performed either at the base station or at the mobile terminal (Sun et al. 2005; Gustafsson and Gunnarsson 2005). This, however, is conditioned upon full synchronization between the two, as outlined in Chap. 2. In LTE cellular communication systems the BS and the MT are not well synchronized, therefore TOA measurements have to be used differentially by TDOA. This requires either that the participating base stations are synchronized between each other or that their synchronization offsets

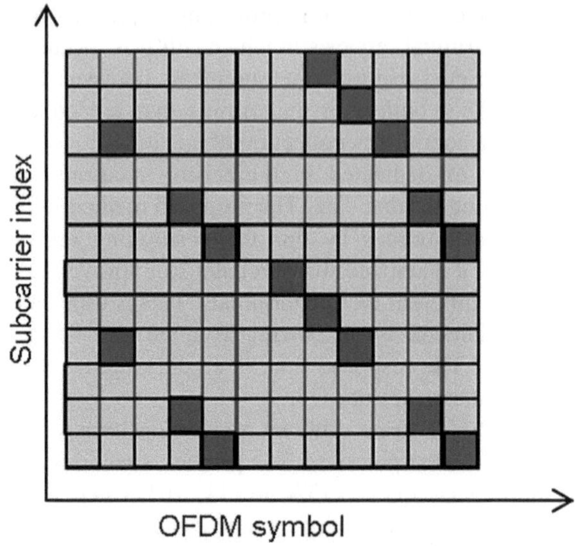

Fig. 5.9 Part of an LTE frame with positioning reference signals in dark (TS 36.211, 2011)

are known via LMUs. U-TDOA is a network-based method, as measurements are taken at the base station (the U stands for uplink) in a dedicated mode. When the MT is nearby, the transmit power of the MT is weak due to power control on the uplink. A drawback is that neighboring base stations receive low signal power. Notwithstanding, the advantage of U-TDOA is that any mobile terminal can be supported independently of its own localization capabilities; these capabilities are provided by the network. In LTE, U-TDOA is under discussion for Release 11 (not yet finalized in early 2012).

5.4 Conclusions

Table 5.3 shows an overview of the different techniques comparing the potential positioning accuracy versus response time (time elapsed for the system to provide the estimated position). The response time of the available location information depends on the amount of additional data that is requested. For example, the network and the protocol that a cellular system uses may gather and distribute relevant data helping to position the mobile terminal.

The simplest and quickest solution to determine location is the attempt of mapping the Cellular identity (CID) to a dedicated position. The positioning uncertainty is related to the cell area size and, if multiple cell IDs can be combined, to reduce ambiguity. To collect additional information to improve the positioning accuracy increases the response time. Such procedures are Enhanced-CID and Adaptive Enhanced Cell ID (AECID). AECID requires even a priori data of the signal strength and initiates additional time-based measurements if the

Table 5.3 Summary of the different techniques and the expected performance (Ericsson White Paper 2011)

Positioning Method	Limitations (environmental and computational)	Positioning performance	Response time	Mobile terminal impact
Cell-ID	No	(Very) low (depends on cell size)	Very low	None
Enhanced cellular ID	No	Medium	Low	Feedback of timing advance
E-cellular-ID + AoA	Rich multipath complicates estimation because of AoA estimation	Medium	Low (AoA estimation)	Feedback from the base station
A-GPS (GNSS)	Indoor reception is weak	High	Medium to high	High: GPS (GNSS) receiver must be available
RF fingerprinting	Indoor lack of measurements	Medium–high (depends on environment and data available)	Low	Database or accurate model
Adaptive enhanced cellular ID (AECID)	No	Medium (depends on additional data)	Medium	Feedback from the base station
U(Uplink)TDOA	Multipath	Medium–high	Medium	Medium (additional measured and communicated from the base station)
O(Observed)TDOA	Multipath	Medium–high	Medium	Low (observed for handover)

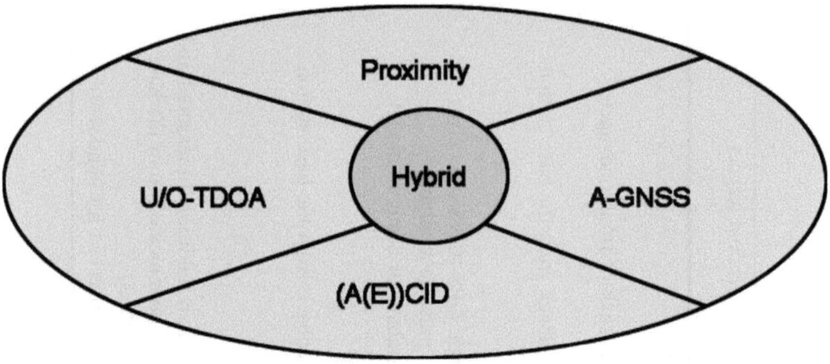

Fig. 5.10 Different positioning solutions that are available and could be jointly used

confidence of the provided RSS based measurement data is low. The time- or ranging-based methods, such observed TDOA or uplink TDOA, require numerous links that need to be scheduled, coordinated and finally processed. Therefore, their response time is significantly larger.

Figure 5.10 shows the different techniques that have been developed and integrated over the last 15 years in cellular mobile radio systems. Several publications (Mensing et al. 2010; Ericsson White Paper 2011; Wigren 2007) draw the conclusion that not a single solution is sufficient to offer reliable positioning information in all environments. When focused on indoor environments, the accuracy performance of proximity methods is linked to the cell size. Adaptive methods, such as AECID, improve accuracy by considering fingerprinting data. This, however, comes with additional computational costs as more data needs to be merged so that response time increases. A-GNSS methods improve positioning by GNSS, but the urban-canyon and indoor environments prove very challenging for GNSS receivers. Therefore, in these environments, supplementary methods are often required to successfully position the mobile terminal. Uplink or Observed TDOA are well understood and accepted now in the recent release of LTE. The additional positioning reference signals as part of the communication signal (and even additional symbols used in the broader channels) indicate that future cellular mobile radio systems will be able to fulfill the requirements that the FCC already demanded in 1996. Finally, Table 5.3 summarizes the different techniques, the expected accuracy performance, the response time, and the impact on the mobile terminals.

References

3GPP, TS 25.305 Universal mobile telecommunications system (UMTS); stage 2 functional specification of user equipment (UE) positioning in UTRAN. Technical specification, version 10.0.0 (2011)

3GPP, TS 36.211 (2011) LTE; Evolved universal terrestrial radio access (E-UTRA); physical channels and modulation. Technical specification, version 10.1.0 (2011)

V. Chandrasekhar, J. Andrews, A. Gatherer, Femtocell networks: a survey. Commun. Mag. 59–67(2008)

C. Drane, M. Macnaughtan, C. Scott, Positioning GSM telephones. Commun. Mag. 46–54(1998)

Ericsson white paper. Positioning with LTE(2011), http://www.ericsson.com/res/docs/whitepapers/WP-LTE-positioning.pdf. Accessed 20 April 2012

FCC, Amending the definition of interconnected VoIP service in Section 9.3 of the commission's rules wireless E-911 location accuracy requirements E-911 requirements for IP-enabled service providers (2011a), http://transition.fcc.gov/Daily_Releases/Daily_Business/2011/db0713/FCC-11-107A1.pdf. Accessed 13 July 2011

FCC, Wireless E-911 location accuracy requirement (2011b), http://hraunfoss.fcc.gov/edocs_public/attachmatch/DA-11-1125A1.pdf. Accessed 28 June 2011

FCC, Wireless E-911 location accuracy requirement—second report and order (2010), http://transition.fcc.gov/Daily_Releases/Daily_Business/2010/db1018/FCC-10-176A1.pdf. Accessed 23 Sept 2010

Foursquare labs Inc. Foursquare application (2012). https://foursquare.com/about/new?from=hp

F. Gustafsson, F. Gunnarsson, Mobile positioning using wireless networks: possibilities and fundamental limitations based on available wireless network measurements. Signal Process. Mag. 41–53 (2005)

S. Kim, Y. Jeong, C. Lee, Interference-cancellation-based IPDL method for position location in WCDMA systems. IEEE Trans. Veh. Technol. 117–126 (2005)

B. Ludden, L. Lopes, Cellular based location technologies for UMTS: a comparison between IPDL and TA-IPDL, *in Proceedings of the 51st Vehicular Technology Conference*, Tokyo, 2000, ed. by IEEE, pp. 1348–1353

J. Medbo, I. Siomina, A. Kangas, J. Furuskog, Propagation channel impact on LTE positioning accuracy: a study based on real measurements of observed time difference of arrival, in *Proceedings of the Personal, Indoor and Mobile Radio Communications* (PIMRC 2009), IEEE 2009

C. Mensing, S. Sand, A. Dammann, W. Utschick, Interference-aware location estimation in cellular OFDM communications systems, in *Proceedings of the IEEE International Conference on Communications 2009 (ICC '09)*

C. Mensing, S. Sand, A. Dammann, Hybrid data fusion and tracking for positioning with GNSS and 3GPP-LTE. Int. J. Navig. Observ. (2010)

M. Monnerat, AGNSS standardization: the path to success in location-based services. Inside GNSS, August 2008

OET BULLETIN No. 71: guidelines for testing and verifying the accuracy of wireless E-911 location systems(2000), http://transition.fcc.gov/Bureaus/Engineering_Technology/Documents/bulletins/oet71/oet71.pdf. Accessed 12 April 2011

Open mobile alliance. OMA secure user plane location V2.0 (2012), http://www.openmobilealliance.org/technical/release_program/supl_v2_0.aspx

Open mobile alliance (2012), http://www.openmobilealliance.org/

J.H. Reed, K.J. Krizman, B.D. Woerner, T.S. Rappaport, An overview of the challenges and progress in meeting the E-911 requirement for location service. Commun. Mag. 30–37 (1998)

Skyhook Inc. Skyhook–location apps. (2012), http://www.skyhookwireless.com/locationapps/. Accessed 26 April 2012

R. Stross, Someday, store coupons may tap you on the shoulder (2010), http://www.nytimes.com/2010/12/26/business/26digi.html. Accesses 25 Dec 2010

G. Sun, J. Chen, W. Guo, K.J.R. Liu, Signal processing techniques in network-aided positioning: a survey of state-of-the-art positioning designs. IEEE Signal Process. Mag. 12–23 (2005)

T. Wigren, Adaptive enhanced cell-ID fingerprinting localization by clustering of precise position measurements. IEEE Trans. Veh. Technol. 3199–3209 (2007)

Wikitude, Wikitude.me (2012), http://www.wikitude.me/w4/wme/map.jsp

Y. Zhao, Standardisation of mobile phone positioning for 3G systems. Commun. Mag. 108–116 (2002)

Chapter 6
Cooperative Localization in Wireless Sensor Networks: Centralized Algorithms

The basic localization techniques known as triangulation and angulation were introduced in Chap. 2. In two-dimensional triangulation the location of a mobile device is computed by measuring its range from at least three base stations with known coordinates. Analogously, in two-dimensional angulation the mobile's location is computed from the arrival angle from at least two stations. While these techniques are practical in Global Positioning Systems or cellular networks, in some networks connectivity of some nodes to even two base stations cannot be guaranteed. A prime example is in wireless sensor networks (WSNs). Because wireless sensors often have a deployment life of months or even years, battery conservation is critical to their operation. This prescribes transmitting infrequently and over short distances. To address the latter, nodes communicate between each other via short, multihop links to stations external to the network (Perkins 2001). The coordinates of the base stations are either hardwired or—because in most cases they are installed outside—can be determined through GPS. In contrast, many WSN applications—such as military or in emergency response—require on-the-fly setup, meaning sensor positions cannot be hardwired and, since sensors are battery constrained, they may not have sufficient power to receive GPS signals. As such, sensors must cooperate in order to extrapolate their locations through multihop links to the stations. This is the basis of cooperative localization.

Cooperative localization was first proposed in Japan to acquire real-time positioning information on mobile robots (Kurazume et al. 1994). Today this concept has been applied not only to WSNs, but also more recently introduced in heterogeneous communication networks. The heterogeneity of today's wireless communication networks can be seen as an additional challenge in localization. Current research aims at porting WSN positioning algorithms into communication networks. For example, in Frattasi (2007) user cooperation was exploited in a least squares framework where cellular and ad-hoc links are combined in a single module of the system. Instead in Figueiras (2008), common Bayesian filtering, namely Kalman filtering, is used for combining short- and long-range links.

C. Gentile et al., *Geolocation Techniques*, DOI: 10.1007/978-1-4614-1836-8_6,

In Cui et al. (2007), the authors proposed a mathematical formulation based on the absolute position obtained by the cellular system followed by a routine optimization that uses the information from the short-range or peer-to-peer links.

The cooperative localization problem can be stated formally as follows. Let the network be composed of two types of nodes: n_A anchor nodes (or anchors), whose locations are known, and n_S sensor nodes (or sensors), whose locations are unknown, for a total of $n = n_A + n_S$ nodes. For simplicity, let the nodes lie in the two-dimensional plane such that node i has location $\boldsymbol{x}_i \in \mathcal{R}^2$ indexed through $i = 1\ldots n_A$ for the anchors and $i = n_A + 1\ldots n$ for the sensors. Let the set N contain all pairs of neighboring nodes, i.e. nodes between which a link exists: $(i,j),\ i<j;\ \|\boldsymbol{x}_i - \boldsymbol{x}_j\| < R$, where $\|\cdot\|$ is the Euclidean distance and the network parameter R is the maximum communication range of the nodes, otherwise known as the *radio range*. The complementary set $\overline{N}$ contains all pairs of non-neighboring nodes: $(i,j),\ i<j;\ \|\boldsymbol{x}_i - \boldsymbol{x}_j\| \geq R$. The measured distance $\hat{d}_{i,j}$ between neighboring nodes i and j is obtained through either one of the received-signal-strength or time-of-arrival techniques introduced in Chap. 2. By processing the anchor locations together with the measured distances, the solution to the problem yields the unknown locations of the sensor nodes in the network.

Centralized cooperative algorithms can guarantee optimal localization results because all the network data is available at a single processing unit. The disadvantage, however, is that this involves relaying information across a large network: from the sensors to the processing unit. If the transmission delays are significant, the processed data may be obsolete upon reception, limiting the algorithms' scalability. Alternatively, local distributed processing at the sensors can more easily maintain network updates in the presence of dynamic links and mobility—an added advantage is shared computational load—but this comes at the price of suboptimal localization results. The choice of centralized or distributed algorithms will depend on the requirements of the application considered. In this chapter, we survey a number of centralized algorithms. Distributed algorithms are treated in Chap. 7.

6.1 Multilateration

The term *multilateration* is derived from the same geometrical principle as triangulation or angulation, but is intended for any constellation of anchor and sensor nodes in the network. This means that even when a sensor lacks direct connectivity to anchor nodes, multilateration techniques can still recover the sensor's location through its indirect connectivity to the anchors via other sensor nodes in the network. The specific network constellation and in particular the anchor–sensor ratio of the nodes will ultimately dictate the attainable degree of localization accuracy.

6.1.1 Atomic Multilateration

The first centralized cooperative localization algorithm considered in this chapter is known as Atomic Multilateration (Savvides et al. 2001). It is one of the earliest and best known algorithms in the field. The multilateriation is atomic in that sense each sensor operates on a small scale—only with its neighboring nodes—if they can furnish the requisite information to determine the sensor's location. The algorithm essentially draws on the same set of equations from (2.3)–(2.6), however applied to sensor networks. The underlying principle is that the measured distance $\hat{d}_{i0}^2$ between a sensor 0 and a neighboring anchor i is related to the sensors locations, $\boldsymbol{x}_0$ and $\boldsymbol{x}_i$ respectively, through the square Euclidean norm as

$$\|\boldsymbol{x}_i - \boldsymbol{x}_0\|^2 = \hat{d}_{i0}^2. \tag{6.1}$$

Expanding the norm in the two-dimensional coordinate space for anchor i and another anchor j also neighboring sensor 0 yields the following quadratic system of equations, each representing a circle:

$$\begin{cases} (x_i - x_0)^2 + (y_i - y_0)^2 = \hat{d}_{i0}^2 \\ (x_j - x_0)^2 + (y_j - y_0)^2 = \hat{d}_{j0}^2. \end{cases} \tag{6.2}$$

By expanding the equations further and subsequently subtracting the second from the first, the difference can be expressed as a linear equation

$$a_{ij,x}x_0 + a_{ij,y}y_0 = b_{ij}, \tag{6.3}$$

where

$$\begin{cases} a_{ij,x} = 2(x_i - x_j) \\ a_{ij,y} = 2(y_i - y_j) \\ b_{ij} = (x_i^2 - x_j^2) + (y_i^2 - y_j^2) - (\hat{d}_{i0}^2 - \hat{d}_{j0}^2). \end{cases} \tag{6.4}$$

As illustrated in Fig. 6.1, the linear difference equation in (6.3) represents the line defined by the two points at which the two circles meet. It can be also written in matrix form as

$$\boldsymbol{A} \cdot \boldsymbol{x}_0 = \boldsymbol{b}, \tag{6.5}$$

where each row of matrix $\boldsymbol{A}$ and vector b corresponds to a pair of anchors (i,j) which both neighbor sensor 0. If the sensor node has $\overline{n}_A$ neighboring anchors, then there will be $\binom{\overline{n}_A}{2}$ rows. Note in (6.4) that the two columns of $\boldsymbol{A}$ are indexed as x and y. For the case $\overline{n}_A = 3$, the system in (6.5) can be resolved to a unique solution for $\boldsymbol{x}_0$, unless the anchors are collinear, for which the system is underdetermined. Even if the anchors are quasi-collinear, a case which is referred

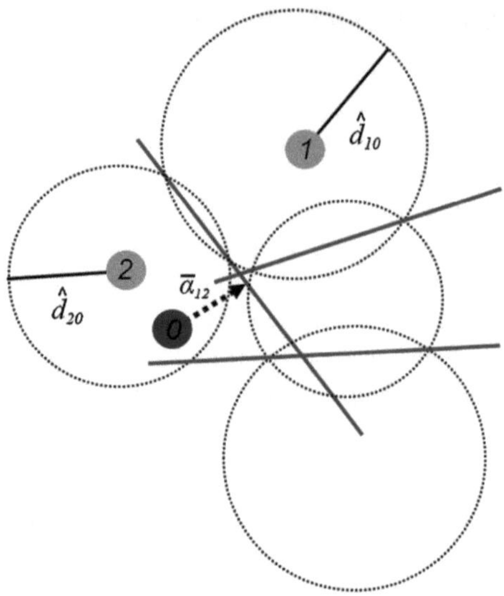

Fig. 6.1 Atomic Multilateration. The linear residual error, $\overline{\alpha}_{12}$, is the distance between the estimated sensor location (*dark*) and the line which intersects the two circles associated with anchors 1 and 2 (*blue*)

to as the Geometric Dilution of Precision (Langley 1999) introduced in Chap. 2, numerical issues may arise. For the case $\overline{n}_A > 3$ non-collinear anchors, the system will be overdetermined if the measured distances contain errors, meaning that the $\overline{n}_A$ circles, or equivalently the $\binom{\overline{n}_A}{2}$ lines, will not intersect at a unique point.

In order to identify a unique position for the sensor, a linear residual error is defined for each linear difference equation:

$$\overline{\alpha}_{ij} = a_{ij,x}x_0 + a_{ij,y}y_0 - b_{ij}. \tag{6.6}$$

The linear residual error is the distance between $\boldsymbol{x}_0$ and the line in (6.3). Then the least squares solution, which minimizes the sum of square linear residual errors:

$$\sum_{\forall(i,j),\left\{\begin{array}{l}(i,0)\in N\\(j,0)\in N\end{array}\right.} \overline{\alpha}_{ij}^2 \tag{6.7}$$

is given by

$$\boldsymbol{x}_0 = (\boldsymbol{A}^T\boldsymbol{A})^{-1}\boldsymbol{A}^T\boldsymbol{b}. \tag{6.8}$$

Figure 6.1 shows the estimated sensor location, $\boldsymbol{x}_0$, and the linear residual error, $\overline{\alpha}_{12}$, associated with the pair of anchor nodes, $\boldsymbol{x}_1$ and $\boldsymbol{x}_2$. (The anchors associated with the other two circles and the residuals associated with the other two lines are suppressed to avoid clutter).

Through triangulation, the locations of all sensor nodes in the network with three or more neighboring anchors can be determined. Once their locations are known, they become *virtual* anchor nodes themselves. In turn they can enable a neighboring sensor with less than three actual anchor nodes to determine its own location. Through this iterative process, the locations of more and more sensors become known. The larger the number of neighboring anchors, the more robust is a sensor node to distance measurement error. As such, at each step in the iteration, the location of the unknown sensor with the largest number of neighboring nodes is determined. Since the unknown sensor may be used as a virtual anchor in later steps, this serves to mitigate the propagation of location error.

6.1.2 Collaborative Multilateration

For all sensors in the network with three or more neighboring anchors (actual or virtual), Atomic Multilateration can be applied. However, if during the iterative process a sensor cannot meet this condition, it means that the sensor cannot gather the minimum data necessary from its neighboring nodes to solve the simplified system of linear equations in 6.5. As an alternative, the sensors can resort to an algorithm defined in Savvides et al. (2002) known as Collaborative Multilateration. By extending its communication reach from single-hop neighbors to multihop neighbors (i.e. neighbors of neighbors) as well, the sensor can then gather the data necessary to solve a system of original (i.e. nonlinearized) quadratic equations derived from 6.1.

The system of equations can be represented graphically by a subdivision of the network which we refer to as a subnetwork. Each anchor in the subnetwork corresponds to a set of known coordinates in the system while each sensor corresponds to a set of unknown coordinates; likewise, each anchor–sensor or sensor–sensor link in the subnetwork corresponds to an equation in the system. If the system has a unique solution, it can be solved using gradient descent or some other nonlinear technique which minimizes the sum of square residual errors between the estimated and measured distances for all links in the subnetwork. The residual error between nodes i and j is defined as

$$\alpha_{ij} = \sqrt{(x_i - x_j)^2 + (y_i - y_j)^2} - \hat{d}_{ij}. \tag{6.9}$$

At a given step, a subnetwork is originated at an unknown sensor node in the network and is expanded upon by adding single-hop neighbors and multihop neighbors in succession, and all associated links. During the expansion, the number of equations included in the system increases with every link added; however the number of unknowns also increases with every sensor added. The expansion continues until the system yields a unique solution for all the sensor nodes included. Conditions for uniqueness are discussed in the following. Because the search space is non-convex, the solution found may be suboptimal. As such, in

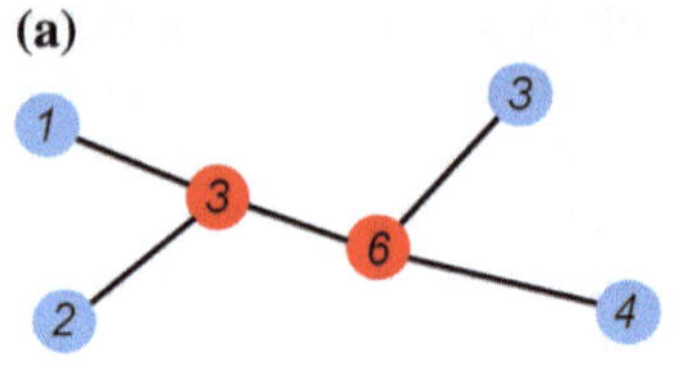

Fig. 6.2 Collaborative Multilateration. Given the network topology in plot (**a**), the locations of the two sensor nodes (*red*) can be identified uniquely from the four anchor nodes (*blue*). The network in plot (**b**), however, suffers from the mirror ambiguity: sensor 6 could be located at either of the two locations shown

order to reduce the search space, the system of equations should be kept as small as possible. Hence, the smallest subnetwork is sought.

A system of quadratic equations has a unique solution if all the nodes in the subnetwork are said to be *participating*—a term coined by the authors in Savvides et al. (2002). All anchors are participating by default. Participating sensors, rather, are determined recursively. The sensors in the subnetwork are labeled by the number, or degree, of participating neighbors they have. Initially, only the anchors are participating. If the degree of a sensor exceeds 2, it is designated as a participating node; also, if two neighboring sensors both have a degree of 2, they are both designated as participating nodes. This ensures that each sensor has at least three participating neighbors, thus enabling Collaborative Multilateration.

Figure 6.2a shows a subnetwork with six nodes: four anchors (*blue*) and two sensors (*red*). At initialization, the sensors both have degree 2—as related to the number of anchor connections each has. In addition, since the two sensors are mutual neighbors, they are designated as participating nodes at the next recursion, making all the nodes in the subnetwork participating. In contrast, Fig. 6.2b shows a subnetwork with five nodes: three anchors and two sensors. At initialization, the degree of sensor 3 is 2 and the degree of sensor 6 is 1. Because the degree of sensor 6 is only 1, despite the fact that the sensors are mutual neighbors, they cannot increase their degrees further. This means that the subnetwork does not have a unique solution. In practice, sensor 6 is subject to the *mirror ambiguity* because the topology of the subnetwork does not provide sufficient information to resolve its location to one of the two positions shown. More about the mirror ambiguity is discussed in Sect. 6.4.2.

6.2 Convex Optimization

The basic multilateration techniques introduced in the previous subsection, although implementable in a centralized fashion, are designed for local processing at the sensor nodes. A sensor gathers primitives—either the measured distance or the measured angle[1]—between neighboring single-hop or multihop nodes in order to determine its location. By virtue of the local processing, the techniques lend to distributed algorithms as well. Techniques based on global optimization for centralized processing, rather, view the network as a whole.

The general approach in optimization problems is to define an objective function—either to minimize or to maximize—and associated constraints on the function variables. The constraints delineate the variable space over which the algorithm can search for the optimal solution. When applied to network localization, typically the variables are the sensor locations and the function to minimize is the error between the measured primitives and the primitives given by the variables. Because the optimization is global, the measured primitives from the whole network are fed to the algorithm which processes them in a centralized fashion. Due to measurement error, when processing them collectively, they will tend to contradict each other. By applying geometrical constraints on the network, which are often convex, the algorithm resolves the contradictions such that the sensor locations are compliant with the physical world. In this subsection, we investigate the sorts of geometrical constraints which can be applied on the network.

6.2.1 Distance Constraints

Localization in WSNs was first posed as a convex optimization problem in Doherty et al. (2001)—convex because the search space over the function variables is convex. Convex optimization is appealing because efficient methods exist to solve for the variables and the solution it yields is optimal. The localization problem is formulated as a convex program (Bazaraa et al. 1990), which in this application is defined as an ensemble of convex geometrical constraints on the sensor locations coupled with an objective function to optimize. If a solution exists—meaning that if the constraints do not contradict each other—then the sensors will lie in the *feasible* solution space, which is just the space formed by the intersection of all the constraints. The optimal solution then lies within this feasible solution space.

The first geometrical constraint that we consider is given by the definition of neighboring nodes provided in the chapter introduction. The constraint translates to an upper bound on the distance between neighboring nodes, i.e. $\left\|\boldsymbol{x}_i - \boldsymbol{x}_j\right\|^2 \leq R^2$, where R is the radio range. This constraint can also be expressed as a linear matrix inequality:

[1] Although described in Sect. 6.1 only for measured range, the multilateration techniques can be readily extended to measured angle-of-arrival by applying the appropriate angulation Eqs. 2.14–2.15 instead of the triangulation Eqs. 2.3–2.6.

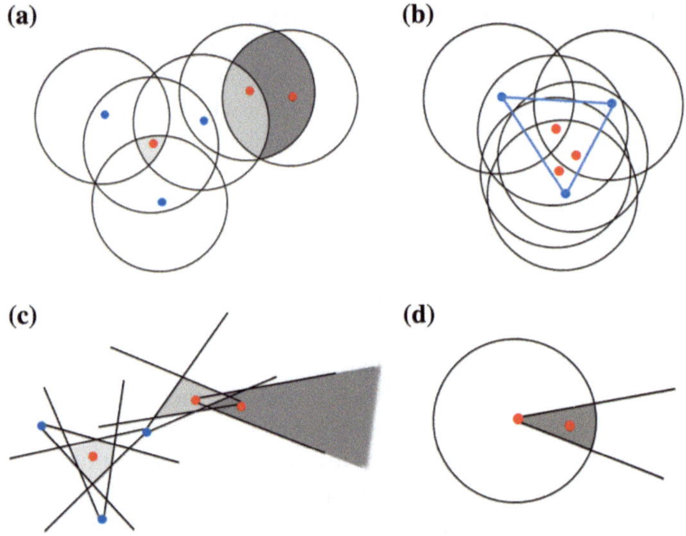

Fig. 6.3 Convex optimization. The convex feasible space for each of the sensor nodes (*red*) is shaded in gray. The anchor nodes are colored blue. Plots (**a**), (**b**) illustrate convex distance constraints on neighboring nodes in the network while plot (**c**) illustrates convex angle constraints. Plot (**d**) illustrates combined distance and angle constraints

$$\begin{bmatrix} I_2 R & \boldsymbol{x}_i - \boldsymbol{x}_j \\ (\boldsymbol{x}_i - \boldsymbol{x}_j)^T & R \end{bmatrix} \geq 0, \tag{6.10}$$

where I_2 is the 2 × 2 identity matrix. The convex area corresponding to the constraint is the area common to the interior of the two circles centered at $\boldsymbol{x}_i$ and $\boldsymbol{x}_j$, both with radius R. Then the feasible solution space for a sensor node is the intersection of the convex areas formed with each of its neighboring nodes. Figure 6.3a illustrates the individual feasible spaces—in different shades of gray—for the three sensor nodes (*red*) in an example network. (The anchor nodes are shown in blue.) Since no objective function is specified, the sensor may equivalently lie anywhere within its respective feasible space.

The shortcoming with this set of limited constraints is that the neighboring nodes lack a complementary lower bound on the distances between them. As a result, the feasible space will collapse within the convex hull of the anchor nodes in the network, yet satisfying all the upper bounds. Figure 6.3b shows such a solution with the convex hull connecting the three anchors—a stark contrast from the correct solution in Fig. 6.3a. Hence, the method generates acceptable results only if all the sensors are actually located within the convex hull of the anchors. In the next section, an approach to incorporate lower bounds to deal with this shortcoming is explained.

6.2.2 Angular Constraints

If the sensor nodes are equipped with steerable directional antennas, a difference class of convex constraints based on angle, rather than distance, can be considered. The measured angle between neighboring nodes i and j from (2.14) is repeated here for convenience as

$$\theta_{ij} = \tan^{-1}\left(\frac{y_j - y_i}{x_j - x_i}\right). \tag{6.11}$$

Assuming that the antennas have a sector-shaped radiation pattern with beam width θ_w, then $\boldsymbol{x}_j$ will lie within the convex area enclosed by the two planes circumscribed by angles $\theta_{ij} - \frac{\theta_w}{2}$ and $\theta_{ij} + \frac{\theta_w}{2}$. The two planes intersect at $\boldsymbol{x}_i$, as shown in Fig. 6.3c. This conical area corresponds to the intersection of the following upper and lower bounds, which can be applied to the sensor locations:

$$\begin{aligned} y_j - y_i &\geq \tan\left(\theta_{ij} - \frac{\theta_w}{2}\right)(x_j - x_i) \\ y_j - y_i &\leq \tan\left(\theta_{ij} + \frac{\theta_w}{2}\right)(x_j - x_i). \end{aligned} \tag{6.12}$$

Since the cone corresponding to each pair of neighboring nodes is unbounded, the sensor locations are only loosely confined, yielding poor results. Rather combining the angular constraints with the distance constraint bounds the conical area, as shown in Fig. 6.3d, drastically improving the results.

6.3 Semi-Definite Programming

The distance constraints considered in Sect. 6.2.1 stem simply from the condition of two neighboring nodes being able to communicate, thereby lying within radio range of each other. In this section, in addition, we assume that the nodes are equipped with ranging capabilities such that a measured distance, $\hat{d}_{ij}$, between neighboring nodes is also available. As such, tighter distance constraints can be applied to the optimization program. These constraints are derived from (6.1), however, in order to account for measurement errors in the system, the equality is written as $\left\|\boldsymbol{x}_i - \boldsymbol{x}_j\right\|^2 = \hat{d}_{ij}^2 + \tilde{\alpha}_{ij}$, where the $\tilde{\alpha}_{ij}$ is denoted as the residual error. The objective function of the program is to minimize, subject to the distance constraints of the network, the absolute residual $\left|\tilde{\alpha}_{ij}\right|$ such that the estimated square distance $\left\|\boldsymbol{x}_i - \boldsymbol{x}_j\right\|^2$ is as close as possible to the measured square distance for all neighboring nodes.

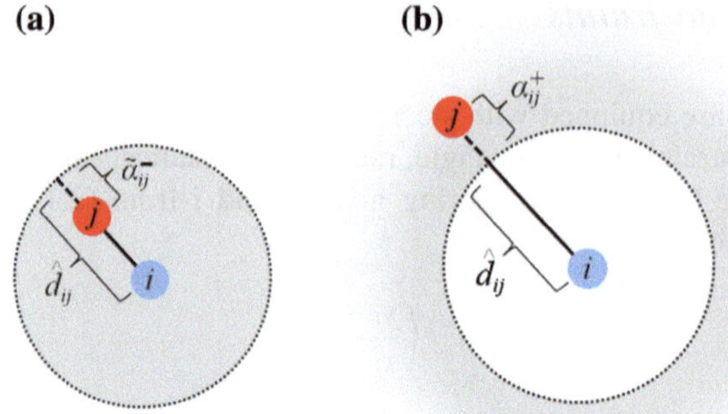

Fig. 6.4 The distance constraint. Plot (**a**) illustrates the upper bound on the distance constraint. The feasible space of the sensor node (*red*) lies inside the circle circumscribed by the measured distance, $\hat{d}_{ij}$, from the anchor node (*blue*). This space is convex. Plot (**b**) illustrates the lower bound for which the feasible space lies outside the circle. This space is non-convex

The optimization program stated above can be written mathematically as:

$$\begin{aligned} \min \quad & \sum_{(i,j)\in N} \left|\tilde{\alpha}_{ij}\right| \\ \text{subject to} \quad & \left\|\boldsymbol{x}_i - \boldsymbol{x}_j\right\|^2 = \hat{d}_{ij}^2 + \tilde{\alpha}_{ij}, \forall (i,j) \in N \end{aligned} \tag{6.13}$$

It can be rewritten equivalently in standard form (i.e. without the absolute value signs) by decomposing the absolute residual into positive $\tilde{\alpha}_{ij}^+$ and negative $\tilde{\alpha}_{ij}^-$ residuals such that $\tilde{\alpha}_{ij} = \tilde{\alpha}_{ij}^+ - \tilde{\alpha}_{ij}^-$ and $\left|\tilde{\alpha}_{ij}\right| = \tilde{\alpha}_{ij}^+ + \tilde{\alpha}_{ij}^-$ (Bazaraa et al. 1990):

$$\begin{aligned} \min \quad & \sum_{(i,j)\in N} \tilde{\alpha}_{ij}^+ + \tilde{\alpha}_{ij}^- \\ \text{subject to} \quad & \left.\begin{aligned} \left\|\boldsymbol{x}_i - \boldsymbol{x}_j\right\|^2 = \hat{d}_{ij}^2 + \tilde{\alpha}_{ij}^+ - \tilde{\alpha}_{ij}^- \\ \tilde{\alpha}_{ij}^+ \geq 0 \\ \tilde{\alpha}_{ij}^- \geq 0 \end{aligned}\right\}, \quad \forall (i,j) \in N \end{aligned} \tag{6.14}$$

Because the objective is to minimize the pairwise residuals and because they are both subject to non-negativity constraints, only the positive or the negative counterpart will be nonzero in the solution.

Now observe that the equality constraint in (6.14) can be decomposed into the intersection of two inequalities, $\left\|\boldsymbol{x}_i - \boldsymbol{x}_j\right\|^2 \leq \hat{d}_{ij}^2$ and $\left\|\boldsymbol{x}_i - \boldsymbol{x}_j\right\|^2 \geq \hat{d}_{ij}^2$. If $\left\|\boldsymbol{x}_i - \boldsymbol{x}_j\right\|^2 \leq \hat{d}_{ij}^2$, then $\tilde{\alpha}_{ij}^- \geq 0$ and the positive residual is binding (i.e. $\tilde{\alpha}_{ij}^+ = 0$). As explained in Sect. 6.2.1, the upper bound can be expressed as a convex constraint. The associated convex area is shaded in gray in Fig. 6.4a. Conversely, if $\left\|\boldsymbol{x}_i - \boldsymbol{x}_j\right\|^2 \geq \hat{d}_{ij}^2$, then $\tilde{\alpha}_{ij}^+ \geq 0$ and the negative residual is binding (i.e. $\tilde{\alpha}_{ij}^- = 0$). The lower bound, however, is non-convex, as seen through the associated shaded area in

Fig. 6.4b. Since the lower bound is part of constraints set, the optimization program as a whole is also non-convex. In the remainder of this section we describe an approach from Biswas et al. (2006) that relaxes the constraints such that the program is rendered convex.

The first step in the approach is to rewrite the square estimated distance in (6.14) as

$$\left\|\boldsymbol{x}_i - \boldsymbol{x}_j\right\|^2 = \boldsymbol{e}_{ij}^T \boldsymbol{X}^T \boldsymbol{X} \boldsymbol{e}_{ij}, \tag{6.15}$$

where $\boldsymbol{X} = [\boldsymbol{x}_1 \boldsymbol{x}_2 .. .\boldsymbol{x}_n]$ and $\boldsymbol{e}_{ij}$ is an $n \times 1$ vector whose entries are all 0, except for entry i which has value 1 and entry j which has value -1. When the measured distances are error-free, all the residuals are equal to zero. It follows that the objective function is also equal to zero and the solution is given by $\boldsymbol{Y} = \boldsymbol{X}^T \boldsymbol{X}$. In this case, both the upper and lower bounds associated with each distance constraint are binding. The nodes can be visualized as forming a rigid structure in the two-dimensional plane with the measured distances fitting perfectly in between. Since in practice the measured distances are erroneous, either the upper or the lower bound will be violated, resulting in $\boldsymbol{Y} \neq \boldsymbol{X}^T \boldsymbol{X}$. In particular, as explained above, when the lower bound is active, the feasible search space becomes non-convex.

This is dealt with by relaxing the optimization program to a *semi-definite* program, i.e. by substituting $\boldsymbol{Y} = \boldsymbol{X}^T \boldsymbol{X}$ with $\boldsymbol{Y} \geq \boldsymbol{X}^T \boldsymbol{X}$:

$$\begin{aligned} &\min && \sum_{(i,j)\in N} \tilde{\alpha}_{ij}^{+} + \tilde{\alpha}_{ij}^{-} \\ &\text{subject to} && \left.\begin{aligned} \boldsymbol{e}_{ij}^T \boldsymbol{Y} \boldsymbol{e}_{ij} = \hat{d}_{ij}^2 + \tilde{\alpha}_{ij}^{+} - \tilde{\alpha}_{ij}^{-} \\ \tilde{\alpha}_{ij}^{+} \geq 0 \\ \tilde{\alpha}_{ij}^{-} \geq 0 \end{aligned}\right\}, \quad \forall (i,j) \in N. \\ &&& \boldsymbol{Y} \geq \boldsymbol{X}^T \boldsymbol{X} \end{aligned} \tag{6.16}$$

By relaxing the program to a semi-definite program, the search is restricted to a convex space, specifically to within a spectrahedron. The program can then be solved through convex optimization methods. This step is tantamount to relaxing the rigid structure of nodes by allowing them to assume dimensions outside the two-dimensional plane. The nodes will do so when the measured distances are erroneous such that the distances can fit in between the node locations, minimizing the objective function. A solution in R^2 is then provided by discarding the dimensions in the solution space which lie above the plane, effectively projecting the sensor locations down onto it. However, this creates an effect similar to the one observed in the previous section, i.e. the sensor locations tend to collapse to the center of the network.

Two techniques are proposed in the same (Biswas et al. 2006) in order to refine the solution. The first is to add a regularization term to the objective function in (6.16) of the form:

$$-\lambda \sum_{(i,j)\in N} \left\|\boldsymbol{x}_i - \boldsymbol{x}_j\right\|^2. \tag{6.17}$$

By increasing the heuristic parameter λ more and more from 0, the solution will return sensor locations which are further and further apart from each other in order to minimize the objective function. Although this does achieve the desired effect of spreading the sensors away from the network center, the solution proves very sensitive to the choice of λ. Details on the selection of λ are provided in the paper. The second technique is to continue the minimization process, however of the original objective function in (6.14) (i.e. before relaxation) through a gradient-descent search using the solution given from the semi-definite program with the regularized objective function as the initialization point.

Finally, note that since the angular constraints in (6.12) are linear, if angle-of-arrival measurements are available, they can be included directly in the semi-definite program.

6.4 Linear Programming

A linear program is an optimization program with a linear objective function and linear constraints. Because they are linear, the program is convex by default. As such, efficient complex optimization methods can be applied. In this subsection a linear program to solve for the locations of the sensor nodes is presented. By applying linear geometrical constraints as opposed to the conical ones described in the previous section, the associated linear program is also convex. The advantage of this approach is that the original geometrical constraints need not be relaxed. Rather, they can be applied as is, resulting in a solution which is inherently compliant with the physical world. An additional advantage of this approach is that linear programs bear smaller computational complexity than semi-definite programs.

6.4.1 Triangle Inequality Constraints

Instead of relaxing the constraints of the semi-definite program in (6.14), (Gentile 2007) proposes applying a different set of geometrical constraints while retaining an equivalent objective function.[2] The paper exploits the triangular structure of the network by imposing the triangle inequality on the link distances. The problem solved can be stated precisely as follows:

[2] Here the objective function minimizes the absolute residuals $|\alpha_{ij}|$ between the measured and estimated distances while in (6.13) it is the absolute residuals $|\tilde{\alpha}_{ij}|$ between the measured and estimated *square* distances which are minimized.

$$\begin{aligned} \min \quad & \sum_{(i,j)\in N} \left|\alpha_{ij}\right| \\ \text{subject to} \quad & \left.\begin{aligned} d_{ij}+d_{jk} &\geq d_{ik} \\ d_{ij}+d_{ik} &\geq d_{jk} \\ d_{jk}+d_{ik} &\geq d_{ij} \end{aligned}\right\}, \quad \forall (i,j,k)\in M, \end{aligned} \tag{6.18}$$

where $d_{ij} = \hat{d}_{ij} + \alpha_{ij}$ and the set M contains all triplets of nodes which form a triangle in the network: $(i,j,k) \in M, (i,j) \in N; (j,k) \in N; (i,k) \in N$. As in (6.16), the problem can be rewritten in standard form by replacing the absolute sign with positive and negative residuals:

$$\begin{aligned} \min \quad & \sum_{(i,j)\in N} \alpha_{ij}^{+} + \alpha_{ij}^{-} \\ \text{subject to} \quad & \left.\begin{aligned} d_{ij}+d_{jk} &\geq d_{ik} \\ d_{ij}+d_{ik} &\geq d_{jk} \\ d_{jk}+d_{ik} &\geq d_{ij} \end{aligned}\right\}, \quad \forall (i,j,k)\in M. \\ & \left.\begin{aligned} \alpha_{ij}^{+} \\ \alpha_{ij}^{-} \end{aligned}\right\}, \quad \forall (i,j)\in N \end{aligned} \tag{6.19}$$

where $\alpha_{ij} = \alpha_{ij}^{+} - \alpha_{ij}^{-}$ and $\left|\alpha_{ij}\right| = \alpha_{ij}^{+} + \alpha_{ij}^{-}$. However, in contrast to (6.16), the solution to the linear program above does not directly yield the sensor locations; instead, it simply yields the estimated link distances. Hence the complete algorithm requires an a posteriori *location reconstruction* stage to furnish the sensor locations.

6.4.2 Location Reconstruction

Provided the estimated distances for all the links in the network from (6.19), in the location reconstruction stage the unknown sensor locations can be determined from the anchor nodes. Accordingly, the stage is originated at any two anchor nodes sharing a neighboring sensor node, as in Fig. 6.5a. Given the anchor locations $\boldsymbol{x}_1 = (x_1, y_1)$ and $\boldsymbol{x}_2 = (x_2, y_2)$ (and the associated distance d_{12} between them), together with d_{13} and d_{23}, by exploiting the Law of Cosines in 6.20(a), the unknown location $\boldsymbol{x}_3 = (x_3, y_3)$ (with respect to the reference coordinate system centered at $\boldsymbol{x}_1$) is furnished from the following set of equations (Capkun et al. 2001):

$$\begin{aligned} &(a) \quad \theta = \cos^{-1}\left(\frac{d_{12}^2 + d_{23}^2 - d_{13}^2}{2d_{12}d_{23}}\right) \\ &(b) \quad \begin{bmatrix} x_3' \\ y_3' \end{bmatrix} = \begin{bmatrix} x_1 \\ y_1 \end{bmatrix} + \begin{bmatrix} d_{23} & \cos\theta \\ s\cdot d_{23} & \sin\theta \end{bmatrix} \\ &(c) \quad \phi = \tan^{-1}\left(\tfrac{y_2-y_1}{x_2-x_1}\right) \\ &(d) \quad \begin{bmatrix} x_3 \\ y_3 \end{bmatrix} = \begin{bmatrix} x_1 \\ y_1 \end{bmatrix} + \begin{bmatrix} \cos\phi & -\sin\phi \\ \sin\phi & \cos\phi \end{bmatrix} \begin{bmatrix} x_3' - x_1 \\ y_3' - y_1 \end{bmatrix} \end{aligned} \tag{6.20}$$

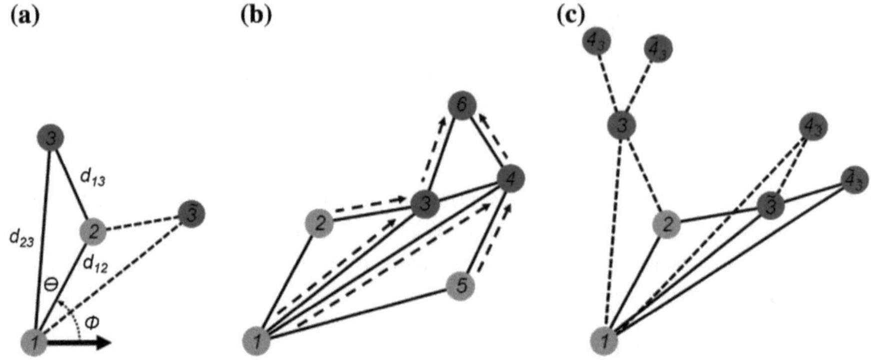

Fig. 6.5 Stages of location reconstruction. **a** Reconstructing sensor location $\boldsymbol{x}_3$ through the three estimated distances (d_{12}, d_{23}, d_{23}) of the triangle and the anchor locations $\boldsymbol{x}_1$ and $\boldsymbol{x}_2$. **b** Location propagation in the network from two known nodes to an unknown node is sequential, following the direction of the arrows. **c** The four possible locations for sensor 4, resulting from the mirror ambiguity

Through this process the two anchor nodes at $\boldsymbol{x}_1$ and $\boldsymbol{x}_2$ are said to "propagate" their locations to the unknown sensor node. From the set of equations, the data provided actually furnishes two candidate locations for the sensor—each mirrored about the line common to $\boldsymbol{x}_1$ and $\boldsymbol{x}_2$. In reference to Fig. 6.5a, those candidates are $\boldsymbol{x}_3$ for $s = 1$ and $\boldsymbol{x}_{\tilde{3}}$ for $s = -1$. As in the example in Fig. 6.2b from Sect. 6.1.2, this is known as the mirror ambiguity. This ambiguity arises from the use of only two anchor nodes to determine a two-dimensional location. How to resolve the mirror ambiguity is discussed next.

Once the location of a sensor node is known, it can serve with another sensor (or anchor) to determine the location of yet another unknown sensor neighboring them both. This is done sequentially. For instance, tracing the arrows in Fig. 6.5b, $\boldsymbol{x}_1$ and $\boldsymbol{x}_2$ propagate their locations to $\boldsymbol{x}_3$, $\boldsymbol{x}_2$ and $\boldsymbol{x}_3$ propagate their locations to $\boldsymbol{x}_4$, and $\boldsymbol{x}_3$ and $\boldsymbol{x}_4$ propagate their locations to $\boldsymbol{x}_6$. If, however, sensor 3's mirror ambiguity cannot be resolved after the first propagation, both candidate locations $\boldsymbol{x}_3$ and $\boldsymbol{x}_{\tilde{3}}$ shown in Fig. 6.5c are retained; information received subsequently through its neighbors about sensor locations yet to be discovered will enable resolution. Specifically, note that each propagation step from the origin potentially doubles the number of candidate locations. After the first step, there are two candidate locations for sensor 3; after the second step, Fig. 6.5c displays the four candidate locations $\boldsymbol{x}_{4_3}$, $\boldsymbol{x}_{\tilde{4}_3}$, $\boldsymbol{x}_{4_{\tilde{3}}}$, $\boldsymbol{x}_{\tilde{4}_{\tilde{3}}}$ for sensor 4. Rather than double these candidate locations for sensor 6 yet further to eight in the third step network redundancy is exploited to dismiss candidates $\boldsymbol{x}_{\tilde{4}_3}$ and $\boldsymbol{x}_{4_{\tilde{3}}}$—both place sensors 1 and 4 within radio range of each other even though they are not actually neighbors (since they cannot communicate with each other). This mechanism suppresses the exponential growth of candidates. Now tracing a different path in Fig. 6.5b, anchors 2 and 5 propagate their locations to the two mirror locations for sensor 4. Only one, however, will coincide with the true location $\boldsymbol{x}_{\tilde{4}_{\tilde{3}}}$; thereby the other

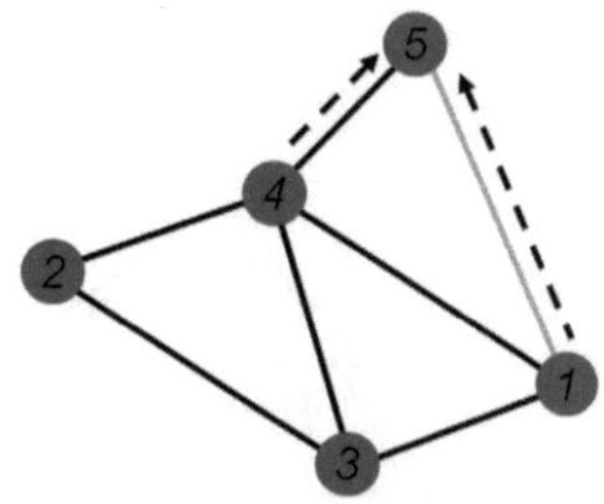

Fig. 6.6 Artificial links. Sensor 5 is connected to the network through sensor 4 only. To enable location propagation to itself, sensor 5 must be connected to at least two nodes. So the artificial link shown in gray is added between sensors 1 and 5

remaining candidate, $\boldsymbol{x}_{4_3}$, can be dismissed. Resolution of sensor 4's location in turn enables sensor 3 to determine its unique location as $\boldsymbol{x}_{\bar{3}}$.

In order to reconstruct the location of a node, the node requires at least two connections to the network. Consider sensor 5 in Fig. 6.6, which has only one link to the network. In this case, an *artificial link* (gray) can be inserted from a non-neighboring node (i.e. $\boldsymbol{x}_1$). This enables location propagation to $\boldsymbol{x}_5$ from $\boldsymbol{x}_1$ and $\boldsymbol{x}_4$. As any other link, an artificial link between $\boldsymbol{x}_i$ and $\boldsymbol{x}_j$, $(i,j) \in \overline{N}$ generates a set of new triangles in the network and so appears in the corresponding set of triangle inequality constraints in (6.19). However, lacking a measured distance for it, an artificial link is less constrained than a normal link. In practice only the inequality $d_{ij} = \hat{d}_{ij} + \alpha_{ij}^{+} \geq R$ can be exploited. So, rather than arbitrarily minimize the positive residual α_{ij}^{+} in the objective function, $\hat{d}_{ij} = R$ is set and the positive bounding constraint $\alpha_{ij}^{+} \geq 0$ alone is included in the linear program ($\alpha_{ij}^{-} = 0$ since $d_{\mathrm{ij}} \nless R$).

A node completely disconnected from the network cannot gather any location information except that it lies beyond the radio range R of all other nodes in the network. Hence no deterministic method exists to compute its location with any meaningful accuracy.

6.4.3 Anchor Nodes

In this subsection, we discuss how to incorporate network anchor nodes into the linear program. For a pair of anchors, this is accomplishing by including the constraints associated with all triangles formed between the pair and any sensor neighboring them both. The residual of the link distance between the anchor pair is set to zero since the distance is known. In general network topologies, however, especially in those with low anchor density, no direct connection may exist between a single sensor node and any two anchors; instead, multihop connections must be considered—in particular, the subnetwork of all nodes along the minimum multihop route between the two anchors. An example subnetwork for the minimum multihop route $\boldsymbol{x}_1 \rightarrow \boldsymbol{x}_3 \rightarrow \boldsymbol{x}_4 \rightarrow \boldsymbol{x}_5 \rightarrow \boldsymbol{x}_2$ is shown in Fig. 6.7a. Highlighted is the set of triangles formed along the route. Although measured values may be available, the link distances (lengths) of the triangle sides must be estimated through a preprocessing step in order to ensure geometrical consistency.

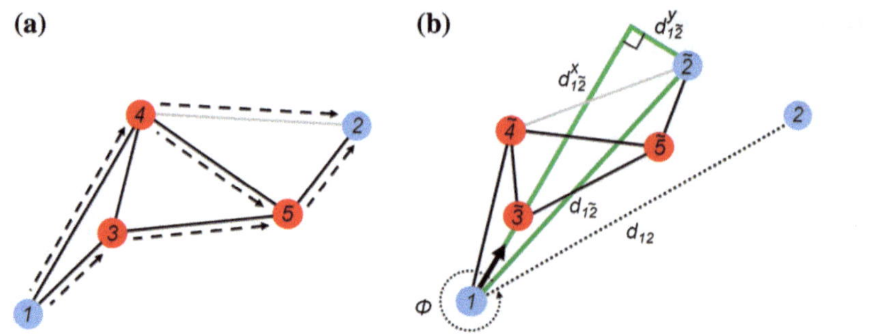

Fig. 6.7 Location propagation between a pair of non-neighboring anchor nodes (*light*) in the network. **a** A subnetwork of nodes along the minimum multihop route between the two anchors. **b** Once propagation takes place in the relative coordinate system, the system has to be scaled and rotated so to align with the anchor nodes

In the first phase of the preprocessing step, a separate linear program including all the triangles in the subnetwork is formulated. Thereafter, location is propagated between the two anchors. In the example subnetwork in Fig. 6.7a, propagation follows the arrows from $\boldsymbol{x}_1$ and $\boldsymbol{x}_3$ to $\boldsymbol{x}_4$, from $\boldsymbol{x}_3$ and $\boldsymbol{x}_4$ to $\boldsymbol{x}_5$ and from $\boldsymbol{x}_4$ and $\boldsymbol{x}_5$ to $\boldsymbol{x}_2$. Notice that an artificial link must be added to enable this propagation path. In contrast to location propagation from two anchor nodes described previously, here propagation originates from a single known anchor node ($\boldsymbol{x}_1$) and an unknown sensor node ($\boldsymbol{x}_3$). Since $\boldsymbol{x}_3$ is unknown, a relative coordinate system oriented along the line between $\boldsymbol{x}_1$ and $\boldsymbol{x}_3$ and centered at $\boldsymbol{x}_1$ is established, as illustrated in Fig. 6.7b. Now $\boldsymbol{x}_1$ and the location of sensor 3 in the relative coordinate system, denoted as $\boldsymbol{x}_{\tilde{3}}$, can be propagated to $\boldsymbol{x}_{\tilde{4}}, \boldsymbol{x}_{\tilde{5}}$, and $\boldsymbol{x}_{\tilde{2}}$, also in the relative coordinate system. The locations can be expressed in terms of the rectangular components (d^x, d^y) of the distances $d = \sqrt{d^{x^2} + d^{y^2}}$ between the nodes on the multihop route. The rectangular components can be computed pairwise through (6.20). The locations follow from the components in sequence as: $(x_{\tilde{3}}, y_{\tilde{3}}) = (x_1 + d_{1\tilde{3}}, 0)$, $(x_{\tilde{4}}, y_{\tilde{4}}) = \left(x_{\tilde{3}} + d^x_{\tilde{3}\tilde{4}}, y_{\tilde{3}} + d^y_{\tilde{3}\tilde{4}}\right)$, $(x_{\tilde{5}}, y_{\tilde{5}}) = \left(x_{\tilde{4}} + d^x_{\tilde{4}\tilde{5}}, y_{\tilde{4}} - d^y_{\tilde{4}\tilde{5}}\right)$, and $(x_{\tilde{2}}, y_{\tilde{2}}) = \left(x_{\tilde{5}} + d^x_{\tilde{2}\tilde{5}}, y_{\tilde{5}} + d^y_{\tilde{2}\tilde{5}}\right)$, or compactly as

$$\begin{bmatrix} x_{\tilde{2}} \\ y_{\tilde{2}} \end{bmatrix} = \begin{bmatrix} x_1 + d^x_{1\tilde{2}} \\ y_1 - d^y_{1\tilde{2}} \end{bmatrix}, \tag{6.21}$$

where

$$d^x_{1\tilde{2}} = d^x_{1\tilde{3}} + d^x_{\tilde{3}\tilde{4}} + d^x_{\tilde{4}\tilde{5}} + d^x_{\tilde{2}\tilde{5}} \text{ and } d^y_{1\tilde{2}} = 0 + d^y_{\tilde{3}\tilde{4}} - d^y_{\tilde{4}\tilde{5}} + d^y_{\tilde{2}\tilde{5}}.$$

The last phase of the preprocessing step is to convert the relative coordinates to absolute coordinates through the transformation below, which is given by scaling and rotating $\boldsymbol{x}_{\tilde{2}}$ such that it aligns with $\boldsymbol{x}_2$, as shown in Fig. 6.7b:

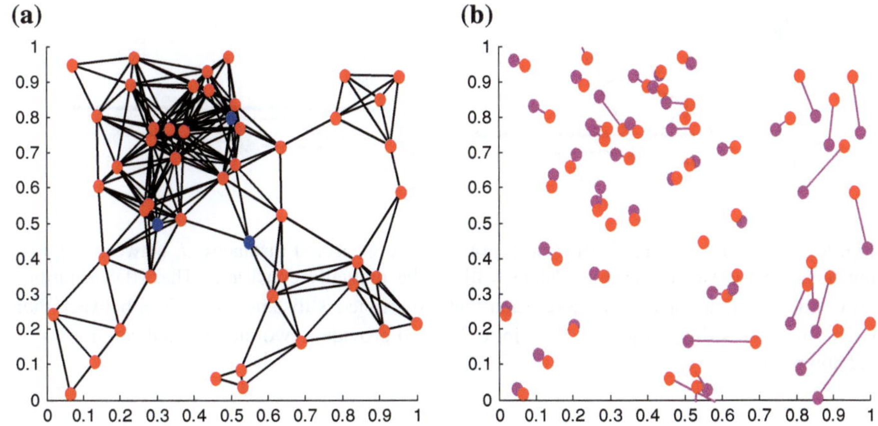

Fig. 6.8 An example network with three anchors and 50 sensors. **a** The links between neighboring nodes in the network. **b** The ground-truth locations of the sensors are shown in *red*, the estimated locations are shown in *magenta*, and the location errors between the two are also shown in *magenta*

$$\begin{bmatrix} x_i \\ y_i \end{bmatrix} = \frac{d_{12}}{d_{1\tilde{2}}} \begin{bmatrix} \cos\phi & -\sin\phi \\ \sin\phi & \cos\phi \end{bmatrix} \begin{bmatrix} x_{\tilde{i}} - x_1 \\ y_{\tilde{i}} - y_1 \end{bmatrix}, i = 2, 3, 4, 5. \tag{6.22}$$

Since $\boldsymbol{x}_{\tilde{2}}$ and $\boldsymbol{x}_2$ are known, the values of d_{12} and ϕ can be calculated, and $d_{1\tilde{2}} = \sqrt{{d^x_{1\tilde{2}}}^2 + {d^y_{1\tilde{2}}}^2}$. The same transformation is used for the other nodes in the relative coordinate system. This completes the preprocessing step.

The disadvantage of this method is that the linear program itself does not inherently contain the location variables in it; as a result, it does not generate a unique solution for them. Therefore, in the reconstruction stage, the location of a sensor node will depend on the propagation path chosen. It has been shown experimentally that by choosing the shortest propagation path to the sensor node, requiring the least number of link distances—each of the distances being independently subject to error—the location error is minimized.

Figure 6.8a displays an example network with three anchor nodes (*blue*) and 50 sensor nodes (*red*) deployed in a 1 × 1 normalized area. The connections between the nodes for radio range $R = 0.25$ appear in black. For simulation purposes, zero-mean Gaussian noise with 0.1 standard deviation was added to the ground-truth link distances. The locations of the unknown sensors were then estimated through the linear programing method. Figure 6.8b shows the sensor locations reconstructed from the algorithm (*magenta*). Also shown in *magenta* are the locations errors between the ground-truth and reconstructed locations. The average location error for this simulation is 0.0432. Notice that the farther the sensors are from the anchor nodes, the greater the error. This is due to the accumulation of error in propagation.

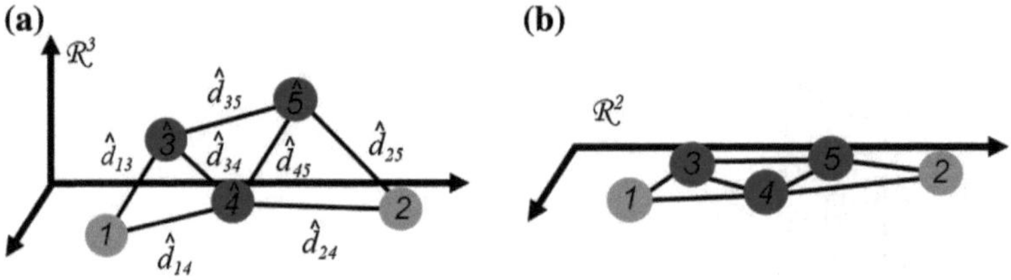

Fig. 6.9 Multidimensional Scaling. **a** In practice, the measured distances, $\hat{d}$, between neighboring nodes in the two-dimensional plane will not be mutually consistent. The MDS solution is provided in a higher space ($\mathcal{R}^3$ here) such that the erroneous distances can fit in between them. **b** The solution is then projected down into $\mathcal{R}^2$ so to provide a two-dimensional solution for the unknown sensors

6.5 Multidimensional Scaling

Multidimensional Scaling (MDS) falls within another class of techniques to estimate the unknown locations of sensor nodes in a network. It was first applied to cooperative localization in Shang et al. (2003). While it does not involve convex optimization, it shares some common aspects with the semi-definite programing technique described in Sect. 6.3. Specifically, when the measured distances are not mutually consistent, the solution is provided in a higher dimensional space such that the erroneous distances have "additional room to fit" in between the sensor locations. The final solution is determined by projecting the sensor locations down onto the two-dimensional plane through principal component analysis. This is illustrated in Fig. 6.9. The types of MDS range from classic, in which the measured distances are presumed to be deterministic, to non-deterministic, in which they are represented through a probability distribution, to weighted, in which a different importance is assigned to each dimension (Costa et al. 2006), to other varieties (Xiang and Hongyuan 2004). Here, we introduce classic MDS.

6.5.1 Principal Component Analysis

The origin of MDS stems from the Law of Cosines in 6.20(a), which relates the lengths of the three sides of a triangle $\Delta\hat{\boldsymbol{x}}_i\hat{\boldsymbol{x}}_0\hat{\boldsymbol{x}}_j$ to the lengths of any two of its sides and the interior angle between the two sides. For convenience, the Law of Cosines is rewritten here as:

$$\frac{1}{2}\left(\hat{d}_{i0}^2 + \hat{d}_{j0}^2 - \hat{d}_{ij}^2\right) = \hat{d}_{i0}\hat{d}_{j0}\cos\theta_{i0j} = \left(\hat{\boldsymbol{x}}_i - \hat{\boldsymbol{x}}_0\right)^T\left(\hat{\boldsymbol{x}}_j - \hat{\boldsymbol{x}}_0\right). \qquad (6.23)$$

The coordinate vectors $\hat{\boldsymbol{x}}$ above are dependent only on the measured distances and are not the *projected* values outputted from the MDS algorithm. Accordingly, we refer to them as the *measured* coordinated vectors denoted by the hat symbol.

Now observe that the right side of (6.23) is simply the inner product between $\hat{\boldsymbol{x}}_i$ and $\hat{\boldsymbol{x}}_j$ with respect to the origin, or center, of the coordinate system, $\hat{\boldsymbol{x}}_0$ (which we can assume to be arbitrary for the time being). As such, the equation associates the measured square distance, $\hat{d}_{ij}^2$, to the measured *centralized* coordinates which we define as $\hat{\boldsymbol{x}}_i = (\hat{\boldsymbol{x}}_i - \hat{\boldsymbol{x}}_0)$ and $\hat{\boldsymbol{x}}_j = \left(\hat{\boldsymbol{x}}_j - \hat{\boldsymbol{x}}_0\right)$. This forms the basis of the technique. The equation can be written compactly by defining further the centralized inner product, $\hat{b}_{ij}$, as

$$\hat{b}_{ij} = \frac{1}{2}\left(\hat{d}_{i0}^2 + \hat{d}_{j0}^2 - \hat{d}_{ij}^2\right) = \hat{\boldsymbol{x}}_i'^T \hat{\boldsymbol{x}}_j', \tag{6.24}$$

or in matrix form as

$$\hat{B} = \hat{\boldsymbol{X}}'^T \hat{\boldsymbol{X}}', \tag{6.25}$$

where $\hat{B}$ is an $n \times n$ matrix of values $\hat{b}_{ij}$ between all nodes i and j in the network.

Principal component analysis on $\hat{B}$ is performed through the Eigendecomposition (Golub et al. 1996) such that the matrix can be factorized as

$$\hat{B} = V\hat{\Lambda}V^T, \tag{6.26}$$

where V is $n x n$ orthonormal matrix of the eigenvectors of $\hat{B}$ and $\hat{\Lambda}$ is the diagonal matrix composed from the n square eigenvalues of $\hat{B}$. Let Λ be a copy of matrix $\hat{\Lambda}$, however with the smallest $(n-2)$ eigenvalues set to zero for $i = 3 \ldots n$. Then the centralized inner product matrix can be projected from the n-dimensional space spanned by its eigenvectors onto the two-dimensional plane as

$$B = V\Lambda V^T. \tag{6.27}$$

By defining X' as the projected centralized coordinates in the plane, it follows from Eq. 6.27—by reversing the factorization step from Eq. 6.25 to Eq. 6.26—that B can be expressed as

$$B = X'^T X'. \tag{6.28}$$

Then X' can be recovered by equating Eqs. 6.27 and 6.28, yielding

$$X' = V\Lambda^{\frac{1}{2}}. \tag{6.29}$$

The residual error α between the measured and projected centralized coordinates is given by the norm:

$$\begin{aligned}\alpha &= \left\|\hat{X}' - X'\right\| = \left\|V\hat{\Lambda}^{\frac{1}{2}} - V\Lambda^{\frac{1}{2}}\right\| = \left\|V\left(\hat{\Lambda}^{\frac{1}{2}} - \Lambda^{\frac{1}{2}}\right)\right\| \\ &= \left\|\hat{\Lambda}^{\frac{1}{2}} - \Lambda^{\frac{1}{2}}\right\| = \sqrt{\sum_{i=1}^{n}\left(\hat{\lambda}_i - \lambda_i\right)^2} = \sqrt{\sum_{i=3}^{n}\hat{\lambda}_i^2}.\end{aligned} \tag{6.30}$$

Since V is orthonormal, multiplying another matrix by V does not vary the norm of that matrix. Hence V can be removed from (6.30). The norm is then written out explicitly as the square root of the sum of square differences between the elements of the two diagonal matrices. Since the first two elements of the matrices are identical, the norm reduces simply to the square root of the last $(n-2)$ terms of the sum. If the measured distances are mutually consistent, then the projected centralized coordinates can be represented exactly in the two-dimensional plane with zero residual error. However, when this is not the case, the factorization in (6.26) given by the Eigendecomposition is such that the residual error is minimized, meaning that X' provides the best approximation of $\hat{X}$ in the two-dimensional space.

6.5.2 Computing the Centralized Inner Product Matrix

Note that while the link distances, $\hat{d}_{ij}$, between two neighboring nodes i and j in (6.24) can be measured—because the origin of the coordinate system, $\hat{\boldsymbol{x}}_0$, does not correspond to the location of an actual node—the distances, $\hat{d}_{i0}$ and $\hat{d}_{j0}$, between the nodes and the origin cannot. Hence this equation cannot be used directly to compute the centralized inner products, $\hat{b}_{ij}$. Rather in this subsection we turn to a method to do so.

First, let the origin of the coordination system be defined as the centroid of the node coordinates:

$$\hat{\boldsymbol{x}}_0 = \sum_{i=1}^{n} \hat{\boldsymbol{x}}_i. \tag{6.31}$$

Substituting this identity into (6.24) implies

$$\left.\begin{array}{l} \sum_{i=1}^{n} \hat{b}_{ij} = 0, \forall j \\ \sum_{j=1}^{n} \hat{b}_{ij} = 0, \forall j \\ \sum_{i=1}^{n}\sum_{j=1}^{n} \hat{b}_{ij} = 0 \end{array}\right\}. \tag{6.32}$$

Now observe that

$$\hat{d}_{ij}^2 = \left(\hat{\boldsymbol{x}}_i - \hat{\boldsymbol{x}}_j\right)^T \left(\hat{\boldsymbol{x}}_i - \hat{\boldsymbol{x}}_j\right) = \hat{b}_{ii} + \hat{b}_{jj} - 2\hat{b}_{ij}. \tag{6.33}$$

Applying further each of the equations in Eqs. 6.32–6.33 yields the respective equations for each of the three:

$$\left.\begin{aligned} \frac{1}{n}\sum_{i=1}^{n}\hat{d}_{ij}^2 &= \frac{1}{n}\sum_{i=1}^{n}\hat{b}_{ii} + \hat{b}_{jj}, \forall j \\ \frac{1}{n}\sum_{j=1}^{n}\hat{d}_{ij}^2 &= \hat{b}_{ii} + \frac{1}{n}\sum_{j=1}^{n}\hat{b}_{jj}, \forall i \\ \frac{1}{n^2}\sum_{i=1}^{n}\sum_{j=1}^{n}\hat{d}_{ij}^2 &= \frac{2}{n}\sum_{i=1}^{n}\hat{b}_{ii} \end{aligned}\right\}. \tag{6.34}$$

Finally, by combining the three equations in a linear system, $\hat{b}_{ij}$ can be solved for as

$$\hat{b}_{ij} = -\frac{1}{2}\left(\hat{d}_{ij}^2 - \frac{1}{n}\sum_{k=1}^{n}\hat{d}_{ik}^2 - \frac{1}{n}\sum_{k=1}^{n}\hat{d}_{jk}^2 + \frac{1}{n^2}\sum_{k=1}^{n}\sum_{l=1}^{n}\hat{d}_{kl}^2\right) \tag{6.35}$$

or in matrix form as

$$\hat{B} = -\frac{1}{2}C\hat{D}^2C, \tag{6.36}$$

where $\hat{D}$ is $n \times n$ matrix of measured distances $\hat{d}_{ij}$ between all nodes i and j and $C = \boldsymbol{I} - \frac{1}{n}\boldsymbol{ee}^T$ is the $n \times n$ matrix known as the *centering matrix*, with $\boldsymbol{I}$ denoting the identity matrix and $\mathbf{e}$ the $n \times 1$ vector of 1s. The measured square distance matrix $\hat{D}^2$ is symmetric positive semi-definite and the centering matrix directly follows from the selection of $\hat{\boldsymbol{x}}_0$ in (6.31). The selection is such that the symmetric positive semi-definite property for $\hat{B}$ is preserved, ensuring that the diagonal matrix, Λ, has only non-negative values, which is a requirement in order for the Eigendecomposition to be used for the factorization in (6.26).

The last step in MDS is to transform the projected coordinates, X', into the final estimates for the sensor locations, provided the known anchor locations of the network. The transformation involves translation, scaling, and rotation, and also reflection to cope with the mirror ambiguity described in Sect. 6.4.2. This is similar in principle to the mapping in (6.22). The transformation is found by minimizing the sum of square errors between the projected anchor locations and the actual locations of the corresponding anchors in the transformed coordinate system. Details of this technique are provided in Horn et al. (1988).

The MDS technique has several drawbacks—the main one being the high computational complexity, $O(n^3)$, of the Eigendecomposition. Another drawback is that the technique assumes that the measured distances between all nodes in the network are available for processing. This, however, is seldom the case, especially in large networks for which the radio range is limited. In the more frequent case, a preprocessing step is necessary in order to generate $\hat{B}$. The algorithm used in this step to compute the distance between non-neighboring nodes may vary, and so in turn will the results generated from the MDS technique. A common approach is to

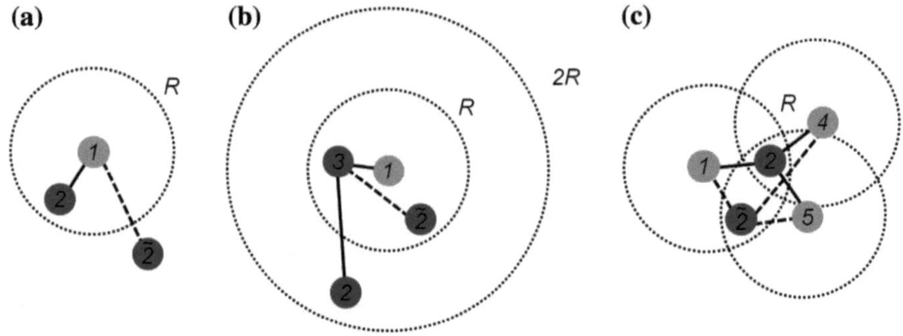

Fig. 6.10 Monte Carlo Localization. **a** Location x_2 is consistent with the true observation that sensor 2 neighbors anchor 1 ($x_{\tilde{2}}$ is not consistent with this observation). **b** Location x_2 is consistent with the true observation that sensor 2 neighbors sensor 3 but not anchor 1. **c** Location x_2 is consistent with the true observation that sensor 2 neighbors anchors 1, 4, and 5

use the shortest multihop distance between any two nodes, which can be computed through Dijkstra's Algorithm or a variant thereof (Cormen et al. 2001).

6.6 Monte Carlo Localization

In this last section of centralized algorithms we describe Monte Carlo Localization (Hu and Evans 2004), which is a technique that can be applied when multiple observations of a sensor's location—recorded over time—are available for processing. The technique was devised for mobile networks in which some or all of the nodes—both the sensors and the anchors—are in motion. However, Rudafshani and Datta (2007) propose an alternative implementation for static networks as well. In static networks, the observations can be exploited to improve performance in the presence of variable channel conditions. The problem is solved in the framework of a first-order Markov process which was described in Sect. 4.3.2. The Markov process generates the posterior probability, $p(\boldsymbol{x}_i|\boldsymbol{o}^t,\ldots,\boldsymbol{o}^0)$, that a given sensor in the network lies at some candidate position, $\boldsymbol{x}_i$, given observations $\{\boldsymbol{o}^t,\ldots,\boldsymbol{o}^0\}$ at discrete time steps from initialization at $t = 0$.

In Monte Carlo Localization, a sensor node can acknowledge two types of observations. The first type of observation is whether the sensor can wirelessly communicate with an anchor node; the observation value can either be true or false. If the observation at time $t, \boldsymbol{o}^t$, is true and if candidate position for the sensor, $\boldsymbol{x}_i$, is consistent with the observation—that is—if its geometrical location lies within the radio range, R, of the anchor, then the likelihood function $p(\boldsymbol{o}^t = \text{true}|\boldsymbol{x}_i) = 1$ and 0 otherwise. Position 2 in Fig. 6.10a displays an example of a true observation. On the other hand, if the observation is false (position $\tilde{2}$), then by the same token $p(\boldsymbol{o}^t = \text{false}|\boldsymbol{x}_i) = 1$ if the candidate position does not lie within range of the anchor and 0 otherwise. In mobile networks, because the nodes are in motion, observations vary continuously and, of the two, the true observation clearly provides more useful localization information—when a sensor moves within an anchor's range, valuable information can be gathered. As mentioned earlier, in static networks observations can vary not because of node mobility but because of variable channel conditions (i.e. fluctuations in the value of the radio range, R). As such, sensors not neighboring any anchor nodes can only make false observations. Hence, their localization accuracy is limited.

The second type of observation, although providing weaker localization information, deals with this limitation. It considers the configuration for which a sensor node and an anchor node are mutual neighbors of another sensor, but are not neighbors themselves. The type of observation is whether the sensor node—which cannot directly communicate with the anchor—can at least communicate with the neighboring sensor; again, the observation value can be either true or false. If the observation is true and, in addition, if the candidate position for the sensor lies within $2R$ of the anchor but not within range of the anchor itself, then the likelihood $p(\boldsymbol{o}^t = \text{true}|\boldsymbol{x}_i) = 1$ and 0 otherwise. An example of a true observation (position 2) is displayed in Fig. 6.10b. On the other hand, if the observation is false (position $\tilde{2}$), then the likelihood is computed vice versa.

Within the same Markov framework, the transition probabilities govern the motion dynamics of the network. Specifically, assuming that the nodes cannot move faster than a maximum displacement $D_{\max}$ between time steps, if candidate location $\boldsymbol{x}_i$ is within the maximum displacement of candidate location $\boldsymbol{x}_{\tilde{i}}$, then the transition probability $p(\boldsymbol{x}_i|\boldsymbol{x}_{\tilde{i}}) = 1$ and 0 otherwise.

Now with the likelihood function and the transition probabilities defined, the Monte Carlo Localization algorithm can be outlined. The algorithm is initialized by randomly generating a set of candidate locations $\boldsymbol{x}_i, i = 1 \ldots n$ for each of the sensor nodes in the deployment area. Then, the probability that a given sensor is at candidate location $\boldsymbol{x}_i$ at time $t = 0$ is set to $p(\boldsymbol{x}_i|\boldsymbol{o}^0) = \frac{1}{n}$, where $\boldsymbol{o}^0$ denotes that no observations have yet been made. At the next time step—and in general at time step t—a new set of candidate locations for each sensor node is generated at random once more. The posterior probability of a new candidate location, $\mathbf{x}_i$, is computed recursively from the previous set at time $t - 1$ through the first-order Markov process [see (4.32)] as

$$p(\boldsymbol{x}_i|\boldsymbol{o}^t, \ldots, \boldsymbol{o}^0) = \eta^t \cdot p(\boldsymbol{o}^t|\boldsymbol{x}_i) \sum_{\tilde{i}=1}^{n} p(\boldsymbol{x}_i|\boldsymbol{x}_{\tilde{i}}) \cdot p(\boldsymbol{x}_{\tilde{i}}|\boldsymbol{o}^{t-1}, \ldots, \boldsymbol{o}^0). \tag{6.37}$$

In static networks, depending on the network constellation and the sensor–anchor ratio in the network, the posteriors will converge to steady-state values after a number of iterations. Then either Maximum Likelihood Estimation or a weighted average of these probabilities is implemented in order to estimate the location of the sensors (Eqs. 4.18 and 4.19, respectively). In mobile networks, rather, the localization algorithm will track the sensor locations through the dynamic posterior probabilities. The sensor locations can be estimated at any step through the same methods for static networks, but, depending on the degree of mobility and the time sampling rate, the posteriors may not converge. As in Sect. 4.3.2, particle filtering is implemented to limit the size of the candidate location sets.

Improvements to the original Monte Carlo Localization algorithm are described in (Baggio and Langendoen 2008). The major improvement arises from the enhanced observation type: instead of observations based on a single anchor—which confines the sensor location to within the anchor's radio range—by using the intersection of radio ranges from multiple anchors, the location of the sensor is confined to a smaller area. This enhanced observation type delivers both greater precision and faster convergence. An example of a true observation (position 2) is displayed in Fig. 6.10c. The intersection area for the three anchors is shaded in gray. The figure also displays a false observation (position $\tilde{2}$).

References

A. Baggio, K. Langendoen, Monte Carlo localization for mobile wireless sensor networks. Ad Hoc Networks Elsevier **6**, 718–733 (2008)

M.S. Bazaraa, J.J. Jarvis, H.D. Sherali, *Linear Programming and Sensor Network Flows*, 2nd edn. (Wiley, New York, 1990)

P. Biswas et al., Semi definite programming approaches for sensor network localization with noisy distance measurements. IEEE: Trans. Autom. Sci. Eng. **4**(3), 360–371 (2006)

S. Capkun, M. Hamdi, J-P. Hubaux, GPS-free positioning in mobile ad-hoc networks. Hawaii Conference on Systems Sciences IEEE, 255–264 (2001)

J.A. Costa, N. Patwari, A.O. Hero, Distributed weighted-multidimensional scaling for node localization in sensor networks. ACM. Trans. Sens. Netw. **1**(2), 39–64 (2006)

Q. Cui et al, A Novel Location Model for 4G Mobile Communication Networks. Vehicular Technology Conference, Fall: IEEE (2007)

L. Doherty, K.S.J. Pister, L. El Ghaoui, Convex Position Estimation in Wireless Sensor Networks Conference INFOCOM. IEEE, pp. 1655–1663 (2001)

J. Figueiras, Accuracy Enhancements for Positioning of Mobile Devices in Wireless Communication Networks [Report]: Ph.D. Dissertation, Aalborg University, Jan (2008)

S. Frattasi, Link Layer Techniques Enabling Cooperation in Fourth Generation Wireless Networks Report: Ph.D. Dissertation, Aalborg University, Alalborg, Sept (2007)

C. Gentile, Distributed sensor location through linear programming with triangle inequality constraints. IEEE: Trans. Wireless Commun. **7**(6), 2572–2581 (2007)

G.H. Golub, C.F. Van Loan, Matrix Computations, 3rd edn. (The John Hopkins University Press, Baltimore, Maryland, 1996).

B.K.P. Horn, H. Hilden, S. Negahdaripour, Closed-form solution of absolute orientation using orthonormal matrices. J. Opt. Soc. Am. A. **5**, 1127–1135 (1988)

L. Hu, D. Evans, Localization for Mobile Sensor Networks. Conference on Mobile Computing and Networking. ACM, pp. 45–57 (2004)

R. Kurazume, S. Nagata, S. Hirose, Cooperative positioning with robots. IEEE Conf. on Robots and Automation, **2**, 1250–1257 (1994)

R.B. Langley, Dilution of Precision. GPS World **10**(5), 52–59 (1999). Questex Media Group, LLC

T.H. Cormen, C.E. Leiserson, Rivest, R.L., Stein, C [1990]. *Introduction to Algorithms*, 2nd edn. (MIT Press and McGraw-Hill, Cambridge, 2001)

C.E. Perkins, Ad Hoc Networking, (Addison-Wesley, 2001)

M. Rudafshani, S. Datta, Localization in Sensor Networks. Conference in Information Processing in Sensor Networks. ACM, pp. 51–60 (2007)

A. Savvides, C.C. Han, C.C. Strivastava, Dynamic Fine-Grained Localization in Ad-Hoc Networks of Sensors. International Conference on Mobile Computing and Networking ACM, 166–179 (2001)

A. Savvides, H. Park, M.B. Srivastava, The Bits and Flops of the N-hop Multilateration Primitive Conference. International Workshop on Wireless Sensor Networks and Applications: ACM, pp. 112–121 (2002)

Y. Shang. et al. Localization from Mere Connectivity. International Symposium on Mobile Ad Hoc Networking and Computing. ACM, pp. 201–212 (2003)

J. Xiang, Z. Hongyuan, Sensor Positioning in Wireless Ad-hoc Sensor Networks using Multi-Dimensional Scaling Conference. IEEE: INFOCOM, pp. 2652–2661 (2004)

Chapter 7
Cooperative Localization in Wireless Sensor Networks: Distributed Algorithms

Chapter 6 introduces centralized algorithms for cooperative positioning. In this chapter, we focus on distributed algorithms applied to cooperative positioning. Distributed algorithms differ on where and when information is processed. Centralized algorithms first collect all potential information at a central unit and then process the data (Figueiras 2008). Compared to this serial methodology, the distributed algorithms process data in parallel at different units. Furthermore, distributed algorithms rely on the connections between geographically distributed sensor nodes for the mutual exchange of information.

In the past only low-cost sensors were available that collected measurement data, such as round-trip TOA and RSS as discussed in Chap. 2. These low-cost sensors could not perform calculations turning raw data into derived quantities and statistics, such as received signal strength and so a centralized approach was necessary. Improvements in low power processing capability allow us to consider distributed implementations.

Although networks with low-cost sensors, especially passive sensors, have some benefits over networks using distributed methodologies; distributed algorithms are often preferable for many reasons (Sahinoglu and Gezici 2010; Wymeersch et al. 2009). One major concern with a centralized approach is that using a central unit for all processing introduces a single point of failure, while distributed methods spread the risk of failure across multiple nodes. Other issues are more logistical. For example, in large sensor networks interference among large numbers of sensor nodes can prevent accurate and reliable communication with a central unit. Estimated positions need to be communicated back to the sensor nodes. Therefore, the energy efficiency of all sensor nodes is reduced by repeated communication with a central unit. Additionally, large networks require a local coordinator that selects or censors transmissions, preferably at the transmitter, to reduce interference and to exploit the diversity by using only the most reasonable links in transmitted signals. Network latency that results from forwarding measurement information via time-consuming multihop links to the

C. Gentile et al., *Geolocation Techniques*, DOI: 10.1007/978-1-4614-1836-8_7,

central unit hampers the ability of large centralized networks to handle moving sensor nodes in a timely fashion. Distributed algorithms handle location changes of sensor nodes much better, and network latency is minimized. In this chapter, we describe the following:

In Sect. 7.1, we present the Cramer–Rao lower bound (CRLB) for cooperative positioning and compare it to the non-cooperative positioning CRLB. The CRLB is the minimum variance unbiased estimator of the positioning accuracy independent of the used positioning algorithm. The CRLB for cooperative positioning can be written as a sum of two parts, where one part is the non-cooperative part (relying only on information from anchors with known position), and the second is the cooperative part (that adds information passed between sensor nodes). There are different applications for the CRLB, such as it can be used to theoretically assess the performance and also allows comparing an implementation or an algorithm against the CRLB as benchmark. Other usages are described in Sect. 7.4 where it is used as criteria to allocate links. Finally, the distributed approach of cooperative positioning allows reducing the complexity of the positioning CRLB.

In Sect. 7.2, a general two-phase framework for distributed positioning algorithms is presented. Passing of only very basic information is described in order to convey how position information propagates through the network. We distinguish several categories of information exchange between nodes that are characterized by the complex operations used to gather information. For example, such information simply states that nodes are connected, while more complex data are ranging estimates, and even more complex data provide position estimates.

In Sect. 7.3, we describe methods for sharing more complete information between nodes including ranging or position estimate error distributions. The importance of this information is that it allows one to better understand how errors propagate through the network. Both the correctness of the shared positioning information and the topology of the wireless sensor network are important to understand network location errors. This motivated researchers to apply belief propagation (Yedidia et al. 2003) using a message passing algorithm.

In Sect. 7.4, we describe methods for link selection. Wireless communication networks with a high density of communication devices require coordination to avoid interference and increase the energy efficiency of the network. For cooperative positioning the approach is to either support other nodes by creating a link, e.g., for round-trip-delay measurements, or support from other nodes, such as sharing positioning information. It is an open question whether a link to another node is worth using to improve its own positioning accuracy. As in communication networks any interference causes errors. Various methods allow predicting the quality of a ranging link, depending on bandwidth, signal power, and other quantities. Censoring links at the transmitter reduces interference and allows coordinating limited resources, such as ranging links between sensor nodes. However, this requires a priori knowledge about the usefulness of the link. An alternative approach is censoring at the receiver at the cost of spending unhelpful resources, like spectrum and energy.

7.1 Theoretical Bounds for Centralized and Distributed Cooperative Versus Non-Cooperative Positioning

In the following, we show the difference between cooperative and non-cooperative positioning based on the evaluation of the Cramer–Rao lower bound (CRLB) (Savvides et al. 2005; Alsindi and Pahlavan 2008; Penna et al. 2010; Zhang 2011). The positioning CRLB is commonly used to assess the performance theoretically and independently of any estimator of a positioning system.

The derivation applies to cooperative positioning independent of the algorithm for solving the positioning problem. However, for large-scale sensor networks, it is reasonable to assume that each node is only connected to a subset of network nodes, i.e., only nodes which are located within the radio range R. A centralized approach in this case is complex and a distributed approach is may be preferred. The CRLB states that the variance of any unbiased estimator is at least as high as the inverse of the Fisher information (Kay 1993). We assume that a parameter, such as the distance r between two nodes (a sensor node S and an anchor node A), is estimated from (e.g. time-based) ranging measurements t_{SA} according to the probability density function $p(r, t_{\mathrm{SA}})$ and the estimated distance $\hat{r}$. The variance $\mathrm{var}(\hat{r})$ is then bounded by the inverse of the Fisher information:

$$\mathrm{var}(\hat{r}) = CRLB(r) \geq \frac{1}{I(r)} \tag{7.1}$$

where the Fisher information $I(r)$is defined as

$$I(r) = -E\left[\left(\frac{\partial^2 \ln(p(r, t_{\mathrm{SA}}))}{\partial r^2}\right)\right] \tag{7.2}$$

and E denotes the expectation value and $\ln p(r, t_{\mathrm{SA}})$ is the likelihood function.

Now, cooperative positioning considers—besides the links between the sensor node to the n_A anchors—also the links between the n_S mobile sensor nodes to each other. Figures 1.5 and 1.6 show the differences of a non-cooperative and a cooperative scenario. The non-cooperative scenario only uses links to anchor nodes or base stations with known positions. The cooperative scenario additionally integrates radio links between sensor nodes. This could additionally mean that some sensor nodes have no direct access to any anchor node at all. The sensor nodes have a non-perfect or unknown location estimate. Different authors (Larsson 2004; Savides et al. 2005; Alsindi and Pahlavan 2008; Penna et al. 2010) derived the Fisher information matrix (FIM) and the Cramer–Rao lower bound (as the inverse of the FIM) for the non-cooperative and the cooperative case. Larsson (2004) applied the CRLB for a homogeneous sensor network for positioning using anchors as well as sensor nodes for a cooperative system. The author considered the geometrical constellation of all nodes by using the GDOP [see Chap. 2 and Eq. (2.6)]. Furthermore, he distinguished between perfect synchronization between the nodes

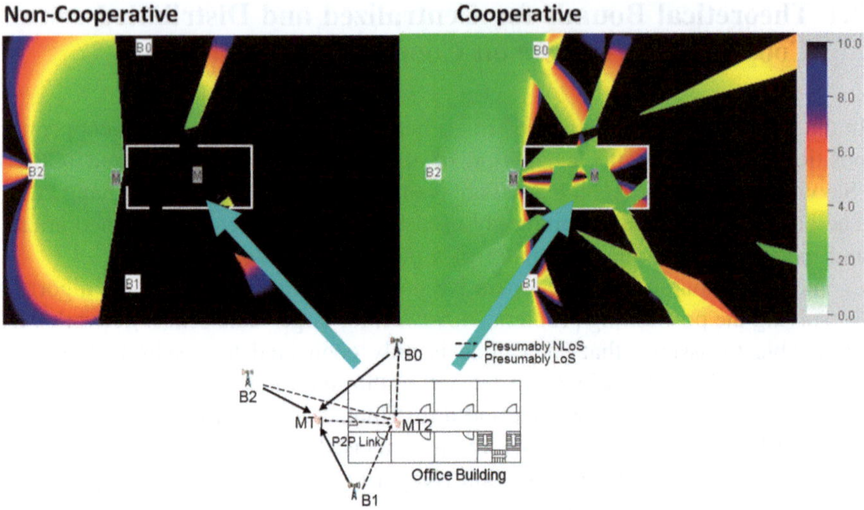

Fig. 7.1 Comparison of the TDOA CRLB of the RMSE for a non-cooperative versus a centralized cooperative positioning system for a setup with anchors (B0, B1, B2) exterior to a small office building as shown. The color bar on right reflects the root-mean-square error in meters. The performance indoors is significantly enhanced by using the additional ranging links of the sensor node (*MT*) via the cooperative links (Online tool: http://www.kn-s.dlr.de/positioning/cooperative.php)

using absolute timing information as well as synchronization uncertainties with non-absolute timing information.

The authors of Savvides et al. (2005) introduced the CRLB for sensor nodes with unknown location and static anchor nodes with uncertain location in a wireless sensor network. The CRLB was used to analyze the effects of network density to provide guidelines for deploying a network. The authors (Alsindi and Pahlavan 2008) defined a generalized CRLB based on additional a priori information about the wireless channel conditions between the cooperating nodes. The channel conditions were assigned to three categories. The first category defined the channel as line-of-sight with free-space path loss. The second category defined that the signal is significantly attenuated through an object but is still received and therefore detectable without causing an extra ranging bias. The third category defined a non-line-of-sight channel that is fully blocked and results in a significant bias for time-ranging methods. The three categories were used as a priori information to generalize the CRLB. The generalized CRLB was applied for optimizing the network. Penna et al. (2010) combined the GNSS system that acts similar to the anchors in wireless sensor network with terrestrial links. The terrestrial links are peer-to-peer links between the sensor nodes. The derived CRLB considered the GNSS system if available and neighboring nodes that cooperated with the other nodes. The cooperation between nodes supported a reasonable performance even in cases where not enough satellites are available. A single node only performed relative ranging with cooperative nodes and without any satellite links.

In the following, we derive the positioning CRLB for non-cooperative and cooperative positioning system. Furthermore, we distinguish how a distributed and a centralized CRLB could be calculated. Two different measurements are considered based on the position of anchor ($\mathbf{x}_{\mathrm{A}}$) and sensor nodes ($\mathbf{x}_{\mathrm{S}}$). The estimated distance between a sensor node S and an anchor node A is:

$$\hat{d}_{\mathrm{SA}} = ||\mathbf{x}_{\mathrm{A}} - \mathbf{x}_{\mathrm{S}}|| + \alpha_{\mathrm{AS}}, \tag{7.3}$$

where α_{AS} is the measurement noise of the link with variance σ^2_{AS}. A non-cooperative positioning system uses only the links between the anchor nodes n_{A} and the sensor nodes n_{S}. For the n_{S} sensor nodes in a non-cooperative positioning system, the FIM of the global parameter vector ($\mathbf{x} = [x_1, \ldots, x_n]$) is a block-diagonal matrix with the FIM of each sensor node along the diagonals. The noise variance of the ranging link between a sensor and an anchor are both symmetric. For the non-cooperative case, considering all n_{S} sensor nodes the global non-cooperative FIM is a block-diagonal matrix:

$$I_{\mathrm{non-coop}}(d) = \begin{bmatrix} I_1 & \cdots & 0 \\ \vdots & \ddots & \vdots \\ 0 & \cdots & I_{n_{\mathrm{S}}} \end{bmatrix}. \tag{7.4}$$

In Eq. 7.1, we defined that the inverse of $I_{\mathrm{non-coop}}(r)$ is the CRLB, and therefore the $CRLB = tr\left(I_{\mathrm{non-coop}}\right)^{-1}$, where $tr(I_{\mathrm{non-coop}})$ is the trace of a matrix $I_{\mathrm{non-coop}}$.

Similarly as in Eq. 7.3 the estimated distance between sensor nodes i at position $\mathbf{x}_{\mathrm{si}}$ and sensor node j at position $\mathbf{x}_{\mathrm{sj}}$, is:

$$\hat{\mathrm{d}}_{\mathrm{si,sj}} = ||\mathbf{x}_{\mathrm{si}} - \mathbf{x}_{\mathrm{sj}}|| + \alpha_{\mathrm{si,sj}}. \tag{7.5}$$

where $\alpha_{\mathrm{si,sj}}$ is the measurement noise of the link with variance $\sigma^2_{\mathrm{si,sj}}$ that is Gaussian distributed. We presume the measurement noise does not depend on the choice of the measuring sensor node (either node i or j).

For the cooperative case, we state the FIM is of the following form:

$$I = I_{\mathrm{non-coop}} + I_{\mathrm{coop}}. \tag{7.6}$$

The second part of the FIM I_{coop} describes the part that is based on the additional information resulting from the cooperative links between the sensor nodes i and j.

$$I_{\mathrm{coop}}(r) = -E\left\{ \begin{bmatrix} H_{11} & \cdots & H_{n_s 1} \\ \vdots & \ddots & \vdots \\ H_{n_s 1} & \cdots & H_{n_s n_s} \end{bmatrix} \Delta_{\mathrm{coop}}(r) \right\}, \tag{7.7}$$

where $\Delta_{\mathrm{coop}}(r)$ is the log-likelihood function and the cross-Hessian matrices $H_{i,j}$ are defined based on $r = [d_{1,i}, \ldots, d_{\mathrm{D},i}]$ where D is the dimension (in our case 2):

$$H_{ij}(r) = \begin{bmatrix} \frac{\partial^2}{\partial d_{1,i} \partial d_{1,j}} & \cdots & \frac{\partial^2}{\partial d_{1,i} \partial d_{\mathrm{D},j}} \\ \vdots & \ddots & \vdots \\ \frac{\partial^2}{\partial d_{\mathrm{D},i} \partial d_{1,j}} & \cdots & \frac{\partial^2}{\partial d_{\mathrm{D},i} \partial d_{\mathrm{D},j}} \end{bmatrix}. \tag{7.8}$$

Therefore, the cooperative FIM is a non-diagonal matrix:

$$I_{\mathrm{coop}}(r) = \begin{bmatrix} I'_1 & \cdots & K_{1,n_s} \\ \vdots & \ddots & \vdots \\ K_{n_s,1} & \cdots & I'_{n_s} \end{bmatrix}. \tag{7.9}$$

With the assumption that the sensor node i is in the range of sensor node j:

$$I'_i = \sum \frac{1}{\sigma^2_{i \to j}} A \tag{7.10}$$

$$K_{ij} = \frac{-1}{\sigma^2_{i \to j}} A \tag{7.11}$$

and the unit-length column vector $A = \frac{d_i - d_j}{\|d_i - d_j\|}$ between the sensor node $\mathbf{x}_{\mathrm{s}i}$ and $\mathbf{x}_{\mathrm{s}j}$. In a centralized system, all links between all nodes are known. In a distributed system, only local information is available and used at each node. This is constrained also by the limited range of each sensor node described in Eq. 7.10. For a distributed cooperative system that only uses local information the local FIM reduces to a set of 2×2 matrix as shown in Eq. 7.12 for sensor node i

$$I_{\mathrm{coop,distr.},i}(r_i) = \begin{pmatrix} I_{i,1} & K_{i,2} \\ K_{i,1} & I_{i,2} \end{pmatrix}. \tag{7.12}$$

and for all sensor nodes n_S in Eq. 7.13 we have

$$I_{\mathrm{coop,distr.}}(r) = \begin{bmatrix} I_{\mathrm{coop,distr.},1}(r_1) & \cdots & 0 \\ \vdots & \ddots & \vdots \\ 0 & \cdots & I_{\mathrm{coop,distr.},n_S}(r_{n_S}) \end{bmatrix}. \tag{7.13}$$

If we compare the cooperative FIM of the centralized system Eq. 7.9 and the individual FIM for a sensor node i in Eq. 7.12 and Eq. 7.13 we can see that the positioning CRLB for centralized cooperative system is more complex. The matrix in Eq. 7.9 is a matrix of size $2n_S \times 2n_S$. The distributed positioning FIM is a diagonal matrix of n_S matrices (each with the dimension D). To calculate the inverse of Eq. 7.13 has a lower complexity compared to the calculation of the centralized FIM in Eq. 7.9.

The performance of the cooperative positioning system is better than the non-cooperative positioning system as more information, i.e., the ranging links between sensor nodes, is considered. The performance bounds for cooperative positioning either centralized or distributed are expected to deliver a similar performance. However, the CRLB is a bound that can be expected to be tight if all information is considered correctly or loose if e.g., wrong assumptions are made. Wrong assumptions are e.g., interfering signals that are not Gaussian distributed or the ranging noise is also not Gaussian distributed. Other error sources are the uncertainty of exchanged positioning information—which is especially problematic in the distributed cooperative positioning system as each node calculates its own position. In Sect. 7.4, we show different techniques about how to select the most suitable connection for a sensor node. Such an evaluation based on the CRLB is e.g., presented by Lieckfeld et al. (2008). The reduction in complexity of evaluating the distributed positioning CRLB is helpful in dynamic situations.

We can summarize that the performance of the different systems follows this expectation in a static setup:

$$CRLB_{\text{Non-coop}} > CRLB_{\text{Coop,distributed}} \cong CRLB_{\text{Coop,centralized}}$$

In Fig. 7.1, we compare the performance of cooperative vs. non-cooperative positioning by using the TDoA CRLB to calculate the lower bound of the root-mean-square error (RMSE) in meters. The non-cooperative scenario used links between the sensor node indoors (MT2) and the anchors (B0, B1 and B2) outdoors. The cooperative scenario, adding only one sensor node (MT1) with non-perfect positioning information to the ranging links of MT2, results in a significant improvement of the performance indoors. Note that, the plotted colors represent the RMSE performance of a sensor node in the rectangular zone in meters.

To conclude the performance can be assessed for distributed and centralized cooperative positioning by the CRLB. However, the calculations are much more complex compared to the non-cooperative positioning due to the non-diagonal matrix itself.

7.2 Distributed Positioning Algorithms

In Sect. 7.1, we derived the CRLB for cooperative positioning based on ranging measurements and compared it to the non-cooperative positioning bound. In this section, we summarize early distributed algorithms for positioning. As discussed in Chap. 6, two different types of sensor nodes are used: Anchor nodes, which know their own position perfectly, and sensor nodes, which do not know their own position at all, or only approximately. However, any sensor node could become a temporary anchor node (e.g. in case they are static for a limited time) when positioning information from neighboring sensor nodes and anchors allow them to determine their positions with high accuracy.

We consider a network with n_A anchor nodes (or anchors) and n_S sensor nodes (or sensors), for a total of $n = n_A + n_S$ nodes. For simplicity, we assume the nodes lie on a 2D plane such that node i has location $\mathbf{x}_i \in \mathcal{R}^2$ indexed through $i = 1, \ldots, n_A$ for the anchors and $i = n_A + 1, \ldots, n$ for the sensors. The set N contains all pairs of nodes connected by links $(i,j), i<j; \left\|\mathbf{x}_i - \mathbf{x}_j\right\| < R$, where $\|\cdot\|$ is the Euclidean distance and the network parameter R is the maximum radio coverage range of the nodes. The complementary set $\overline{N}$ contains all pairs of nodes not connected by links: $(i,j), i<j; \left\|\mathbf{x}_i - \mathbf{x}_j\right\| \geq R$. The measured distance $\hat{d}_{ij}$ between neighboring nodes i and j is obtained through one of the received signal strength (RSS) or time-of-arrival (ToA) techniques introduced in Chap. 2.

Distributed algorithms have a startup phase to gather rough position estimates, and a refinement phase to improve the estimates. We describe two different distributed methods to estimate the positions of sensor nodes. When enough anchors are in range and all nodes perform ranging and estimate distances between sensor nodes, these techniques can be extended by atomic multilateration (Sect. 6.1.1) and collaborative multilateration (Sect. 6.1.2).

One of the first publications (Savarese et al. 2001) on distributed algorithms for positioning described an ad hoc wireless sensor network. The wireless sensor network contains two types of nodes. There are n_A anchor nodes that act as reference points and know their own position, and there are n_S sensor nodes that are not aware of their positions. The authors argue that received signal strength (RSS) is the preferred method of measuring the distance between nodes as angle of arrival requiring multiple antennas at the nodes is too complex for a simple wireless system. Time-based methods, such as ToA (or TDoA) require a (partial) synchronous network. Received signal strength method is supported by two properties of the wireless sensor network: (a) Dense interconnectivity leading to redundancy in the range measurements; (b) Limited mobility which accounts for long observation times to remove fast-fading effects through integration.

The authors distinguish two phases in the positioning process, a startup and a refinement phase. In the startup phase, the anchor nodes broadcast their own position together with a hop counter across a limited range. Sensor nodes that are close by receive this message, rebroadcast the position information, and increase the hop counter number by one. The sensor nodes may receive multiple messages, and therefore check for the lowest hop counter to ensure conformity with the distance model. Initially, the distance between an anchor node and a neighboring sensor node is estimated from signal strength measurements received by either node. In the refinement phase for each hop, the distance model assumes an average distance that is repeatedly refined at the anchor nodes. The sensor nodes infer the distance to anchors using the first distance estimate between anchor and sensor node together with the total number of hops. Anchor nodes receive hop counts with position information from other anchor nodes. The total distance calculated from different positions of the anchor nodes and the hop counter is the common denominator to calibrate the distance model of each hop. The sensor nodes triangulate when position information from at least three anchor nodes (to avoid ambiguity) is available in a 2D plane. The weakness of simple distance calculations based on hops is the distance model. It works well for isotropic networks, but errors are large for anisotropic network structures.

The ad hoc positioning algorithm (Niculescu and Nath 2001) proposed to position the nodes with exact locations by extending the coverage of GPS using the ad hoc network with fixed nodes having known positions (called landmarks—we keep calling them anchor nodes in the following). This algorithm assumes that nodes (anchors and sensors) have a very limited communication range that only links direct neighbors with each other to maintain low power consumption. The proposed scheme used anchor nodes that are connected to GPS information to calculate absolute position information of the sensor nodes. The ad hoc positioning algorithm intends to reduce the network traffic compared to Savarese et al. (2001). Therefore, anchors do not update the network topology when receiving messages from other anchors and the system copes with dynamically changing positions of the sensor nodes.

The simplistic distance calculations were also addressed by Niculescu and Nath (2003). They exploited the regular network structure topology with an additional rule. The regular network topology shows that messages reach all sensor nodes hop-by-hop (Niculescu and Nath 2003). Figure 7.2 depicts such a network at three succeeding time steps (*T*=1, *T*=2, and *T*=3). The network consists of anchor nodes (green (circled A)) and sensor nodes (orange and blue (circled S)). The sensor nodes differ by having no position estimates (blue circles) or by having their own estimates (orange circles). Sensor nodes are capable of calculating their own position if they collect information from at least three different anchor nodes, and only then the sensor nodes forward the location message from the anchor nodes to their neighboring sensor nodes. Collecting the messages from the different anchors can occur at a single time step such as *T*=1. When not enough (at least three) anchor messages are received at the same time instance, anchor messages over multiple time steps are needed. At *T*=2, the orange sensor nodes receive forwarded anchor information through the left orange sensor nodes, and consider the anchor information already received at *T*=1. The information forwarded earlier is shown at *T*=2 by the dashed line. The bold dashed represents the used information from the sensor node. At *T*=3, the right sensor node (the only one left in blue at *T*=2) estimates its position based on the messages received at *T*=2 and *T*=3. The participating sensor nodes forward the position information from the anchors together with distance information represented, for example, by the number of hops, or by true distances based on the ranging between participating nodes.

The distributed methods presented were the first proposed for (ad hoc) wireless sensor networks. They do not consider any error propagation on erroneously estimated ranges or positions. However, this omission was already recognized as an open problem especially as the ranging errors were as high as 90 % of the distances between nodes.

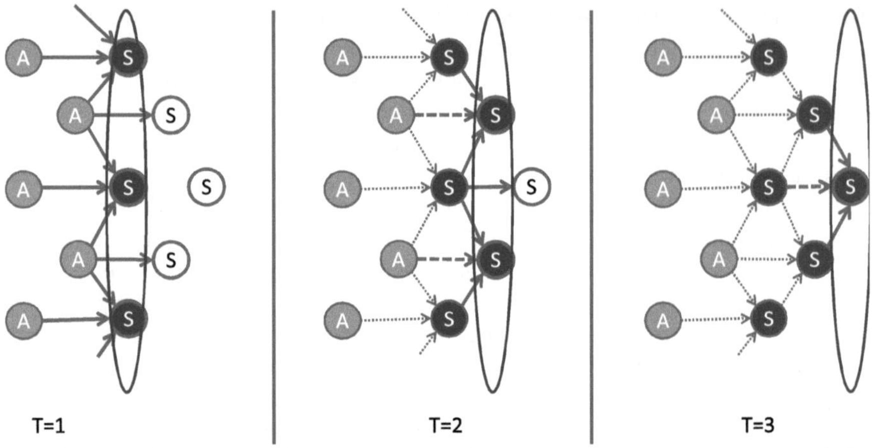

Fig. 7.2 Controlled flooding of messages: At time instance $T = 1$ the anchors (*green circled A*) broadcast their messages to their neighboring sensor nodes. At $T = 2$, all sensor nodes (*orange or blue circled S*) which received at least three messages before (*orange circled S*) broadcast the received anchor messages to their neighboring sensor nodes. Finally, at $T = 3$ the last sensor receives anchor information through the sensor nodes that calculated their position at $T = 2$

7.3 Distributed Network Error Propagation

Section 7.2 presented distributed algorithms that are based on ranging measurements using received signal strength or hop counts. The overall performance of the results in the cited papers showed an improvement, but it was also reported that some sensor nodes performed worse. The propagation of erroneous measurements was not considered. In this section, we present techniques that are based on passing positioning estimates (messages) to neighboring sensor nodes together with estimates of their reliability to help to reduce error propagation.

Message passing is rather new compared to the original concepts that build on time-based, signal strength based ranging. Message passing techniques build on distributed methods to infer from shared knowledge (Wymeersch et al. 2009). The link between sensors is used to exchange relevant positioning data. This exchanged data could contain more than the estimated position of the sensor node itself. It may include a time stamp, uncertainty information about the own estimate (parameterized, or estimated samples), noise estimates of the own estimate (uncertainty), ranging information from other neighboring nodes, etc. Figure 7.3 shows an example of sharing information by passing messages between sensor and anchor nodes. The question marks above the mobile sensor nodes represent not only their unknown positions but also the question where the information could be processed—either at the location server (which represents the central unit) or at the sensor node itself.

In the following, we present two message passing techniques, namely belief propagation and nonparametric belief propagation. Both methods differ by how the

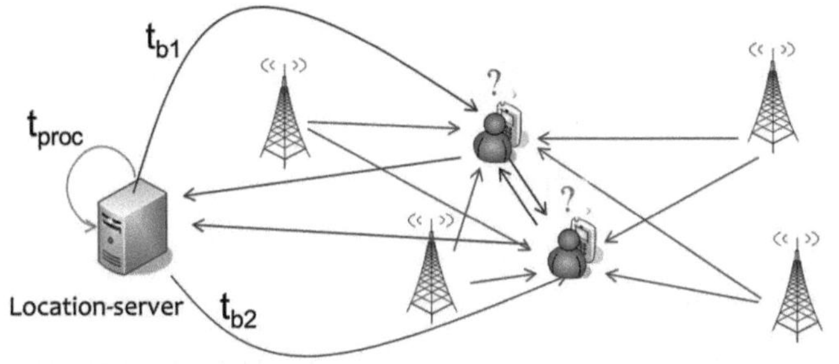

Fig. 7.3 Network structure for centralized (Location-server) cooperative or distributed (each node) cooperative positioning

information is shared. Belief propagation assumes that positioning estimates and their uncertainties could be exchanged based on simple values, such as the mean and the variance, that are based on a statistical distribution, e.g., the Gaussian distribution. However, the distribution function (Gaussian or any other) that could be assumed is not necessarily correct, and therefore nonparametric belief propagation (NBP) methodologies are presented. NBP methods can account for effects that cannot be easily represented by known distributions. Furthermore, as belief propagation relies on conditions that could cause loops, a subsection is devoted to the assessment of the correctness of the beliefs.

7.3.1 Belief propagation

A first example of distributed positioning is based on belief propagation (BP) (Yedidia et al. (2003)) which is a well-known graphical model for inference in statistical physics, artificial intelligence, computer vision, etc. Inference describes how existing knowledge about prior results is used to derive conclusions about the future. Section 6.6 introduced Monte Carlo localization that applies multiple observations, which also change in time, to process them using the posteriori probability. Monte Carlo localization is a probabilistic algorithm that can be applied centrally as described in Sect. 6.6.

BP is an efficient method to localize sensor nodes jointly with uncertainty about the location estimate. MCL represents the posterior probabilities by approximating the desired distribution through a randomly chosen set of weighted particles. MCL can be applied centrally and as the information is collected locally and might be processed at each sensor individually it is also a distributed algorithm. The posterior probability is called the belief $P(y_k|x_{0...k})$, where y_k is the state at time instance k, and $x_{0...k}$ represents the data starting at time instance $t = 0$ up to time instance $t = k$. The data of each time instance contains ranging information and a motion model of the object. The set of weighted particles represents the belief, and

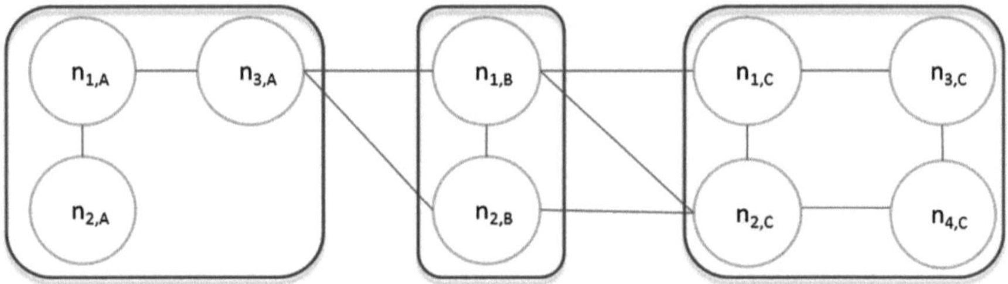

Fig. 7.4 Grouping of sensor nodes in group A (*left*), B (*middle*), C (*right*). The messages are exchanged through the links between the groups

therefore this method is also known as particle filtering (see also for more details in Chaps. 8 and 9 together with inertial (non-radio) sensors).

In their contribution, Ihler et al. (Sudderth et al. 2010; Ihler 2007) compared parametric belief propagation to nonparametric belief propagation. Savic et al. build on the work from Ihler et al. and investigated nonparametric belief propagation based on spanning trees (Savic et al. 2010) for loopy networks. They also collected experimental data to verify their theoretical assessment and their simulations. These publications will be used in the following to explain the concept and the ideas. Assume we have n_s sensor nodes distributed in a planar area spanned over two dimensions with location $\mathbf{x}_i$ (we omit the index S for sensor node in the following to ease the reading. However, where it is needed we will mention the explicit role of the node). The node i measures the distance between itself and node j with probability $P_0(\mathbf{x}_i, \mathbf{x}_j)$ and with this P_0 describes the probability of sensor node i detecting sensor node j. We assume a symmetric link—which is not generally true, e.g., if different frequency bands are used, but simplifies the following derivations. The distance measurement d_{ij} between sensor nodes in Eq. 7.4 is noisy, which is taken into account by $\alpha_{ij} \sim p(\mathbf{x}_i, \mathbf{x}_j)$:

$$d_{ij} = ||\mathbf{x}_i - \mathbf{x}_j|| + \alpha_{ij} \tag{7.14}$$

Furthermore, the sensor node i may use a prior distribution $p_i(\mathbf{x}_i)$ that contains either sufficient information, e.g., if it is an anchor node to resolve ambiguity problems or the information is not informative at all (just to initialize the system). Altogether a joint distribution for all n_s sensor nodes would be

$$p(x_1, \ldots, x_{n_s}, \{d_{ij}\}) = \prod_{(i,j)} p(d_{ij}|\mathbf{x}_i, \mathbf{x}_j) \prod_i p_i(\mathbf{x}_i) \tag{7.15}$$

under the assumption the link between sensor nodes i and j exists. The Eq. 7.15 is like the derivation in Sect. 6.6 about Monte Carlo localization. In the following we show how graphical models can help to enable simple algorithms to merge the factorization of the probability function e.g., in Eq. 7.15. A simple example is shown in Fig. 7.4 where multiple sensor nodes are grouped into three cliques A (left nodes: $n_{1,\mathrm{A}}, n_{2,\mathrm{A}}, n_{3,\mathrm{A}}$), B (middle group of nodes: $n_{1,\mathrm{B}}, n_{1,\mathrm{B}}$) and C (right group of nodes:

$n_{1,\mathrm{C}}, n_{2,\mathrm{C}}, n_{3,\mathrm{C}}, n_{4,\mathrm{C}}$). The collected information in each cliques forwarded based on the distribution. We denote a set of random variables $x_\mathrm{A} \in \mathrm{A},\ x_\mathrm{B} \in \mathrm{B}$ and $x_\mathrm{C} \in \mathrm{C}$ which are conditionally independent. We can use this relationship in the joint distribution:

$$p(x_\mathrm{A}, x_\mathrm{B}, x_\mathrm{C}) = p(x_\mathrm{B})p(x_\mathrm{A}|x_\mathrm{B})p(x_\mathrm{C}|x_\mathrm{B}) \tag{7.16}$$

If we want to obtain relative positions, given only the relative measurements d_{ij}, the sensor location may be solved up to an unknown rotation, translation, and ambiguity of the entire network. However, we would like to obtain the absolute coordinates; so, we need at minimum three anchor nodes (assuming perfect synchronized nodes). The prior distributions of an unknown and anchor node at position $\mathbf{x}_i^a$ are respectively given by:

$$p_i^{\text{unknown}}(\mathbf{x}_i) = \begin{cases} \dfrac{1\,[\mathrm{m}^2]}{\text{areasize}}, \\ 0, \quad \text{otherwise.} \end{cases} \tag{7.17}$$

$$p_i^{\text{anchor}}(\mathbf{x}_i) = \delta(\mathbf{x}_i - \mathbf{x}_i^a) \tag{7.18}$$

If the sensor nodes have no a priori information about their own position a uniform distribution is chosen as initial setup. On the other hand if the unknown nodes may obtain some prior information (e.g., high/low probable locations, dependences between unknown groups of nodes, etc.) it should be reflected in the Eq. 7.15 by biasing the prior distribution accordingly.

For large-scale sensor networks, it is reasonable to assume that only a subset of pairwise distances is available due to limited connectivity, i.e., only between sensors which are located within the same radio range R. A simple model of probability of detection is given by:

$$P_d(\mathbf{x}_i, \mathbf{x}_j) = \begin{cases} 1, & \text{for} \|\mathbf{x}_i - \mathbf{x}_j\| \le R, \\ 0, & \text{otherwise.} \end{cases} \tag{7.19}$$

Empirical approximations for $P_d(\mathbf{x}_i, \mathbf{x}_j)$ could be estimated using real experiments in the deployment area of interest or we may consider channel models for specific environments such as indoor or urban canyon together with information about obstruction of the line-of-sight path. Such information is also investigated in Chap. 3 that proposes several methods to resolve the wireless channel. We have to exchange information between the nodes which are not directly connected. We define a pair of nodes i and k to be one-step neighbors of one another. This pair of nodes observes a pairwise distance of d_{ik}. This is represented in Fig. 7.4 by each clique of sensor nodes A, B, or C. Then, we define two-step neighbors of node k to be all nodes i such that we do not observe the distance d_{ik}, but observe d_{kj} and d_{ji} for some node j. This could be the nodes in Fig. 7.4 between clique A and B, or B and C. The same pattern works for the three-step neighbors, and so forth. These n-step neighbors ($n > 1$) have information about the distance between them.

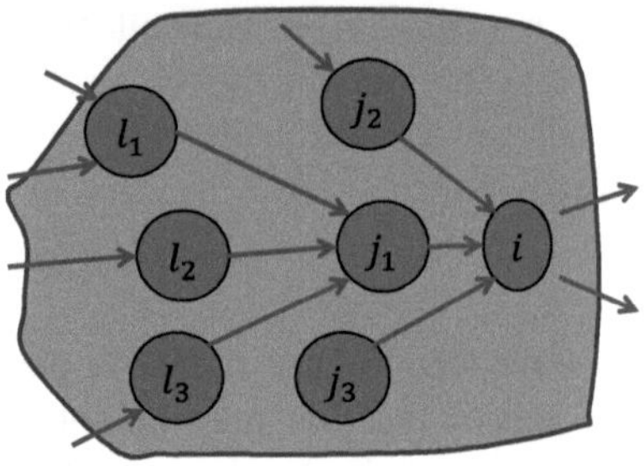

Fig. 7.5 Belief propagation by message passing between sensor nodes

Therefore, two nodes not observing their distance are too far apart. In the following, we will use only one-step and two-step neighbors. The relationship between the graph and joint distribution may be quantified in terms of potential functions ψ which are defined over each of the graph's cliques in:

$$p(x_1,\ldots,x_{N_j}) \propto \prod_{\text{cliques C}} \psi_C(\{x_i : i \in C\}) \tag{7.20}$$

We can define potential functions which can express the joint posterior distribution. This only requires potential functions defined over variables associated with single nodes and pairs of nodes. Single-node potential at each node i and the pairwise potential between nodes i and j, are respectively given by:

$$p(x_1,\ldots,x_{n_k},\{o_{ij}\},\{d_{ij}\}) = \prod_{i,j} p(o_{ij}|\mathbf{x}_i,\mathbf{x}_j) \prod_{i,j} p(d_{ij}|\mathbf{x}_i,\mathbf{x}_j) \prod_i p_i(\mathbf{x}_i) \tag{7.21}$$

In Eq. 7.21, we merge Eq. 7.15 with either Eq. 7.19 to take into account the existence of the link between the sensor nodes reflected by o_{ij}. We expect that some nodes have a higher probability to detect nearby neighbors, so the probability of detection P_d could be given. However, uncertainty in the measurement process such as physical barriers, multipath, and interference results in the fact that sometimes, especially in indoor scenarios, nearby sensors may still not be able to observe each other. Moreover, for the noise distribution $p_{\{\alpha\}}$, we choose the Gaussian distribution which represents an approximation of the real scenario. Finally, the joint posterior distribution is given by:

$$p(x_1,\ldots,x_{N_j}|\{o_{ij},d_{ij}\}) \propto \prod_i \psi_i(\mathbf{x}_i) \prod_{i,j} \psi_{ij}(\mathbf{x}_i,\mathbf{x}_j) \tag{7.22}$$

To estimate the posterior marginal distributions for each node i:

$$p(\mathbf{x}_i|o_{jl},d_{jl}) = \int p(x_1,\ldots,x_{N_u}|o_{ij},d_{ij})\mathrm{d}x \tag{7.23}$$

Having defined a graphical model, we can now estimate the sensor locations by applying the belief propagation (BP) algorithm. The form of BP as an iterative local message passing algorithm makes this procedure trivial to distribute among

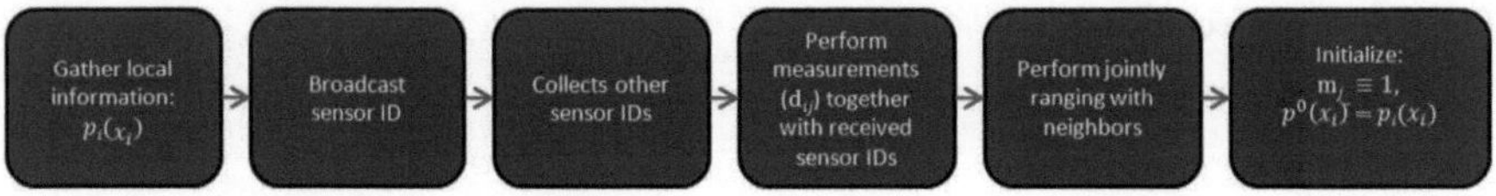

Fig. 7.6 Initialization of the BP algorithm

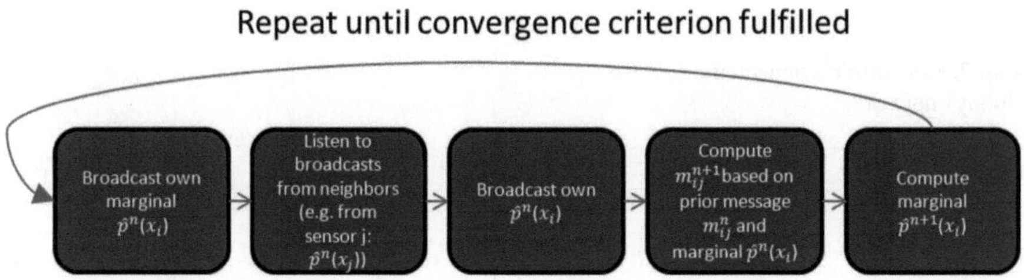

Fig. 7.7 Belief propagation for sensor nodes that is repeated till it fulfills a convergence criterion such as an achieved accuracy uncertainty

the wireless sensor nodes as shown in Fig. 7.5. Figures 7.6 and 7.7 describe the initialization and computation phase of this algorithm.

We apply BP to estimate each sensor's posterior marginal, and use the mean value of this marginal and its associated uncertainty to characterize sensor positions. Each node i computes a quantity known as its belief $M_i^k(\mathbf{x_i})$, which is the posterior marginal distribution of the position $\mathbf{x_i}(\mathbf{x_i^r}, \mathbf{x_i^k})$ at iteration k, by taking a product of its local potential ψ_i with the messages from its set of neighbors G_i (all nodes in range R):

$$M_i^k(\mathbf{x}_i) \propto \psi_i(\mathbf{x}_i) \prod_{j \in G_i} m_{ji}^k(\mathbf{x}_i) \tag{7.24}$$

The messages m_{ij}, from node i to node j, are computed by:

$$m_{ij}^k(\mathbf{x}_j) \propto \int \psi_{ij}(\mathbf{x}_i, \mathbf{x}_j) \frac{M_i^{k-1}(\mathbf{x}_i)}{m_{ji}^{k-1}(\mathbf{x}_i)} \mathrm{d}\mathbf{x}_i \tag{7.25}$$

The messages m_{ij}^k after the kth iteration are exchanged or broadcasted to the neighboring nodes. To obtain distance measurements, each sensor has to broadcast its ID and to listen for others sensor nodes broadcasts. For any received sensor node ID, each sensor (except anchors) has to estimate the distance to it. In case of $d_{ij} \neq d_{ji}$, the sensor has to communicate with its observed neighbors in order to symmetrize the distance measurements such that $d_{ij} = d_{ji} = \frac{d_{ij}+d_{ji}}{2}$. For tree-like network structures, the number of iterations should be at most the length of the longest path in the tree. However, it usually runs until all unknown nodes obtain information from a minimum of three non-collinear anchor nodes.

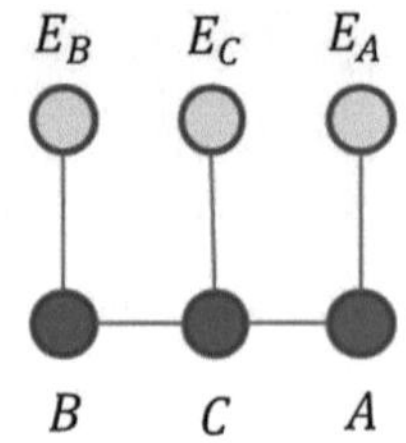

Fig. 7.8 Single connected network

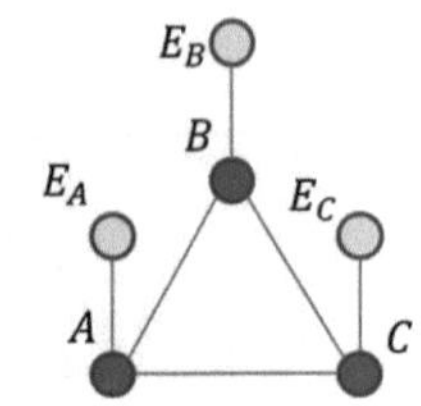

Fig. 7.9 Multiple connected (loopy) network

7.3.2 Correctness of Belief Propagation: Double Counting Problem

Each node may refer to his own achieved accuracy uncertainty. If it is not known by the exchanging sensor node it could be misinterpreted. Different convergence criteria for the different nodes may cause that the same information is taken into account multiple times as it is differently interpreted by the different nodes. In a network topology that is not known in advance this may provokes loops.

Figure 7.8 shows a single connected network with three unknown nodes (*A*, *B* and *C*) and three anchor nodes (E_A, E_B, E_C) which represent the local evidence. Such evidence is e.g., an illustration of the locally computed position estimate. The message-passing algorithm (belief propagation) can be thought of as a way of communicating local evidences between nodes such that all nodes calculate their beliefs given all the evidence. Node *B* receives a message from node *A*, and forwards this to node *C*. Vice versa node *B* sends a message to *A* from node *C* without any information from *A*. Figure 7.9 shows a multiple connected network with a loop. Node *A* sends information to node *B* and node C sends in the next iteration the same information to node *A*. Here, double counting cannot be avoided.

In order for BP to be successful, we need to correctly count or avoid any double counting (Pearl 1997; Mooij and Kappen 2007; Weiss 2000; Weiss and Freeman 2001)—this is important in a situation in which the same evidence is passed multiple times around the network and mistaken for new evidence. Therefore, double counting is not possible in a single-connected network. When a node receives evidence, it can not receive the same evidence again. However, in a loopy network such as in Fig. 7.9 double counting cannot be avoided. BP could still lead to correct inference if the same evidence is "double counted" in equal amounts. This could be solved by building an unwrapped network that replicates the

Fig. 7.10 Unwrapped network after the first iterations

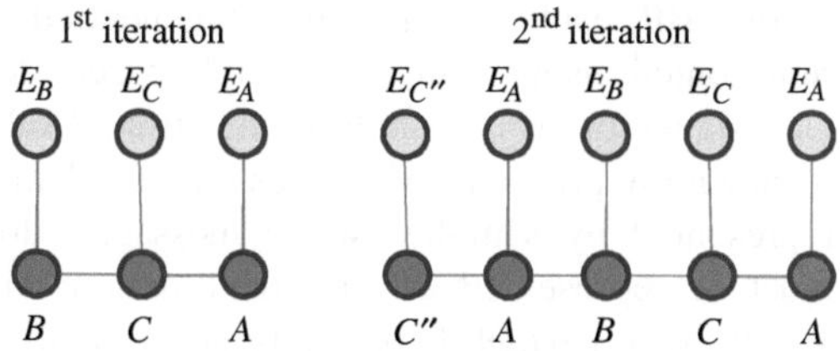

evidence and the transition matrices while preserving the local connectivity of the loopy network. Figure 7.10 shows the first two iterations of an unwrapped loopy network. The message received by node B after several iterations of BP in the loopy network is identical to the final messages received by node B in the unwrapped network. Thus, we can create an infinite network. The unwrapped network, e.g. in Fig. 7.10, is single connected, and therefore BP delivers correct beliefs. To make use of the beliefs requires that the probability distributions that are induced by the loopy problem are similar. In a single-loop network, the similarity is given after several iterations (Fig. 7.10). Generally, BP converges if additional sensor nodes do not change the posterior probability of the already existing nodes in the center.

Another method is proposed by Yedidia et al. to assess the correctness of BP through Bethe approximation. The authors proved that for a single-connected network it is possible to check the correctness, but for loopy networks it is an approximation. The reader is referred to their publication (Yedidia et al. 2005).

7.3.3 Non-parametric Belief Propagation

In the following section, we outline non-parametric belief propagation. In a wireless positioning system not all parameters are Gaussian distributed. Therefore, if we consider that parameters have nonlinear relationships and non-Gaussian uncertainties sensor localization by BP is undesirable. Particle filter (Ristic et al. 2004) based approximation of BP, called nonparametric belief propagation (NBP) (Sudderth et al. 2010; Savic et al. 2010), and enables the application of BP to inference in sensor networks as well. To briefly introduce a particle filter we assume that $\mathbf{x}_i$ is the variable to be estimated and $\mathbf{y}_\mathbf{i}$ is its observation. The sequence $x_0, \ldots, x_N$ is a Markov process and the observations $y_0, \ldots, y_N$ $(i = 1, \ldots, N)$ are independent. We build this on the following approximation $p(x) \approx \sum_{i=1}^{N} w^i \delta(x - x^i)$ and it builds on an iterative update of the measurements (w^i, x^i) recursively using the Bayesian rule $p(\mathbf{x}_i|y_{\{1:i-1\}}) = \int p(\mathbf{x}_i|x_{\{i-1\}})p(x_{\{i-1\}}|y_{\{1:i-1\}})\mathrm{d}x_{\{i-1\}}$.

In NBP, each message that is generated and communicated (broadcasted) is represented using either a sample-based density estimate (e.g. a mixture of Gaussians) or as an analytic function. Both types are needed for the sensor localization problem. Messages observed along the direct links (one-step) are represented by samples, while messages along unobserved links (two-step,...) must be represented as analytic functions since their potentials have the form $1 - P_d(\mathbf{x}_i, \mathbf{x}_j)$ which is typically not normalizable, as e.g., each sensor node may use unknown individual levels for received signal strength measurements. The belief and message update equations are performed using stochastic approximations in two stages: first, drawing samples from the belief $M_i^k(\mathbf{x}_i)$, then using these samples to approximate each outgoing message m_{ij}^k. We discuss each of these steps in the following. Given N weighted samples $\{W_i^{k,l}, X_i^{k,l}\}$ from the belief $M_i^k(\mathbf{x}_i)$ obtained at iteration k, we can compute a Gaussian mixture estimate of the outgoing BP message m_{ij}^k. We first consider the case of observed edges between unknown nodes. The distance measurement d_{ij} provides information about how far sensor j is from sensor i, but no information about its relative direction. To draw a sample of the message $\left(X_{ij}^{l,k+1}\right)$, given the sample X_i^l which represents the position of sensor i, we simply select a direction θ at random, uniformly in the interval $[0, 2\pi]$. We then shift X_i^l in the direction of θ by an amount which represents the estimated distance between nodes j and i $(d_{ij} + \alpha^l)$:

$$\begin{aligned} x_{i,j}^{k,l+1} &= X_i^{l,k} + (d_{ij} + v^l)\left[\sin(\theta^{l,k})\cos(\theta^{l,k})\right], \\ \theta^{l,k} &\sim U[0, 2\pi) \text{ and } \alpha^l \sim p_a \end{aligned} \quad . \tag{7.26}$$

$U[0, 2\pi]$ represents the uniform probability density function in the interval from 0 to 2π, and α^l is the measurement noise with distribution p_a (e.g., Gaussian). We can now calculate the weight of this sample $\left(w_{ij}^{l,k+1}\right)$ using Eq. 7.24, Eq. 7.25, the kernel density estimate (KDE) of potential function ψ, and reasonable approximation of this kernel function with delta impulse:

$$m_{ij}^{k+1} \propto \sum_j w_{ij}^{l,k+1} k_{ij}(\mathbf{x}_j - \mathbf{x}_{ij}^{l,k+1}) \tag{7.27}$$

Using last equation and the fact that this message exists only if there is detection between these two nodes (with probability P_d), the weight of sample $\mathbf{x}_{ij}^{l,k+1}$ is given by:

$$w_{ij}^{l,k+1} = \frac{P_d\left(\mathbf{x}_i^{k,l}, x_j\right) W_i^{l,k}}{m_{ji}^k(\mathbf{x}_i^{k,l})} \tag{7.28}$$

The optimal value for bandwidth h_{ij}^{k+1} could be obtained in a number of possible techniques. The simplest way is to apply the "rule of thumb" estimate (Silverman 1986):

$$h_{ij}^{k+1} = N^{-\frac{1}{3}} \mathrm{var}\left(\mathbf{x}_{ij}^{k+1}\right) \tag{7.29}$$

An important modification to this procedure can be used to improve accuracy and computation cost. Our goal is to accurately estimate belief in the regions of the state space in which it has significant probability mass. A reasonable distribution is one which allows us to accurately estimate the portions of the message m_{ij} which overlap these regions of the state space. Our additional information involves utilizing previous iterations' information to determine the angular direction to each of the neighboring sensors. In particular, we use samples from the marginal distribution computed in the previous iteration to form the relative direction θ:

$$\theta^{l,k} = \arctan\left(X_j^{l,k} - X_j^{l,k}\right), \; k > 1, \; \theta^{k,l} \in [-\pi, \pi] \tag{7.30}$$

Therefore, in the first iteration ($i = 1$) we calculate θ. This additional information increases the accuracy of this algorithm. The next task is to obtain messages from anchor nodes to unknown nodes (only observed edges). It could be done using previous procedure (each sample of the belief would be placed at the known location of the node and weighted by $1/N$), but it will increase computation and communication cost. Only the location of the anchor node $(\mathbf{x}_i^*)$ is used to calculate the analytic form of the message. Therefore, this message is proportional to the potential function which is constant over the iterations and depends only on the location of the unknown node $\mathbf{x}_j$ given by

$$m_{ij}^{k+1}(\mathbf{x}_j) \propto \psi_{ij}(\mathbf{x}_i^*, \mathbf{x}_j). \tag{7.31}$$

The messages along unobserved edges are represented by an analytic function. With the probability of detection P_d, and samples from the belief M_i^k, an estimate of the outgoing message to node j is given by:

$$m_{ij}^{k+1}(\mathbf{x}_j) = 1 - \sum_j W_i^{l,k} P_d(X_i^{l,k}, \mathbf{x}_j) \tag{7.32}$$

Then, the messages from the anchor nodes $\left(W_i^{l,k} = \frac{1}{N}\right)$ are given by

$$m_{ij}^{k+1}(\mathbf{x}_j) = 1 - P_d(x_i^*, x_j) \tag{7.33}$$

To estimate the belief $M_j^{k+1}(\mathbf{x}_j)$ using Eq. 7.23, we draw samples from the product of several Gaussian mixture and analytic messages. It is difficult to draw samples from this product. Therefore, we use a proposal distribution, sum of the Gaussian mixtures, and then reweigh all samples. This procedure is well known as mixture importance sampling. Denote the set of neighbors of j, having observed links to j and not including anchors, by G_j^0, and the set of all neighbors by G_j. In order to draw N samples, we create a collection of k_sN weighted samples (where $k_s \geq 1$ is a parameter of the sampling algorithm) by drawing k_sN/G_j^0 samples from each message m_{ij} with $i \in G_j^0$ and assigning each sample a weight equal to the ratio:

$$W_j^{l,k+1} = \prod_l m_{lj}^{k+1} / \sum_{l \in G_j^0} m_{lj}^{k+1} \tag{7.34}$$

Some of these calculated weights are much larger than the rest, especially after more iterations. This means that any sample-based estimate will be unduly dominated by the influence of a few of the particles, and the estimate could be erroneous. To avoid this, we draw N values independently from collection with probability proportional to their weight, using resampling with replacement. This means that we create N equal-weight samples drawn from the product of all incoming messages. A node is located when a convergence criteria is met, e.g., the Kullback–Leibler (KL) divergence can be used, a common measure of difference between two distributions (Ihler 2007). As we already mentioned, the BP/NBP convergence is not guaranteed in a network with loops (Pearl 1997) or even with convergence, it could provide us less accurate estimates. Savic and Zazo (2009) applied NBP for localization and showed that there was no convergence problem, but the accuracy was sometimes dissatisfying.

Nonparametric belief propagation is the next step compared to belief propagation as it also takes into account non-Gaussian dependencies of different effects. This applies especially for distributed algorithms that will be executed on low-complex sensor nodes. With this, further advantages like energy efficiency and communication overhead can be limited. A key argument would also be the low latency that is especially of interest in dynamic scenarios.

In the next section, we outline link selection algorithms that censor either already at the transmitter or at the receiver. These techniques would play in favor of the presented message passing algorithms to limit their complexity.

7.4 Link Selection

In this section, we outline censoring techniques to reduce the number of wireless links between nodes at the transmitter or at the receiver. Links contain information and are generally beneficial for performance, but significantly increase the complexity. If a minimum number of links exists for a node, it depends on the requested location accuracy to decide on using additional links. Furthermore, a link may interfere with other links to alleviate the estimated performance of the nodes connected by the link. This is especially of interest in dense wireless sensor networks or cellular mobile radio networks in urban or indoor environments such as shopping malls.

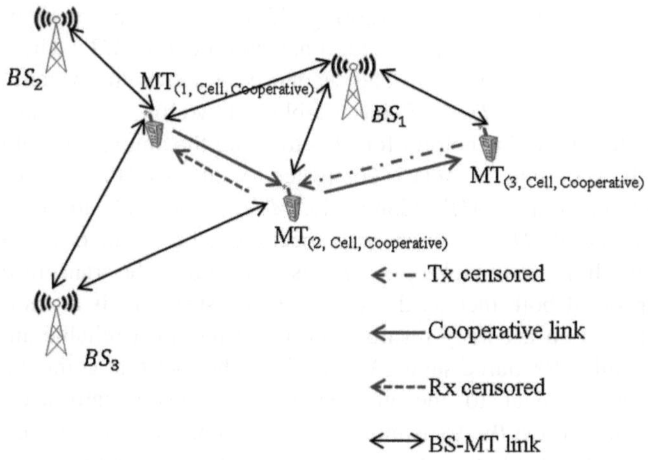

Fig. 7.11 Censoring schemes either on the transmitter (*Tx*) or on the receiver (*Rx*) side

Therefore, we are looking for criteria and methods to perform a wise link selection. The actual impact of location methods in communications, in particular within cooperative schemes, is crucial for feasibility. Recent studies based on event driven packet-oriented tools evaluated the effects of cooperative approaches within WiFi based or IR-UWB based positioning under realistic network deployment and system architectures (e.g., (WHERE project partners 2009)). These studies have pointed out the tight tradeoff that exists between latency and overhead (caused by cooperation) and between positioning accuracy and robustness (demanded by applications). As an example, for a cooperative Wi-Fi positioning system an increase of the number of mobile terminals increases the throughput overhead and measurement delays also increase significantly. Furthermore, it has been illustrated in IR-UWB mesh networks (Denis et al. 2009) that blind (non-exhaustive or unlisted) update scheduling in jointly cooperative distributed iterative positioning and ranging has a negative impact on overhead and latency. Hence, these results clearly emphasize the current needs for more efficient cross-layer strategies that more carefully account for the actual cost of cooperation, while still benefiting from positioning performance enhancements (e.g., from information redundancy, spatial diversity, Euclidean graph rigidity, etc.).

When the target to be localized has more than the minimum sufficient number of reference nodes available, a link selection procedure may be applied either at the transmitter or the receiver. Figure 7.11 depicts a scenario with several base stations that act as anchors, and mobile terminals that act as sensor nodes. The three MTs have different positions and different links to BSs. MT_1 has a connection to three BSs and it is expected that this MT estimates its own position well. MT_2 has only two links to BSs and therefore, benefits from the cooperative link to MT_1. MT_3 has only two links, and only one link to a BS. Das and Wymeersch (2012) use this scenario as their basic scenario to differentiate three different cases.

The first case is called receive censoring: MT_1 receives information from three different BSs and does not require additional support from MT_2. The link $MT_1 \rightarrow MT_2$ is censored at the receiver. The second case shows the benefit for MT_2 of using the reverse link to MT_1. MT_2 has established two links to BSs and a third link to MT_1. The link to $MT_1$is helpful for MT_2 to avoid the ambiguity that occur when using only the BSs. Finally, MT_3 is connected to only one BS, and therefore relies on the additional link to MT_2. Conversly, MT_2 is not likely to benefit from the position estimate of MT_3 which cannot expect a good estimate of its own position. Therefore, the link between $MT_3 \rightarrow MT_2$ is censored at the transmitting node.

The purpose of both increased accuracy and restrictions is to limit the use of resources. Increased accuracy results from using the most reliable and geometrically useful links. Resource saving comes from the fact that computational complexity is proportional to the number of links taken into account during information fusion. On the transmitter side, the improvement for the overall network is more advantageous as transmit signal and processing power are reduced. Also, the additional spectrum becoming available is a limited resource that can be used elsewhere. Therefore, link selection and censoring offer solutions for very dense scenarios. Lieckfeldt et al. (2008) investigated the use of the CRLB at each node to decide if a link should censored or not. Das and Wymeersch (2012) proposed to apply their sum-product (SPAWN) algorithm that uses belief propagation, saying that each node estimates its own belief and uncertainty of its own position. Together with a threshold, the sensor node decides whether the transmit link should be censored or not. This prevents unreliable position estimates to be broadcasted. Therefore, the node is unaware of its neighbors and the authors (Das and Wymeersch 2012) called the scheme neighbor-agnostic transmit censoring. On the other hand, when the node is aware that the broadcast of its own estimate is unhelpful for neighboring nodes, it censors its own transmission. The authors call this neighbor-aware transmit censoring.

In the paper from Lieckfeldt et al. (2008), the geometrical impact has been addressed and compared to the approach of using the closest reference nodes. However, most of these selection schemes have been proposed for centralized localization. Recent approaches address distributed and cooperative scenarios, and the selection criteria are mainly based on theoretical localization performance limits such as CRLB (Denis et al. 2009; Das and Wymeersch 2012; Raulefs et al. 2012). In Denis et al. (2009), unreliable links are successively discarded based on CRLB analysis during the first phase—connectivity based coarse positioning. In this way, resources are saved as the number of packet exchanges in the refined ToA-based ranging phase is reduced.

Other references that apply censoring are in Kaplan (2006a, b). The methods discussed there are based on minimization of mean square error (MSE). Utility of each set consists of N nodes is defined as the reciprocal of the mean square error.

To conclude, censoring is especially of interest in cooperative positioning networks when the density of sensor nodes is high and some sensor nodes will be out of reach to any of the anchor nodes. Furthermore, the complexity is kept low by using distributed algorithms, and the performance is maximized where sensor nodes perform their own calculations. Latter is an important issue for large scale, dynamic networks that cannot be controlled centrally. Therefore, the scalability of such link selection methods together with the positioning algorithms is an important ingredient.

7.4.1 Practical Application: Firefighters

In the following, we report from a publication that presents practical constraints of a cooperative positioning system using an extended Kalman filter (EKF). Chapter 9 presents the EKF in detail and also details about inertial sensors. Wu et al. (2009) considered a scenario of a group of sensor nodes that moves jointly through an unknown territory with limited support of anchor nodes. They considered the case that none, one or two anchor nodes may be available from time to time. The group of sensor nodes exchange regularly information of their inertial sensors, such as motion sensor (accelerometers) and rotation sensors (gyroscopes), and the sensor nodes perform ranging between themselves. The number of participating sensor nodes is n_s and defines the magnitude of the teamwork effect.

The proposed use case is a group of firefighters that enters a building with an unknown structure (a map is unavailable or the structure is changed because of the fire). The control center of the firefighters requires following and maybe controlling the movements of the firefighters. A GPS device is not an adequate solution to estimate the position of the firefighters as GPS not necessarily operates in such environments. Furthermore, in time of need only relative positioning information is maybe sufficient to position a firefighter. Contrary to a wireless sensor system, that performs trilateration and needs four anchors, this system claims anchors are not always mandatory. The authors proposed an EKF as an appropriate algorithm because the real-time requirements prohibit a more complex implementation (constraints are namely battery consumption and weight, and processing power). The EKF algorithm performed well compared to the optimized solution. The authors focused their investigations on three aspects. The size of the team is defined by the number of sensor nodes. The system benefits from the number of sensor nodes as the drift rate reduces by a factor of $1/\sqrt{(n_s)}$. This is defined as the teamwork effect. Furthermore, a second effect that is reported is the "reset effect". The reset effect describes a situation where the position estimates of the sensor nodes improve significantly or abruptly instead of gradually. The conditions that change in such a situation are, e.g., a sensor node connects to another sensor node with higher certainty of its own position estimate. Such a situation is for example the merger of two groups of sensor nodes which cause an abrupt improvement. The third reported effect is named the anchor effect, where adding an anchor significantly reduces the absolute positioning error by constraining ambiguity. Therefore, each time an anchor node entered the scenario the resulting effect is called the anchor effect.

7.5 Conclusions

In this chapter, we presented distributed algorithms that are investigated for cooperative positioning. The Cramer–Rao lower bound is a tool that was considered to estimate the positioning performance based on:

- Geometric constellation of the sensor nodes to each other
- Synchronization uncertainties between the sensor nodes
- Variable conditions of the wireless channel
- Heterogeneous networks (terrestrial communication network with peer-to-peer links and satellite navigation system (e.g. GPS))

Besides ranging between sensor nodes position estimates of the neighboring sensor nodes may be exchanged. Message passing algorithms based on belief propagation (parametric or non-parametric) offer feasible solutions as they naturally fit distributed network architectures. However, for distributed, changing and also unknown network topologies, any shared information could be considered multiple times inadvertently without knowing. The presented correcting method weights such information properly. High density mobile networks without any access coordination suffer from the interfering links. Therefore, for the scalability of positioning algorithms a key ingredient are the presented link selection methods. Local link selection reduces the active links to relevant links. The presented link selection methods use the geometrical constellation between the nodes and the exchanged messages between the nodes.

References

B. Denis, M. Maman, L. Ouvry, On the scheduling of ranging and distributed positioning updates in cooperative IR-UWB networks, in *International Conference on Ultra-Wideband.* IEEE, 2009, pp. 370–375

K. Das, H. Wymeersch, Censoring for Bayesian co-operative positioning in dense wireless networks. IEEE J. Sel. Areas Commun. **30**, 9 (2012) (Special issue on Cooperative Networking—Challenges and Applications)

J. Figueiras, Accuracy Enhancements for Positioning of Mobile Devices in Wireless Communication Networks. Ph.D. Dissertation, Aalborg University, Jan 2008

F. Penna, M.A. Caceres, H. Wymeersch, Cramer-Rao bound for hybrid GNSS-Terrestrial cooperative positioning. Commun. Lett. **14**(11), 1005–1007 (2010)

A. Ihler, Accuracy bounds for belief propagation, in *Proceedings of the Twenty-Third Conference on Uncertainty in Artificial Intelligence,* 2007

J.S. Yedidia, W.T. Freeman, Y. Weiss, in *Understanding belief propagation and its generalizations,* ed. by G. Lakemeyer, B. Nebel Exploring artificial intelligence in the new millennium (Morgan Kaufmann Publishers Inc., CA, 2003)

L.M. Kaplan, in *Global Node Selection for Localization in a Distributed Sensor Network*, ed. by IEEE. Transactions on Aerospace and Electronic Systems, Jan 2006a, pp. 113–135

L.M. Kaplan, in *Local node selection for localization in a distributed sensor network,* ed. by IEEE. Transactions on Aerospace and Electronic Systems, Jan 2006b, pp. 136–146

S.M. Kay, *Fundamentals of Statistical Signal Processing Estimation Theory*, vol 1 (Prentice Hall, NJ, 1993)

E.G. Larsson, Cramer-Rao Bound Analysis of Distributed Positioning in Sensor Networks. *Signal Processing Letters*, Mar 2004, pp. 334–337

D. Lieckfeldt, J. You, D. Timmermann, in *Distributed Selection of References for Localization in Wireless Sensor Networks.* Workshop on Positioning, Navigation and Communication (WPNC), 2008, pp. 31–36

J.M. Mooij, H.J. Kappen, Sufficient conditions for convergence of the sum product algorithm. IEEE Trans. Inf. Theory **53**(12), 4422–4437 (2007)

N. Alsindi, K. Pahlavan, Cooperative localization bounds for indoor ultra-wideband wireless sensor networks. Eurasip. J. Adv. Signal Process. **2008**(1), 852509, (2008). doi10.1155/2008/852509

D. Niculescu, B. Nath, in *Ad hoc positioning system (APS).* Global Telecommunications Conference, 2001 (GLOBECOM '01). IEEE, 2001. pp. 2926–2931

D. Niculescu, B. Nath, DV based positioning in Ad hoc networks. Telecommun. Syst. **22**(1), 267–280 (2003)

J. Pearl, *Probabilistic Reasoning in Intelligent Systems*: *Networks of plausible inference* (Morgan Kaufmann, CA, 1997)

B. Ristic, S. Arulampalam, N. Gordon, *Beyond the Kalman filter—particle filters for tracking applications* (Artech House, London, 2004)

R. Raulefs, S. Zhang, C. Mensing, Bound-based spectrum allocation for cooperative positioning. Trans. Emerging Tel. Tech. (Wiley, 2012). doi:10.1002/ett.2572

Z. Sahinoglu, S. Gezici, in *Enhanced position estimation via node cooperation.* IEEE International Conference on Communications (ICC). IEEE, 2010, pp. 1–6

C. Savarese, J.M. Rabaey, J. Beutel, in *Locationing in distributed Ad-Hoc wireless sensor networks.* International Conference on Acoustics, Speech and Signal Processing (IEEE, Salt Lake City, 2001)

V. Savic, S. Zazo, in *Sensor localization using generalized belief propagation in network with loops.* 17th European Signal Processing Conference EUSIPCO (Glasgow, UK, 2009), pp. 75–79

V. Savic, A. Poblacion, S. Zazo, M. Garcia, Indoor positioning using nonparametric belief propagation based on spanning trees. EURASIP J. Wirel. Commun. Netw. **1**, 1687–1699 (2010)

A. Savvides, W. Garber, R. Moses, M.B. Srivastava, An analysis of error inducing parameters in multihop sensor node localization. IEEE Trans. Mobile Comput. **4**(6), 567–577 (2005)

B.W. Silverman, *Density Estimation for Statistics and Data Analysis* (Chapman and Hall, London, 1986)

E.B. Sudderth, T.I. Alexander, I. Michael, Nonparametric belief propagation. Commun. ACM **53**(10), 95–103 (2010)

Y. Weiss, W.T. Freeman, Correctness of belief propagation in Gaussian graphical models of arbitrary topology. Neural Comput. **13**, 2173–2200 (October 2001)

Y. Weiss, Correctness of local probability propagation in graphical models with loops. Neural Comput. **12**, 1–41 (2000)

WHERE project partners. Deliverable D2.2: Cooperative Positioning (Intermediate Report). 8 Mar 2009. http://www.kn-s.dlr.de/where/public_documents_deliverables.php#

S. Wu, K. Jim, S.-C. Mau, T. Zhao, in *Distributed Multi-Sensor Fusion for Improved Collaborative GPS-denied Navigation.* ION 2009 International Technical Meeting (Anaheim, USA, 2009), pp. 109–123

H. Wymeersch, J. Lien, M.Z. Win, Cooperative localization in wireless networks. Proc. IEEE **97**(2), 427–450 (2009)

J.S. Yedidia, W.T. Feeman, Y. Weiss, Constructing free energy approximations and generalized belief propagation algorithms. IEEE Trans. Inf. Theory **51**(7), 2282–2312 (2005)

S. Zhang, *Distributed Cooperative Positioning for Next Generation Mobile Radio Systems*, master thesis, (Technical University of Munich, Munich, 2011)

Chapter 8
Inertial Systems

The sensors available for tracking systems can be loosely broken into groups based on the measurement reference frame: (1) idiothetic or body reference sensors such as inertial sensors and encoders which measure motion of the body in the body frame; and (2) allothetic sensors or external reference sensors such as magnetic field, processed GPS and pressure which measure heading, 2D location and elevation in the global or Earth frame, and local reference sensors which measure ranging or bearing to local reference points/landmarks (Chaps. 2, 9).

In this chapter, we focus on the use of inertial idiothetic sensors as part of a pedestrian location and tracking system in GPS denied areas. Inertial measurements are differential measurements in the sense that they quantify changes in speed or direction. The two primary types of inertial sensors are accelerometers and gyroscopes. Accelerometers measure instantaneous changes in speed, or equivalently force, and gyroscopes provide a fixed frame of reference with which to measure orientation or equivalently change in direction. As such, any navigation information obtained from the sensor system is used to compute movement relative to the starting location and orientation. A navigation system that estimates current position from a prior known position and measurements of motion (for example, speed and heading) and elapsed time is called a *dead reckoning system.* A serious disadvantage of relying on dead reckoning for navigation is that the errors of the process are cumulative without other externally referenced corrections such as GPS, compass, or landmark-based corrections.

Nevertheless, dead reckoning techniques have been used for decades by the Department of Defense, NASA, and others for sophisticated navigation systems (Titterton and Weston 2004). These systems have relied on very high quality mechanical, fiber optic, or laser sensors with size, weight, power, and cost beyond the range of what is needed for consumer applications.

Table 8.1 gives an idea of the error growth performance requirements for different classes of inertial measurement units (IMUs). The limits on error growth set requirements on the types of sensors that can be used.

The ability to fabricate microelectromechanical systems (MEMS) using semiconductor device fabrication technologies (Ghodssi and Lin 2011) has lead to the

C. Gentile et al., *Geolocation Techniques*, DOI: 10.1007/978-1-4614-1836-8_8,

Table 8.1 Inertial measurement unit classifications

Class	Error growth limit	Gyro type/bias	Accelerometer type/bias
Military grade	1 NM/24 h	Mechanical, electrically suspended gyro (ESG) 0.005°/h	Servo accelerometer <30 μg
Navigation grade	1 NM/h	Ring laser gyro (RLG), fiber optic gyro (FOG) 0.01°/h	Servo or vibrating beam 50 μg
Tactical	<10 NM/h	RLG, FOG 1°/h	Servo, vibrating beam, MEMS 1 mg
Consumer	>10 NM/h	MEMS >1°/h	MEMS >1 mg

development of low size, weight, power, and cost MEMS inertial sensors. Using these consumer grade navigation sensors that have only recently become an option in cell phones, new data is available that can be leveraged to improve location accuracy for on foot personnel in GPS denied or degraded areas. Already cell phone applications include the ability to enhance location using cell carrier location services (cell tower triangulation—Chap. 5) and Wi-Fi (provided by Skyhook and now Apple, Google, and others—Chap. 4).

The motivation for this chapter is to review how the new consumer grade sensors can be used to improve pedestrian tracking. The chapter is organized as follows: First, we discuss some of the limitations of GPS as a sensor for pedestrian tracking. Then, we discuss MEMS inertial sensors and review some of the standard computations that are used with general inertial navigation sensors. Pedestrian tracking is unique relative to vehicle tracking where dynamic models are available and control inputs are known, so next we review specific approaches to implementation of pedestrian navigation systems.

Heading errors are the largest source of error in inertial pedestrian navigation systems so we divert the discussion from inertial sensors alone to include a discussion of the use of magnetic sensors for heading correction. It is important to note that using magnetic sensors for heading correction indoors is not straight forward because of the common occurrence of large magnetic disturbances caused by electrical systems and the building structure itself.

We conclude the chapter with a discussion of accuracy metrics. It is typical to see inertial navigation system error quoted as a percent of distance travelled; however, this can be very misleading as it does not account for heading errors which are typically the largest source of error.

8.1 Limitations of GPS for Pedestrian Tracking

The impression that GPS can be used for accurate personnel location is not true. For example, the GPS systems used in cars, correct locations to known road maps, thus, giving the appearance of high accuracy. Because of this many users are unaware that GPS location data can be quite poor and deteriorate significantly near

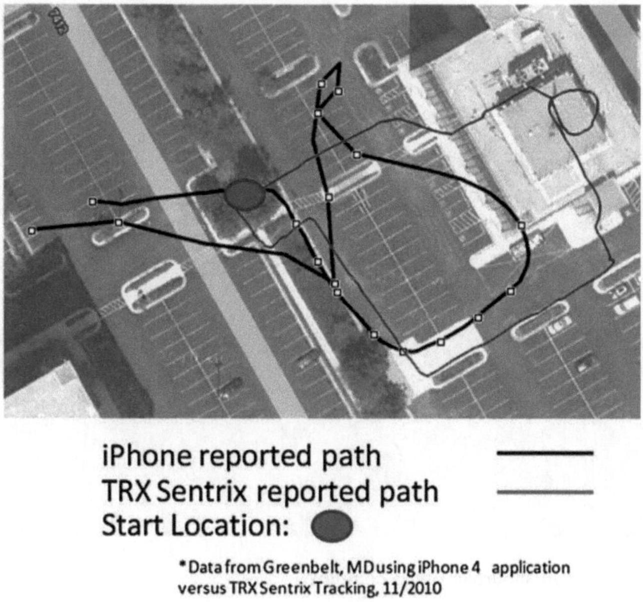

Fig. 8.1 GPS path (*red*) versus fused path estimate (*blue*)

even isolated buildings. Figure 8.1 shows a pedestrian path taken near a relatively isolated building where one might expect accurate GPS location estimates. The blue path, which fuses inertial sensor data (from the TRX waist worn inertial navigation unit–INU), provides an accurate representation of the true path; the red GPS only path shows significant error.

Another issue with GPS is that it does not have consistent performance across time for a given location. Even for a similar path type the errors are a function of the atmospheric conditions and the visible satellites at that time, and they can vary significantly. Examples of several trials taken near our Greenbelt office building with Android cell phone GPS are shown below. Test results are included here to clarify the need for enhancing tracking with other sensors in urban and suburban areas. The course walked is shown in Fig. 8.2. The course was chosen such that a large part of the path was away from the office building where we would expect GPS locations to be "good". Another part is near or inside the office building, where we would expect GPS locations to be poor or nonexistent.

The path error results are computed by comparing each GPS point that comes in with the ground truth location.[1] When GPS data are missing, there is no error reported.

The reported paths for trials 1 and 3 are shown in Figs. 8.3 and 8.4 respectively.

[1] The ground truth location is obtained by pinning the drift compensated inertial path at each of several surveyed marker locations to the surveyed marker location, and then interpolating the inertial tracks between the surveyed marker points. Inherent in this method is an assumption that

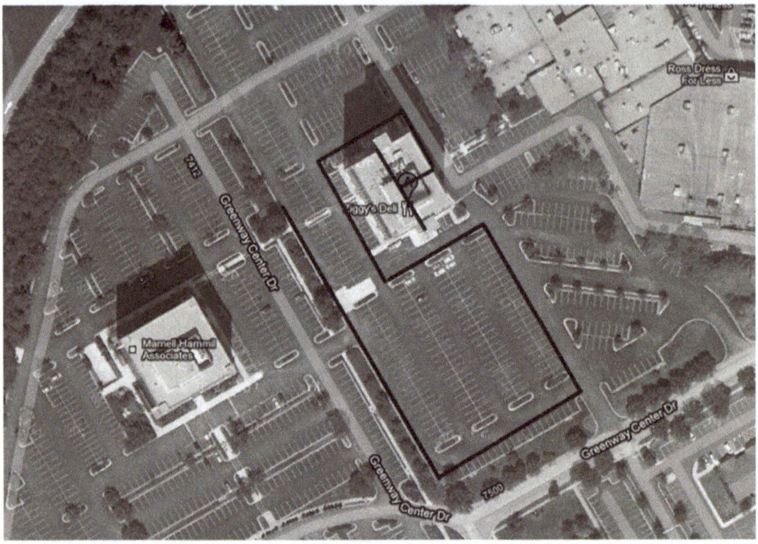

Fig. 8.2 Test course in suburban office complex

Fig. 8.3 Trial 1—4:29 pm GPS path *green*, ground truth *gray*

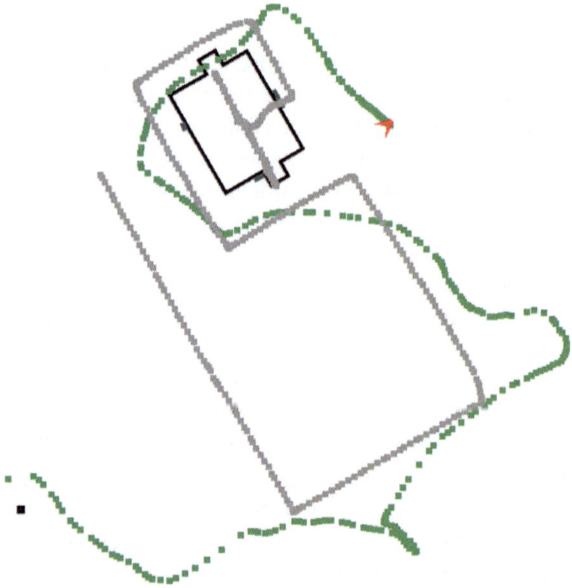

(Footnote 1 continued)
the approximate inertial path shape (after compensation for drift and scaling errors) for the collected data is correct.

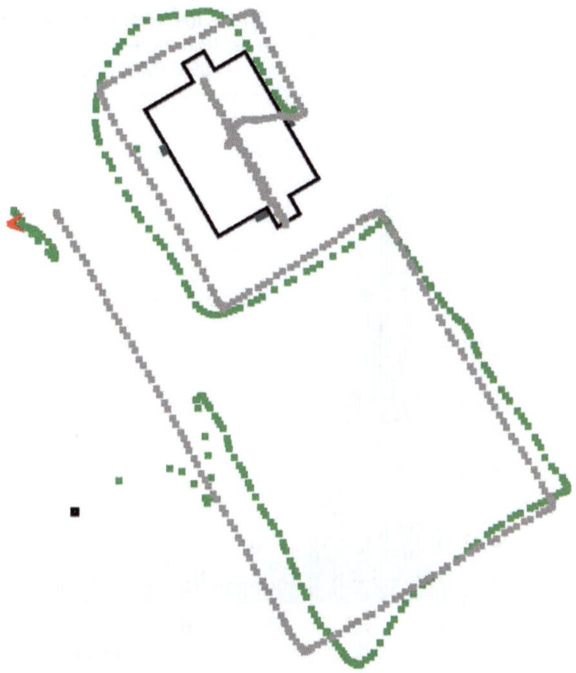

Fig. 8.4 Trial 3—8:04 pm GPS path *green*, ground truth *gray*

The error histograms of four tests at varying times (afternoon to evening) over a single day are shown in the figures below.

In the best case, Trial 3—Fig. 8.7, the expected error is 13.6 m and 50 % CEP[2] is 10 m. Trial 4 error histogram shown in Fig. 8.8, with 22 m expected value but 50 % CEP of 10.7 m, is taken right after Trial 3. Despite the close timing of the tests, the GPS data has an initialization error, finally converging to a better solution with error characteristics similar to Trial 3. Most people expect that their GPS location will be within 10 m of ground truth "the majority of the time". This equates to 50 % CEP that is 10 m. Trials 3 and 4 meet or come close to meeting this performance criterion. Constraining only the CEP allows some fairly large errors to appear without penalty. Notice, Trial 4 has a significant number of errors between 50 and 70 m but a CEP of 10.7 m!

Trials 1 and 2 do not achieve the expected GPS performance of 10 m CEP. Trial 1—Figs. 8.3 and 8.5 also had an initialization error and generally lower quality tracking performance with expected value 33.6 m and CEP 27 m. The worst performance was for Trial 2—Fig. 8.6 with expected error 65.9 m and CEP of 68.6 m. In this case, there is no discernable grouping of the errors. Overall, the

[2] The CEP or circle of equal probability is the radius of a circle whose boundary contains 50 % of the errors. While 50 % is a very common definition for CEP, the circle dimension can be defined for different percentages.

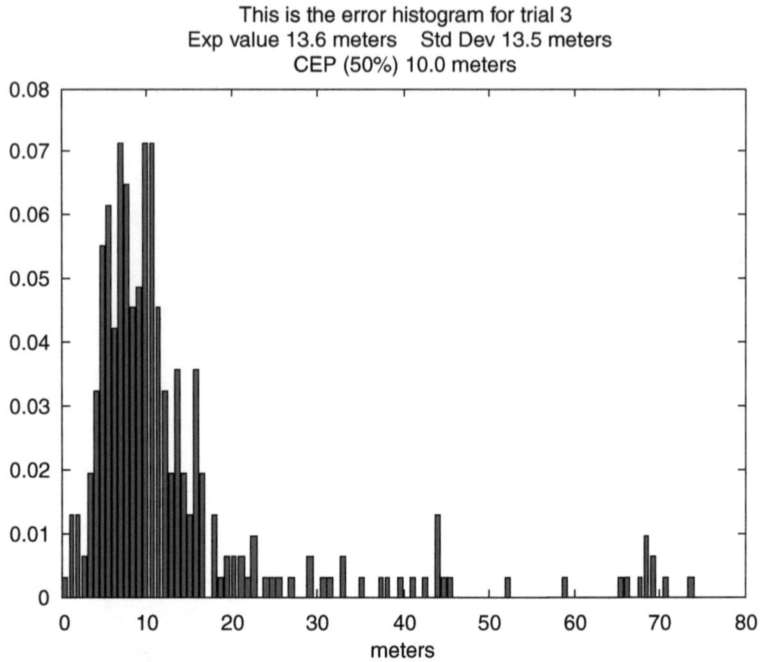

Fig. 8.5 Trial 3—8:04 pm GPS error histograms

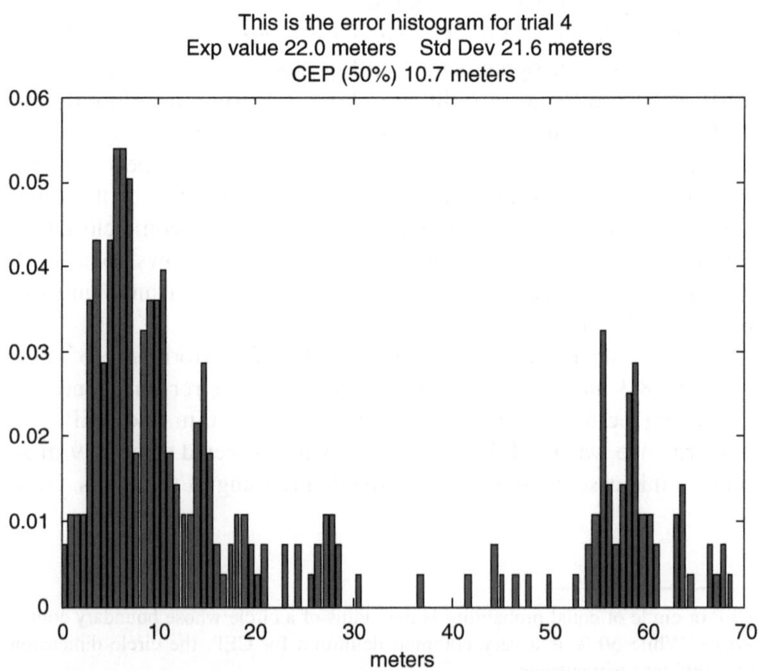

Fig. 8.6 Trial 4—8:10 pm GPS error histograms

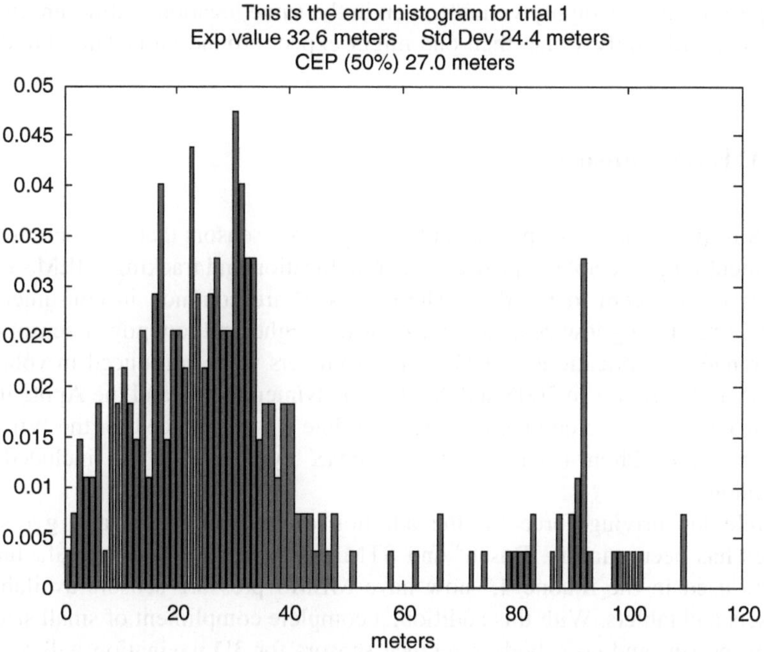

Fig. 8.7 Trial 1—4:29 pm GPS error histograms

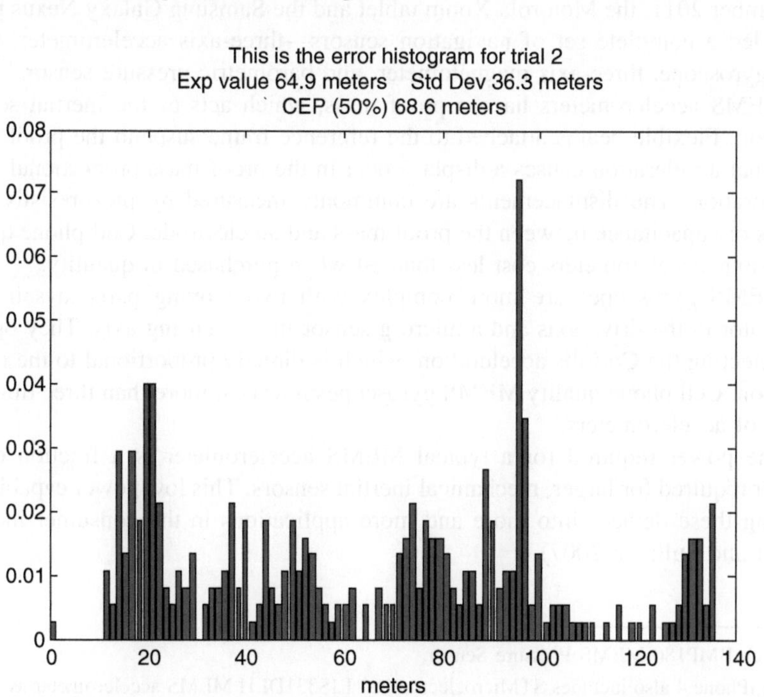

Fig. 8.8 Trial 2—6:28 pm GPS error histograms

GPS performance is highly variable even in the same location at different times of day and significantly worse than one might expect around an isolated building.

8.2 MEMS Sensors

If GPS is questionable for personnel tracking, other sensors that can measure local movement might be able to provide accurate location and tracking. MEMS inertial sensors would seem to be ideal. These sensors are common in consumer electronics, introducing low cost, small size, low weight, and low power sensors for a wide range of applications. MEMS accelerometers were introduced in volume to consumer electronics in 2006 and 2007 in the Nintendo Wii and the Apple iPhone (which also includes a compass). A gyro module was introduced for the Wii and in 2010 and the iPhone4 has the first MEMS gyroscope to be included in a smartphone.

While the driving force for the addition of accelerometers and gyroscopes sensors has been gaming, Bosch[3] and STMicroelectronics, who manufacture the sensors used in the iPhone 4,[4] now have MEMS pressure sensors available for handsets and tablets. With this addition, a complete compliment of small size, low weight, power, and cost, body reference sensors for 3D navigation will be available for cell phones and other consumer electronics (Bouchaud 2011). As of December 2011, the Motorola Xoom tablet and the Samsung Galaxy Nexus phone included a complete set of navigation sensors—three-axis accelerometer, three-axis gyroscope, three-axis magnetometer, and barometric pressure sensor.

MEMS accelerometers have a proof mass, which acts as the inertial sensing element. Flexible beams attached to the reference frame suspend the proof mass. External acceleration causes a displacement in the proof mass proportional to the acceleration. The displacements are commonly measured by piezoresistive elements or capacitance between the proof mass and an electrode. Cell phone quality three-axis accelerometers cost less than $4 when purchased in quantity.

MEMS gyroscopes are more complex with two moving parts, a self-tuned resonator in the drive axis and a micro-g sensor in the sensing axis. They operate by detecting the Coriolis acceleration, which is directly proportional to the rate of rotation. Cell phone quality MEMS gyroscopes now cost more than three times the price of accelerometers.

The power required for a typical MEMS accelerometer is a fraction of the power required for larger, mechanical inertial sensors. This low power capability is driving these devices into more and more applications in the consumer industry (Frost and Sullivan 2007).

---

[3] Bosch BMP180 MEMS Pressure Sensor.

[4] The iPhone 4 also includes STMicroelectronics' LIS331DLH MEMS accelerometer as well as its L3G4200D MEMS digital three-axis gyroscope.

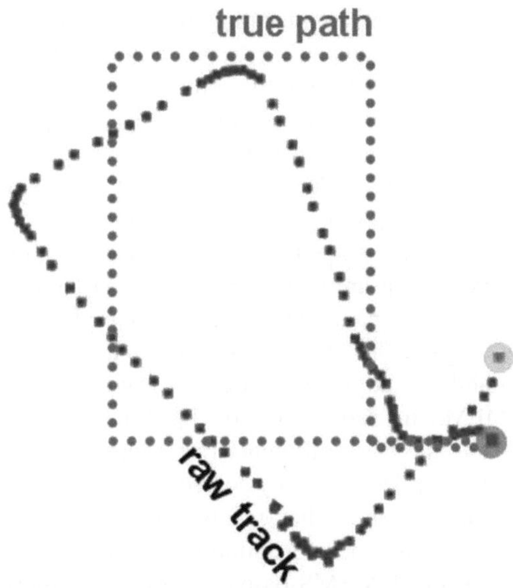

Fig. 8.9 True path traversed is shown versus inertial path estimate with uncompensated bias

Low-cost MEMS sensors make it possible to build small, low power inertial navigation units (INUs); however, MEMS sensors are subject to large inertial drift and other errors, and these must be accounted for in the design and operation of MEMS-based INUs.

Figure 8.9 illustrates of the affect of uncompensated gyro bias on the path estimate. This is referred to as inertial drift.

Performance of inertial sensors is characterized by several standard metrics. Understanding what these mean will help to select the appropriate sensor to meet performance requirements. Bias stability, angular random walk, scale factor accuracy, and dynamic range are all used to quantify the accuracy of the sensors. (Woodman 2007).

Bias (run-to-run bias stability) is the variation in offset of the stationary sensor output value from zero from one power-on cycle to another. This is a static offset, so it can be detected and compensated by online filtering.

In-run bias stability provides a measure of the operational bias variation over a specified period of time, typically around 100 s, in fixed conditions (usually including constant temperature). Bias stability is usually specified as a standard deviation value with units of deg/h. If Bt is the known bias at time t, a bias stability of 1 deg/h over 100 s means that the bias at time (t + 100) s is a random variable with expected value Bt and standard deviation 1 deg/h. The reported bias stability is due to flicker noise in the electronics. (Woodman 2007) In-run bias stability is the most important parameter in determining the accuracy of the resulting navigation solution. Temperature variations and other environmental disturbances can also cause the bias to fluctuate during operations and result in acceleration and/or rate measurements that are different from the true value.

Gyroscope angular random walk, which is caused by thermomechanical noise which fluctuates at a rate much greater than the sampling rate of the sensor, is a secondary factor contributing to the error. (Woodman 2007) This term is a measure of the variation of the averaged output of a stationary gyroscope over time. The scale factor accuracy is also important. It relates to the repeatability of measurements. Varying scale factor can result in two different measurements even when the motion has not changed.

Dynamic range refers to the range of acceleration or angular rate that a sensor is able to measure without saturating. Often there is a tradeoff among sensitivity and other error parameters to get increased dynamic range.

Again, the factors that contribute most to tracking inaccuracy are in-run bias stability and temperature sensitivities. Because of the manufacturing process, MEMS inertial sensors of the same model coming off the same production line may have very different error behaviors. Low-cost MEMs gyroscopes are generally supplied with only coarse factory trimmed offset and scale parameters. Device mounting and other manufacturing procedures can cause these parameters to change, and therefore additional calibration is required by the OEM-manufacturer.

Typical MEMS-based gyroscopes experience high thermal drift that is not accounted for in the bias stability measurements which are taken under fixed conditions. However, high quality MEMS sensors are available that are temperature calibrated over the operational range of the sensors.

The relationship between bias and temperature is often highly nonlinear for MEMs sensors. (Woodman 2007) When using sensors that have not been temperature calibrated, it is likely that runtime drift errors will be primarily caused by temperature variations. Most IMUs contain internal temperature sensors which make it possible to correct for temperature induced bias effects. At a minimum, static offset calibration should be performed on each gyro device in a temperature-controlled chamber. During calibration temperature and rate output are logged. A curve relating temperature and gyro bias can then be determined.

For accelerometers, temperature calibration is generally not necessary.

8.2.1 MEMS Sensors for Navigation

While developing applications using MEMS sensors, navigation algorithms that have worked on other types of sensors may not perform as well. Inertial bias and drift errors in MEMS sensors can lead to large position errors over a relatively short period.

The in-run bias stability numbers for MEMS sensors are significantly higher than their non-MEMS counterparts. For example, mechanical gyrocompasses achieve a bias stability on the order of 1 arcsecond/h (better than 0.00015 deg/h). Ring laser gyros or fiber optic gyros achieve a bias stability on the order of 1 NM/h (around 0.015 deg/h). MEMS sensors achieve bias stability on the order of 15 deg/h (earth's rotation rate) (Al-Sheimy 2009). One direct advantage of the higher accuracy gyro-sensors is that, when at rest on the Earth, they can use measurements of the rotation of the Earth to initialize to true North and thus compute a reference for absolute heading. Because these gyros are not magnetic, this will be possible in conditions where magnetic disturbances are present. For MEMS sensors, the Earth's rotation is in the noise and so they can only measure heading relative to their initial heading without input from an external source such as a compass which would be affected by external magnetic disturbances.

If one integrates the acceleration to arrive at a position, an uncompensated accelerometer bias error will introduce an error proportional to the elapsed time in the velocity estimate and an error proportional to the square of the elapsed time in the position estimate. The best MEMS accelerometers can have an in-run bias stability of 0.05 mg[5] where 1 g $= 9.8$ m/s^2. Uncompensated accelerometer bias at this level for only 1 min can introduce almost a meter of error. Cell phone quality MEMS accelerometers have 1 mg in-run bias stability which could lead to an error of 18 m over 1 min.

Uncompensated gyro bias errors introduce errors in orientation proportional to time t; however these errors are actually more of an issue when they occur in the pitch or roll axis. Uncompensated pitch or roll orientation error will cause a misalignment of the inertial measurement system, and therefore a projection of the gravitational acceleration vector in the wrong direction. Using a small angle approximation, this would lead to a position error of $\frac{1}{6}g\Delta\omega t^3$, where ω is the rotation rate in the pitch or roll axis.

The best temperature calibrated MEMS gyros currently available can have an in-run bias stability of 12 deg/h or 0.003 deg/s. An uncompensated bias of 0.003 deg/s in the pitch or roll axis would produce an error of more than 20 m in a minute! The sensors found in cell phones now are not temperature calibrated and the in-run bias stability is closer to 0.03 deg/s. An uncompensated bias of 0.03 deg/s in the pitch or roll axis would give an error of over 20–30 m in 30 s!

These drift numbers are alarming but demonstrate the worst case drift. Fortunately, additional orientation information is available via the accelerometers when the device is not moving. Assuming the device is at rest on the Earth, it will experience 1 g of acceleration. This constrains the possible device orientations to a plane that fixes the pitch and roll axes with respect to the Earth's frame of reference. Yaw information (earth frame) is not available since yawing the device will not change the direction of its gravity vector. Yaw information can be corrected using a compass *when* good compass data is available.

[5] Run-to-run bias is higher and the assumption is that constant offset can be compensated by filtering methods (Kalman filter or other) but the variation around that bias (in-run bias stability) is harder to compensate.

8.2.2 Inertial Navigation Unit (INU) Orientation Estimation

As described above, errors in orientation can quickly lead to very large errors. In this section, we present several orientation estimators: (1) angular rate-based; (2) accelerometer-based; and (3) combined rate-accelerometer. The last estimator is robust in that it combines the gyroscope and accelerometer estimates in a way that mitigates their inherent limitations.

A quaternion representation for orientation is often used to avoid the singularities in the Euler angle parameterization when pitch approaches ±90°. It is especially important to use a quaternion representation for orientation in body-mounted systems to avoid this singularity.

8.2.2.1 Quaternion Representation of Orientation

Because the orientation estimation is performed in the space of quaternions, Q, it is useful to review the pertinent mathematical properties associated with them. Here, we present only that much information as is necessary for the discussion. A detailed discussion of quaternions may be found in (Kuipers 1999).

A quaternion q is specified by four real values and may be represented as a four dimensional vector

$$q = \begin{pmatrix} q_0 \\ q_1 \\ q_2 \\ q_3 \end{pmatrix}$$

A quaternion can also be thought of as an extension of the complex numbers that have three imaginary components represented with unit axes i, j, and k. In the complex representation the quaternion is written as

$$q = q_0 + iq_1 + jq_2 + kq_3$$

and its conjugate is

$$\bar{q} = q_0 - iq_1 - jq_2 - kq_3$$

The quaternions together with the notion of a multiplicative operator define a mathematical group. This group follows all the normal laws of algebra except that multiplication is not commutative, i.e., in general, $q \cdot p \neq p \cdot q$, for quaternions $p,\ q \in Q$. Multiplication of quaternions is defined by

$$q \cdot u = \begin{pmatrix} q_0 & -q_1 & -q_2 & -q_3 \\ q_1 & q_0 & -q_3 & q_2 \\ q_2 & q_3 & q_0 & -q_1 \\ q_3 & -q_2 & q_1 & q_0 \end{pmatrix} \begin{pmatrix} u_0 \\ u_1 \\ u_2 \\ u_3 \end{pmatrix}$$

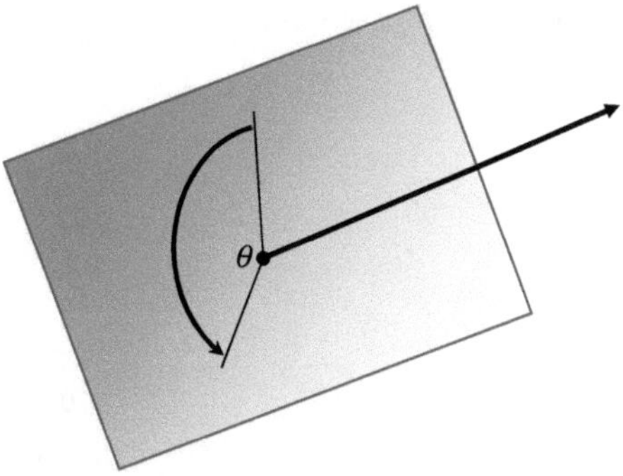

Fig. 8.10 Quaternion rotation angle and surface normal to the plane of rotation

Quaternions represent an orientation by specifying a rotation angle θ and vector $v = [v_1 \quad v_2 \quad v_3]^T$ in three-space representing the surface normal to the plane of rotation (Fig. 8.10).

A quaternion encodes the rotation into four real numbers

$$q(v,\theta) = \begin{pmatrix} \|v\| \cos(\theta/2) \\ v_1 \sin(\theta/2) \\ v_2 \sin(\theta/2) \\ v_3 \sin(\theta/2) \end{pmatrix}$$

Essential to the representation of an orientation is the mapping $T_q : Q \mapsto Q$ defined as

$$T_q u = q \cdot u \cdot \bar{q}$$

where q has unit norm. The resulting $T_q u$ has the interpretation of a rotation in 3D space. The rotation by the angle θ is in the plane that is normal to the vector q. Thus, q represents an orientation in 3D as a rotation and the mapping $T_q u = q \cdot u \cdot \bar{q}$ performs that rotation.

In the sequel, we denote q as the orientation quaternion. This is the fundamental quantity that is to be estimated at each time step of interest. It is important to note that orientation is relative to some starting attitude. In fact, the quaternion q has associated with it a well-defined operator that when applied to rotated points in 3D space, recovers the originating attitude.

8.2.2.2 Gyroscope Orientation Estimate

First, consider an orientation quaternion estimate based on the measurement of three orthogonal gyroscopes yielding measurements $\omega \in R^3$. At any time t, the angular rates $\omega(t)$ given by the gyroscopic measurements govern directly the continuous-time quaternion propagation equation $\dot{q}(t) = \frac{1}{2}\omega(t) \cdot q(t)$, where $\omega(t) = [0 \quad \omega_x(t) \quad \omega_y(t) \quad \omega_z(t)]^T$ and T denotes the transpose. Substituting $t = k\Delta$

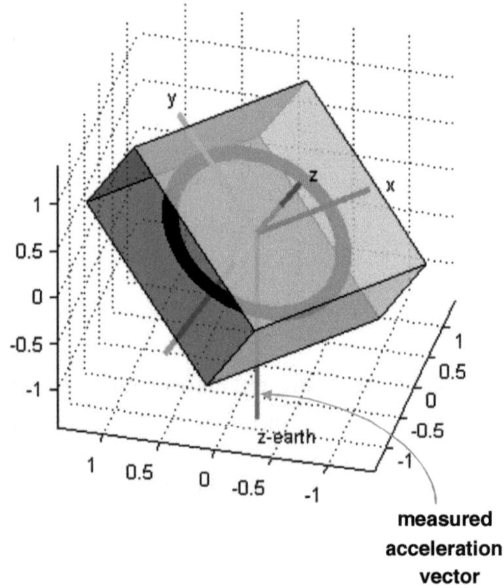

Fig. 8.11 Body frame versus the Earth frame

(where Δ is the fixed intersample interval) gives the discrete-time version of gyroscopic based quaternion estimate as

$$q_{\omega,0} = 1,$$
$$q_{\omega,k+1} = q_{\omega,k} + \frac{1}{2}\Delta \cdot \omega_k \cdot q_{\omega,k}.$$

This estimate is subject to bias errors in the measured angular rates. Any such unmitigated bias will lead to drift of the estimate away from its true value. The estimate is also subject to rate saturation which could cause errors in the estimate.

8.2.2.3 Accelerometer Tilt Estimate

The accelerometer measurements can also be used to obtain a partial orientation estimate—pitch and roll in the Earth reference frame (Fig. 8.11). Consider an orientation quaternion estimate based on the measurement of three orthogonal accelerometers yielding measurements $a \in R^3$. The gravity quaternion is denoted as $g = [0 \quad 0 \quad 0 \quad -1]^T$ which reflects the fact that a three-axis accelerometer not experiencing any non-gravitational related acceleration on the Earth will measure -1 g in the z-direction and zero elsewhere. An object that is oriented so that its z-axis is not coincident with gravity will have nonzero components in the x- or y-directions. Thus, a determination of the quaternion that reorients the measurements so that they are aligned with the z-axis is a pitch and roll estimator.

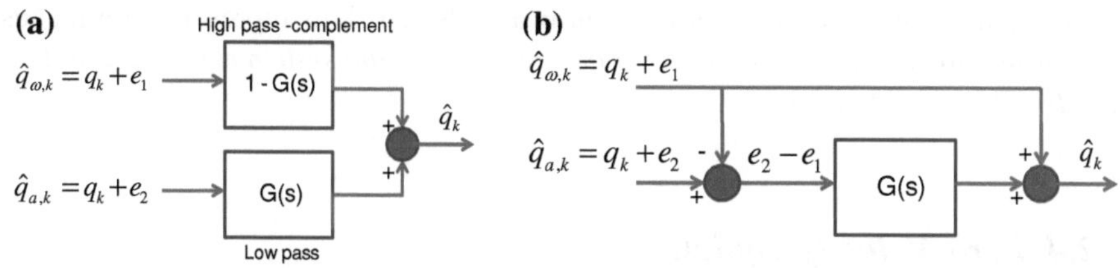

Fig. 8.12 **a** Complementary filter **b** Equivalent representation—filter operates only on the noise

An object that is oriented so that its z-axis is coincident with gravity is said to be in the *Earth frame* of reference. Measurements made with respect to the object are said to be in the *body frame* of reference. It can be shown that the translation between the body frame of reference and the Earth frame of reference can be represented by the quaternion square root of the product of the gravitation quaternion and the conjugate of the measured acceleration quaternion $a_k = \begin{bmatrix} 0 & a_x & a_y & a_z \end{bmatrix}^T$. This leads to the accelerometer estimate

$$q_0 = 1,$$
$$q_{a,k+1} = \sqrt{g \cdot \bar{a}_k}.$$

This estimate is subject to noise and invariant to rotations around the "yaw" axis so the solution is not unique. It is assumed that gravitational acceleration is dominant.

8.2.3 Complementary Filters

The gyroscopes and accelerometers each provide orientation information. This gyro estimate is good over the short term but suffers from bias as well as saturation errors that cannot be compensated without additional information. Assuming zero (or near zero) non-gravitational acceleration, the accelerometer data can be used to obtain a noisy measurement of pitch and roll relative to the Earth frame as described above.

Complementary filters are often used when you have two noisy measurements of the same signal with complementary properties. For example, one sensor provides good information only in the short term (high frequency data and low frequency noise), while the other provides good information over the long term (low frequency data and high frequency noise). A classic example is the combination of gyro rate data (very good short term but drifts over long term) with accelerometer tilt sensor data (very good on average but—not correct during acceleration) for orientation estimation.

A simple estimate would be to send the gyro orientation estimate, $\hat{q}_{\omega,k} = q_k + e_1$, through a high pass filter and the accelerometer tilt estimate, $\hat{q}_{a,k} = q_k + e_2$, through a low pass filter and then add them (Fig. 8.12a). Note that in this case if the errors are zero, $\hat{q}_k = q_k$.

An equivalent representation is given in Fig. 8.12b. This representation makes it clear that the *complementary filter operates only on the system errors and not on the dynamical quantities.*

8.2.4 Zero Velocity Updates

The process of detecting a point in time when the velocity of an object is zero and resetting the integral of acceleration at that point to zero is termed a *zero velocity update*. These updates can be used to limit error growth in inertial tracking devices.

In applications such as surveying when using high quality inertial measurement units, the use of frequent zero velocity updates to remove the effects of gravity and other biases is standard. For example, Applanix makes vehicle and man wearable systems for surveying applications that notifies the user to stop every few minutes for a zero velocity update. The frequency of the updates needed is dependent on the motion type. Because pedestrian motions are less smooth, higher frequency updates (every couple of minutes) are needed when tracking pedestrian motion versus vehicle motion where less frequent updates are typically needed.

If one were to attempt to build a similar system using current MEMS sensors instead of the high quality inertial sensors used in the Applanix system, the zero velocity updates would be needed around once per second making the system impractical. Because of the high frequency of updates needed, while using MEMS sensors in pedestrian applications, sensors are often placed on the foot to leverage the zero velocity periods that occur each time the foot with the sensor is flat on the ground. This is discussed in more detail in the section on foot-mounted Systems on page 234.

8.3 Inertial Systems for Pedestrian Tracking

In robotics and other vehicle tracking applications, Bayesian filtering methods (e.g., Kalman filter, particle filter—see Chap. 9) have been successfully used to help correct drift by fusing inertial measurements with prior knowledge and other sensor measurements. In these applications, system models and control inputs are known and typically measurements from other sensors (wheel encoders, etc.) are available as well. Given knowledge of the system model and control inputs, the location of the robot or vehicle can be predicted (prediction step), and then sensor measurements are used to provide corrections to the estimate to compensate for unmodeled dynamics or other disturbances (update step).

The problem of tracking and locating pedestrians presents a set of challenges that is unique relative to most vehicle tracking problems. Human walking has been studied for more than 50 years using a variety of tools (Onyshko and Winter 1980; Zajac and Neptune et al. 2002; Zajac and Neptune et al. 2003). Most previous

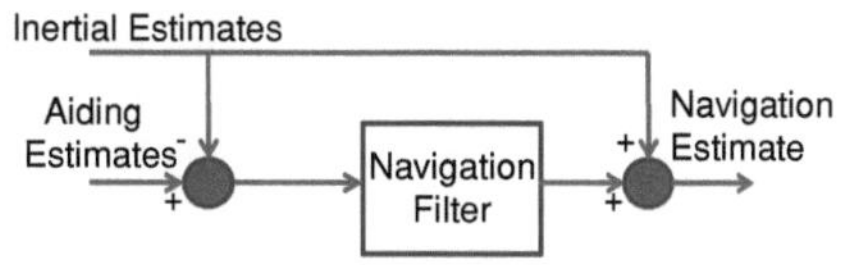

Fig. 8.13 Complementary navigation filter

work has been concerned with "simple" steady-state walking, that is, healthy or infirm individuals moving at a normal pace. While a variety human motion models are available, the inputs that drive those models are hard to identify. Additionally, the effectiveness of detailed motion models is questionable as the basis for non-linear filtering and estimation methods since the tracking system based on measurements at one body location (on the waist or the foot are most common) has low effective observability. Because of these issues, routine application of standard filtering methods using these models does not provide the same type of benefits that they do in robotic systems where known dynamic models and control inputs are the norm.

8.3.1 Classical Filtering Methods

Despite the lack of models and control input information, a common approach has been to insert a model that simply represents sensor model transformations (integrations) from the measurement space (accelerations, angular velocities) to location and orientation, and ignores the human aspect of the motion altogether. But because of the missing model and control information, there is little information for the model prediction step. The physical constraints of human motion limit the movement distance over the sample period, but the direction is unconstrained (without map information—map type corrections are discussed further in Chap. 9). When lower quality inertial sensors are integrated to produce position estimates an independent velocity measure or zero velocity update is needed every few seconds to control drift.

As discussed above, *a complementary filter operates only on the system errors and not on the dynamical quantities such as position and velocity. For this reason, they are useful when system models and control inputs are not known.* For these reasons, it is common in inertial navigation systems to use a complementary filter (Fig. 8.13).

The inertial system is then corrected in accordance with the filter's best estimates of the system errors (Brown 1972; Higgins 1975; Foxlin 1996). When set up in this way, an aided inertial system can be thought of as the inertial system providing the estimated trajectory (or system control inputs), and the aiding sources providing the noisy measurements that allow computation of corrections to the trajectory (Brown 1972).

Because of the poor quality of sensors and the lack of models, another common approach in body-mounted navigation systems has been to essentially ignore

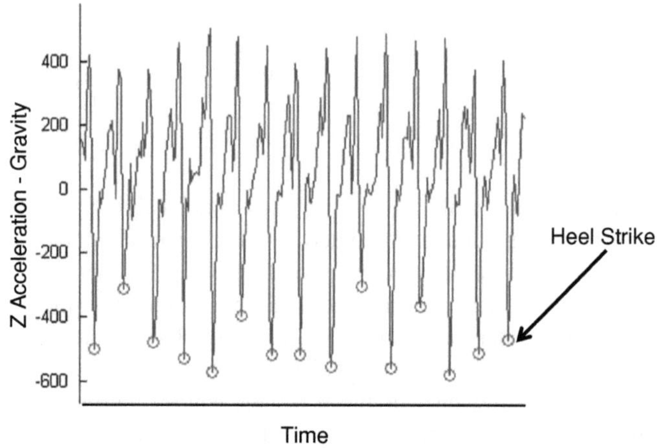

Fig. 8.14 Z-axis accelerometer signal with steps marked

standard filtering methods altogether, and instead, to develop pedometer algorithms that rely on very simple motion models. In pedestrian tracking, the motion models typically referred to in the literature describe classification of motion type (walking, running, crawling…) and step length and frequency (Judd 1997; Funk et al. 2007).

8.3.2 Torso-Mounted Systems

Torso-mounted inertial sensors are typically attached at the waist and centered in the front or in the back of the torso to be closest to the center of gravity where there is less extraneous motion. Other mounting locations, such in a vest pocket are possible, but the mounting location affects the character of the motion signatures so a system may have to be tuned for a specific mounting location. Moving a system designed for waist mounting to another location on the body can cause performance issues.

Waist mounted inertial tracking systems that use MEMS sensors are typically developed as a pedometer-based systems. The simplest of the pedometer type systems detects each step and uses a fixed predefined step length to compute the distance travelled, assuming all motions are walking or running forward (Judd 1997). Because of the computational simplicity, this type of pedometer has relatively low power use. It provides adequate performance for runners and other athletes with an approximately fixed pace attempting to measure their workout distance.

Step detection is a critical function in any pedometer system. Figure 8.14 shows raw z-axis accelerometer data from a waist mounted sensor for a person going up 12 steps. Circles mark each step detected. It is clear from this sample that the peaks

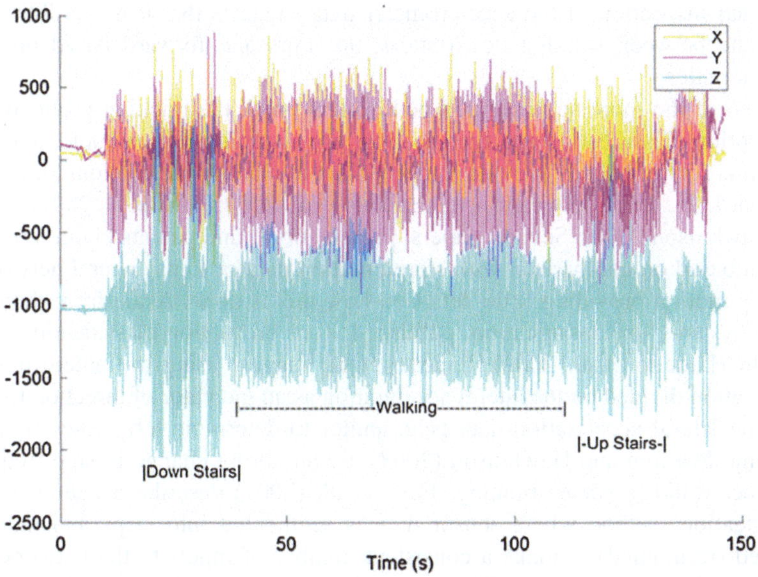

Fig. 8.15 Raw accelerometer signals from X-, Y- and Z-axis during typical pedestrian motions

are not always clean and that there is significant magnitude variation even when performing the same task.

Using accelerometers to monitor and classify human motion has been an area of research since the late 1990s for activity monitoring in medical research, for example see (Bouten et al. 1997; Herren et al. 1999). Wearable computing and navigation applications followed this work to improve basic fixed gait pedometers (Lee and Mase 2001; Bao and Intille 2004). More sophisticated pedometers provide step length estimation based on height, step frequency, and other factors. In general, speed and step length increase when the step frequency increases, and for a given step frequency, step length remains fairly constant (with some distribution about a nominal value). Considering the human body locomotion and its physical restrictions, different methods have been proposed to approximate the step length. Linear models have been derived by fitting a linear combination of step frequency and measured acceleration magnitude to the captured data. Pedometer systems may provide a mechanism to use GPS or other measures to adaptively update the step length estimates (Ladetto 2000; Lee and Mase 2001; Ladetto et al. 2002; Fang et al. 2005; Godha et al. 2006). Chau (2001a, b) presents a review of analytical techniques which have the potential for a step data analysis, including: Fuzzy Logic, statistical, fractal, wavelet, and Artificial Neural Network methods.

Figure 8.15 shows three-axis accelerometer data taken while walking in an office building. In this particular segment, the subject walks down four flights of stairs, down a hallway, and up four flights of stairs.

Visual inspection of the accelerometer data suggests that it is possible to differentiate between walking down stairs, up stairs and forward based on signal characteristics.

Some of the more sophisticated pedometers break the tracking problem down into motion classification and then scaling, not assuming, for example, that every motion is forward. They provide a mechanism to classify the motions as forward, backward, up, down, left, right, etc. (Ladetto et al. 2002; Funk et al. 2007; Soehren and Hawkinson 2008). Several papers, presentations, and patents claim to classify motion based on comparison with stored motion data or to use neural networks to classify motion providing little detail on how this is done. Aside from the use of vision systems for classification, published work on motion classification is limited. In (Ladetto et al. 2002), Ladetto et al. suggest using the antero-posterior acceleration divided by the lateral acceleration as an indicator of direction together with the lateral acceleration data peak angles to determine left versus right side stepping. Soehren and Hawkinson (2008) use an abrupt change in step frequency to detect walking versus running. Funk et al. (2007) describe a neural network classification method where sensor data is segmented into steps and then normalized (resampled) to make a consistent number of inputs to the network independent of step frequency. This method has been used to classify standard pedestrian motions as well as more utilitarian job related motions such as crawling and climbing ladders.

8.3.3 Velocity Sensors

The addition of a velocity measurement can significantly improve position accuracy and enables the use of integration of sensor data combined with standard filtering techniques for waist mounted MEMS inertial systems. As a result, an active area of research is the development of sensors for measuring velocity. Two common approaches for obtaining velocity measurements use image sensors computing optic flow or optical feature movement (Veth 2011) and Doppler velocimeters (Weimann et al. 2007; Hopkins et al. 2009; McCroskey et al. 2010).

Optical systems use the apparent motion of portions of an image between frames and determine relative motion of the camera (Lucas and Kanade 1981; Harris and Stephens 1988; Harris 1992; Shi and Tomasi 1994; Bay et al. 2008; Karvounis 2011; Veth 2011). This requires three basic operations (1) finding features in an image suitable for tracking, (2) matching these features in a subsequent image, and (3) solving for the resulting camera motion.

Using a pinhole camera model, a feature point $p = (x, y)$ in an image taken at focal length f is mapped to the real world 3D feature point $P = (X, Y, Z)$ by the equation

$$p = \frac{fP}{Z}$$

By measuring the velocity of the image feature by its frame to frame movement, the motion of the 3D scene structure can be derived using the following equations:

$$u = \frac{t_z x - t_x f}{Z} + \frac{\omega_x xy}{f} - \omega_y (f + \frac{x^2}{f}) + \omega_z y$$

$$v = \frac{t_z y - t_y f}{Z} + \omega_x (f + \frac{x^2}{f}) - \frac{\omega_y xy}{f} - \omega_z x$$

where (u, v) is the image velocity vector, Z, is the distance to the 3D object which has translational velocity (t_x, t_y, t_z) and rotational velocity$(\omega_x, \omega_y, \omega_z)$ relative to the camera. By tracking feature points across several image frames, the 3D location and motion parameters can be computed (Strelow and Singh 2002; Kolodko and Vlacic 2005).

The main difference between the optic flow and feature tracking approaches is in the method used to select these points of interest (Veth 2011). Feature tracking algorithms select specific feature locations within the image that have significant spatial intensity changes, such as corners, because they have a high probability of being tracked in subsequent images. Feature descriptors that encode information from the localized image feature are used to match features in subsequent images (Lucas and Kanade 1981; Harris and Stephens 1988; Harris 1992; Shi and Tomasi 1994; Bay et al. 2008). Optical flow techniques simply divide the image into a grid of image patches, without regard to the quality of the patch for tracking. Image intensity patterns are compared to determine matches in subsequent images (Barron et al. 1994; Barrows 2011). The optic flow approach is simpler computationally and very amenable to software and hardware optimization (Barrows 2011) but is not as robust for matching the patches. The underlying mathematics of estimating the camera motion is the same in optical flow and feature tracking algorithms. Feature tracking algorithms provide an added benefit that the features can be saved as landmarks and used to correct location when the feature/landmark is revisited. This feature mapping and correction process is accomplished using simultaneous localization and mapping algorithms (Veth 2011) which will be discussed in the Chap. 9.

Leveraging the development of millimeter wave components for automotive radar collision avoidance systems, low cost and low power Doppler velocimeters may soon be available that are able to accurately sense the relative velocity of the torso with respect to the ground on a body-mounted sensor. The Doppler velocimeter works by sending out high frequency (GHz) signals which are reflected by obstacles such as the ground and received again by the sensor. Measuring the Doppler frequency shift, the speed can be determined. Doppler velocimeter measure speed along its sensitive axis so the orientation of the Doppler beam with respect to the navigation axes must be known. Honeywell's 92 GHz Doppler velocimeter performance has been measured with the beam pointed straight ahead at a wall with error less than 1 cm/s and pointed 45° own at a carpet target with error less than 3 cm/s (McCroskey et al. 2010).

Velocity aiding from optical or radar sensors provides a relatively drift-free measurement to bound the drift of inertial sensors; however, the velocity sensors come with added system complexity and additional computational expense. Both optical and radar-based velocity sensors compute subject velocity relative to the environment and both sensors produce accurate estimates when the environment has sufficient visible differentiation or the environmental materials produce sufficient radar returns.

Since the velocity computation is relative to what is in the field-of-view, both sensors are also susceptible to errors caused by people or objects moving into or blocking the field-of-view of the sensor. Random sampling methods such as RANSAC (RANdom SAmple Consensus) can allow the system to eliminate outliers (Fischler and Bolles 1981). The complimentary inertial sensors, which have low drift over very short periods and are unaffected by these types of environmental factors, provide a basis for screening velocity measurements. By comparing each measurement to the navigation system's computed velocity, a statistical algorithm such as a median filter can be used to determine whether to accept the measurement (Weimann et al. 2007; Veth 2011). The purpose of the median filter is for removal of the nonGaussian outliers caused by, for example, incorrectly matched stereo features. The median absolute deviation (MAD) is defined as the median of the absolute deviations from the median:

$$\mathrm{MAD(X)} = \underset{x \in X}{\mathrm{median}}(|x - \mathrm{median(X)}|)$$

As a rule of thumb, the median and MAD are typically not affected by outliers unless more than 50 % of the data are outliers, whereas, the mean and standard deviation could be affected by a single outlier (Hampel et al. 1986). The filter excludes features outside some fixed number of MADs from the median. Theoretically, with any nicely distributed data, a single MAD away from the median in either direction will bound 50 % of the data. With a Gaussian distribution, two MADs will bind about 82 % of the data. Three MADs are nearly equivalent to two standard deviations under a Gaussian distribution, bounding about 95 % of the data (Hampel et al. 1986).

8.3.4 Foot-mounted Systems

The technique of zero velocity updating (ZUPT) has been used to extract high quality velocity information directly from foot-mounted accelerometers for pedestrian tracking (Foxlin 2005; Godha et al. 2006). With appropriately placed sensors, the characteristics of human motion can be used to provide a zero velocity update that allows a very good estimation of velocity over each stride from the acceleration data, minimizing the effects of bias and noise. Consider how a person's feet move during walking. Figure 8.16 illustrates the phases of walking. Each foot alternates between a period of motion when the foot swings forward and a period of

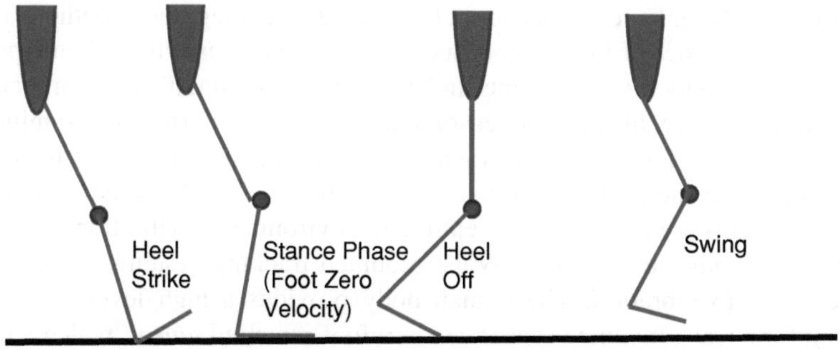

Fig. 8.16 foot-mounted sensors allow frequent zero velocity updates

no motion when the other foot swings forward. The motion is segmented by events on the tracked foot: heel strike, foot flat, heel off, toe off. The foot is considered to be approximately at rest when the foot is flat. So if the sensor is, for example, placed on the foot or in the heel of a shoe, the stationary period of the foot can provide a zero velocity update and allow correction of the velocity over each stride, thus minimizing accumulated error and providing a better position estimate.

In addition, because the precise placement of sensor is known, the sensor's pitch and roll is also known precisely during the stationary period,[6] and thus can be corrected over the stride. The yaw (heading) is the only variable that cannot be corrected by the zero velocity update. Researchers at Carnegie Melon (Laverne et al. 2011) are developing shoe embedded radar with the goal of improving the quality of the velocity update by continuously measuring the velocity of the foot with respect to the ground.

In the same research project (Laverne et al. 2011), Lavern et al. showed that the shock to the IMU in boot mounted systems can cause significant heading errors over time if the heading is not compensated. An obvious complimentary sensor is a magnetic sensor for heading correction. For foot-mounted sensors in pristine outdoor environments without magnetic disturbances, this works well. Indoors, foot-mounted magnetic sensors have issues due to the proximity of the sensor to the floor which is often a source of magnetic disturbances in large buildings where steel infrastructure and reinforced concrete are standard construction practices. Brandt and Phillips (2003) proposed an approach to controlling foot-mounted gyro drift—use of foot-to-foot RF range measurements. Using this method, Lavern et al. were able to compensate the heading errors with the addition of an IMU on each foot and foot to foot ranging.

One advantage of the ZUPT system algorithm is that it can efficiently track the different modes of walking (forward, backward, sidestep) without any additional modeling (Godha et al. 2006). However, while zero velocity updating has been

[6] If the pitch and roll are not known precisely they can be estimated using the accelerometer data directly or by using an extended Kalman filter which takes advantage of the fact that the tilt errors will be correlated with horizontal velocity errors.

shown to significantly reduce errors relative to direct integration methods during walking, other modes of locomotion, such as running and crawling, do not provide the needed zero velocity update time, and thus suffer from significantly larger errors.

Additionally, uncompensated sensor bias and sensor noise have significantly negative effect on foot-mounted systems because the sensor data is being integrated to provide the position estimate over each step. For the same reason, foot-mounted systems are also more susceptible to environmental vibration from heavy machinery, or other disturbance. Waist mounted tracking systems are inherently less sensitive to vibration. The human body provides a high level of vibration damping for a waist mounted sensor versus a foot-mounted sensor, which will pick up the full vibration of the surrounding environment. Additionally, waist-mounted pedometer type sensors recognize the general shape of the signature of the motion and do not rely on double integration of accelerometer data that is highly susceptible to vibration disturbances and other sensor noise.

Another challenge for foot-mounted systems is that people have low tolerance for any foot-mounted systems that might interfere with mobility, including wired interconnections of the boot sensors to the body. While energy harvesting and battery technology continue to improve, technology is not available today to harvest sufficient energy from human motion or thermal sources to power a suite of navigation sensors in the boot.

8.3.5 Cell Phone Systems

The ultimate system for pedestrian tracking would use small light weight sensors embedded in a device that someone would carry with them every day, such as a cell phone. Cell phone sensors will very likely not be worn in a fixed location, such as around the waist (centered back or front), which is a requirement for most pedometer based personnel tracking systems available today. The unknown placement/orientation of the sensors relative to the subject adds complexity since the algorithms must now also determine the direction of a "forward" step relative to the orientation of the sensors.

Cui et al. did an international study to assess the most probable location that cell phone users would carry their phone (Cui et al. 2007). The closest to a waist worn system is a belt clip. A belt clip is the preference for only 13.8 % of men and less than 1 % of women. The pocket is the preferred location phones for men at 60 % but only 16.4 % of women. For women, 61.4 % carry the phone in their purse/bag, whereas only 10 % of men carried their phone this way (Cui et al. 2007). Researchers have begun work on identifying carrying position of cell phones and relative motion direction which are both critical issues in tracking them. (Blanke and Schiele 2008; Steinhoff and Schiele 2009) Blanke and Schiele have published their data sets (http://www.mis.tu-darmstadt.de/datasets) from tests of various subjects carrying IMUs of similar quality to those found in cell phones in their trouser pockets.

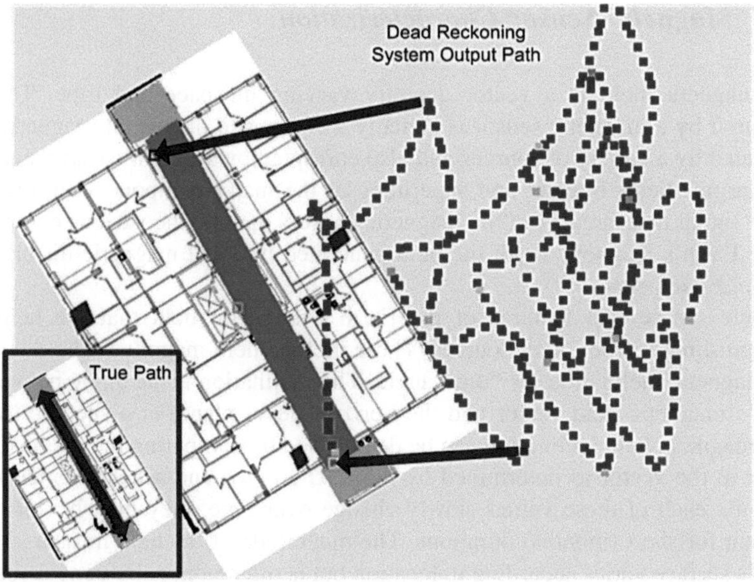

Fig. 8.17 Commercial dead reckoning system demonstrates the need for sophisticated algorithms for inertial and magnetic sensor fusion indoors

8.4 Heading Correction

Heading errors due to gyro drift are a significant cause of errors in inertial systems. While pitch and roll can be corrected during zero velocity updates by realigning with gravity, heading errors (yaw) cannot. A standard approach to correct heading errors is to combine MEMS six degrees of freedom inertial sensors (IMU) with three-axis magnetic sensors to provide yaw correction (Foxlin 2005; Godha et al. 2006; Funk et al. 2007). This approach has challenges delivering highly accurate orientation in environments with magnetic disturbances, some of which may be caused by objects the person is carrying.

Magnetic interference from doorways and beams and other ferrous objects in the building can cause very large nonGaussian disturbances. Figure 8.17 shows an example of a commercial navigation unit designed for outdoor navigation (where magnetic disturbances are typically less of a problem). The navigation unit used a simple implementation of Kalman filter based fusion of magnetic sensor data with inertial data. The magnetic data is fused under the assumption that the magnetic heading disturbances are Gaussian (standard Kalman filter assumption) without any attempt to remove magnetic anomalies. The path walked should reflect five traversals of the hallway but the true path is not recognizable from the output of the dead reckoning system.

The challenge in developing fusion algorithms is to recognize and eliminate poor magnetic readings before the heading errors accumulate.

8.4.1 Magnetic Sensor Characterization

The magnetic field is a vector quantity varying in space and time. The field measured by a magnetic sensor is actually a composite of several magnetic fields generated by a variety of sources and also corrupted by sensor measurement noise. The various fields interact and superimpose; the major components in this composite signal measured from the magnetic sensors include the various components of the Earth's magnetic field, the field generated by local magnetic disturbances, and noise elements.

There are several features of the Earth's magnetic field that are helpful in distinguishing it from other sources. These include field magnitude, field stability, and magnetic inclination, or "dip" angle. The inclination is the angle between the Earth's magnetic field vector and the horizon (level plane) at a specific location (The magnetic field inclination can be determined by comparing the magnetic field vector to the vector as determined by the INU's gyros and accelerometers).

While each of these values slowly change over time, they may be considered constant for short (months) durations. The magnitude of the field measured on the Earth's surface varies according to location but is often nominally considered to be about 0.5 gauss (50 μT). Declination for a given location describes the angle between magnetic and true North. Because this value cannot be measured unless true North is known, it will not be very useful for determining field accuracy but is important for providing the correct heading relative to true North. There are fairly accurate, reliable models describing Earth's magnetic field that can be leveraged for the auto calibration (Maus et al. 2010). These could be easily implemented, for example, as a database of inclination, declination, and field strength by location.

8.4.2 Magnetic Sensor Calibration

More troublesome for accurate angle measurement than the variations in the Earth's magnetic field are the local variations in the field. The local field variations are typical indoors where the building structure and power systems can create magnetic disturbances. Even outdoors disturbances can be caused by nearby buildings, vehicles, power lines, buried pipes, and even the subject's individual things. External disturbances can be handled by developing sensor fusion algorithms as described above. Local disturbances due to fields generated on the sensor board itself can be minimized by calibration.

Asymmetry that is seen in the range and offset of magnetic field values for each device is due to constant interference caused by the PCB layout and nearby parts. In Fig. 8.18, the x and y field outputs are plotted as the sensor is held flat and rotated around the z-axis. Without calibration, the field values are not uniform in peak magnitude in the x- and y-direction, as would be expected and the values are

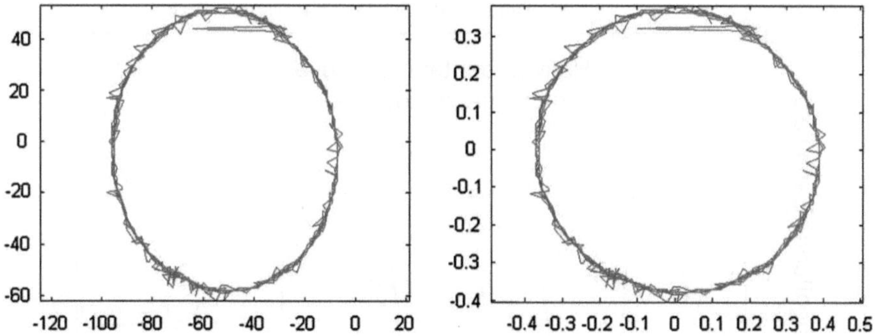

Fig. 8.18 XY magnetic data plot before calibration and after calibration

offset from one another. This causes the xy plot to appear oval and not centered at the origin.

This calibration need only been done once before final delivery of the sensor, and the calibration values can be programed into the device. To determine sensor calibration values, the offset values and range of sensor data must be determined. The calibration should be done where magnetic interference is minimized (e.g. outside away from buildings). For each magnetic field sensor (x, y, and z), we need to determine minimum and maximum values. The INU is rotated about the x, y, and z axes and minimum and maximum values for each are determined. Then calibration parameters are computed by

$$\text{OFFSET} = (\min + \max)/2 \text{ and RANGE} = \text{abs}(\max - \min)$$

Similar procedures must be performed to calibrate to the device on which the sensor is mounted.

In addition, the magnetic field readings can lead to significant heading or azimuth errors if they are not tilt compensated. The tilt compensation requires additional sensor tilt information that can be computed using the three-axis accelerometers. Without continuous tilt compensation, tilt of the sensor can cause significant heading errors. By computing how the tilt errors propagate to the azimuth angle, it can be seen that the heading errors are strongly affected by the azimuth angle itself and they are also affected by the field inclination—the angle the magnetic field makes with the horizontal plane when facing magnetic north (Ladetto et al. 2002).

8.4.3 Inertial Navigation Unit (INU): Compass Fusion

In a INU compass fusion system reported in Funk et al. (2007), it was observed consistently that the mean heading error is generally large when either the magnitude or inclination of the magnetic field is far from the expected values or the

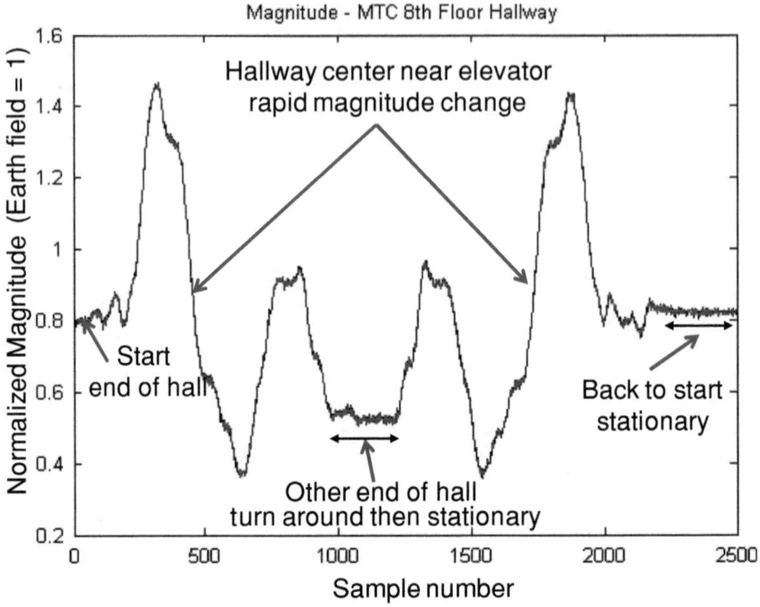

Fig. 8.19 Magnetic field magnitude variation over time

field variance is high. For example, in the hallway in which the data was taken for Fig. 8.17 the magnetic magnitude (Fig. 8.19) and inclination (Fig. 8.20) have high variance and vary significantly from their nominal values of normalized magnitude 1 and inclination 66.5° as a the subject traverses the hallway back and forth once.

Another indicator of the quality of the magnetic data is how well it follows the inertial orientation over the short term.[7] When the magnetic field data is undisturbed, the change in inertial heading should match the change in magnetic field heading. On the other hand, the compass provides an angle with absolute reference in presence of "clean" Earth field. With this in mind, the compass/gyro fusion algorithm is designed to allow the gyroscope to control high frequency angle variations and the compass to control low frequency variations when the compass data reliable.

Using the above indicators of data reliability (magnetic field magnitude, inclination and variance and agreement with inertial heading changes) the feedback algorithm is able to attenuate data that has been affected by magnetic disturbances from the building feedback and minimize output error variance when a person is moving (Funk et al. 2007).

Figure 8.21 illustrates the operation of algorithm where $d\theta_{gyro}$ is the gyro measured angular rate, $\theta_{compass}$ is the measured compass heading, |H| is the magnetic field magnitude, θ_{out} is the fusion algorithm output, and $e = \theta_{compass}(i)-$

[7] Short is relative to the quality of the gyroscope. Refer to Table 8.1 for typical gyro drift rates by class.

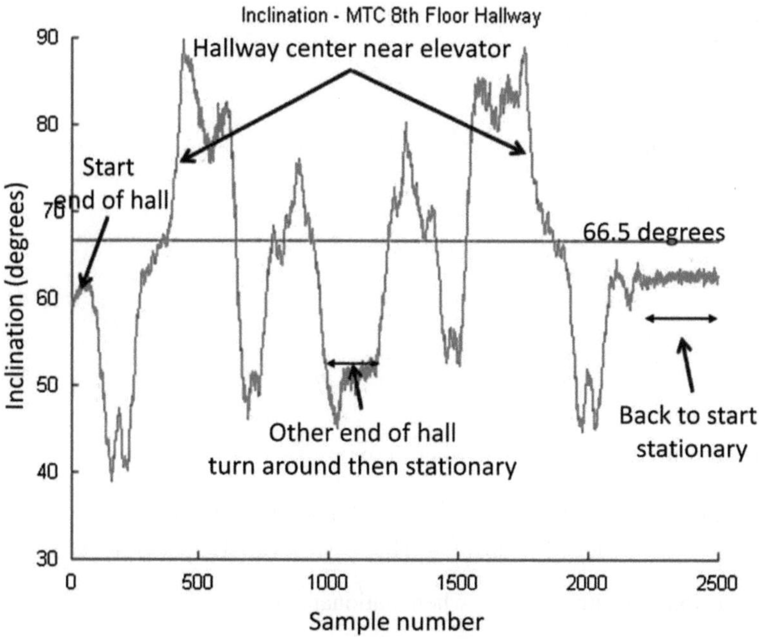

Fig. 8.20 Magnetic field inclination variation over time

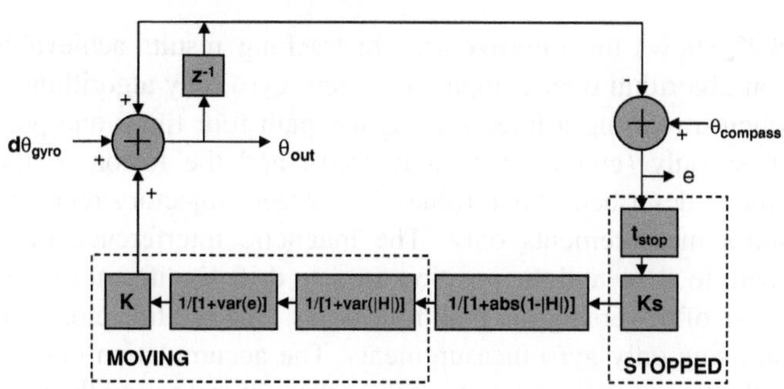

Fig. 8.21 Gyro-compass fusion algorithm

θ_{out}(i−1) is the difference between the compass angle and the fusion angle. The dashed sections are only active when the user is moving or when the user is stopped, respectively. This is because magnetic field variance is not an accurate indicator of compass correctness when the user is stationary. K is the time constant for rate correction when moving and $t_{stop}K_s$ is the time varying constant for rate correction when stopped. The attenuation factors are: 1/[1 + abs(1−|H|)] which attenuates fields that vary from the expected normalized Earth field value of 1, 1/[1 + var(|H|)] which attenuates fields that have high magnitude variation, 1/[1 + var(e)] which attenuates fields that have high error variance. The range for all attenuation factors is 0–1.

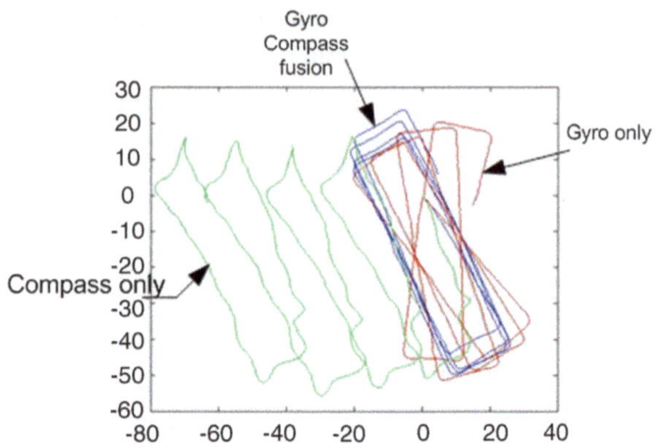

Fig. 8.22 Tracking results from traversing a rectangular path with magnetic distortion–*Green*–compass only, *Red* -gyro only, and *Blue*—compass and gyro

The fusion algorithm output when moving is:

$$\theta_{out}(i) = \theta_{out}(i-1) + d\theta_{gyro} + [\theta_{compass}(i) - \theta_{out}(i-1)] \\ * K/([1 + \text{abs}(1 - |\text{H}|)] * [1 + \text{var}(|\text{H}|)] * [1 + \text{var}(\text{e})])$$

The fusion algorithm output when stationary is:

$$\theta_{out}(i) = \theta_{out}(i-1) + d\theta_{gyro} + [\theta_{compass}(i) - \theta_{out}(i-1)] \\ * t_{stop}K_s/[1 + \text{abs}(1 - |\text{H}|)]$$

Figure 8.22 shows the improvement in tracking results achievable with this simple fusion algorithm over compass only and gyro only algorithms. The data is collected when traversing a fixed rectangular path four times and post processed using compass only (green), gyro only (red), and the fusion of the gyro and compass data as described above (blue). The green trajectory represents the path using compass measurements only. The magnetic interference in the hallway causes the path to skew and the position error to drift significant in the x-direction over the course of traversing the path four times. The red trajectory represents the computation using only gyro measurements. The accumulation of errors with the gyro causes the trajectory to rotate over the course of traversing the path four times so that eventually the heading estimates drift unacceptably. The blue trajectory represents the computation using both gyro and compass data together. As is evident from the figure, the error is significantly reduced in the trajectory with simple feedback control.

8.5 Accuracy Metrics

The accuracy of pedometer based inertial tracking systems is often quoted as a percentage of distance travelled. Quoting error as a percentage of distance travelled can be very misleading; distance travelled metrics tend to underestimate the

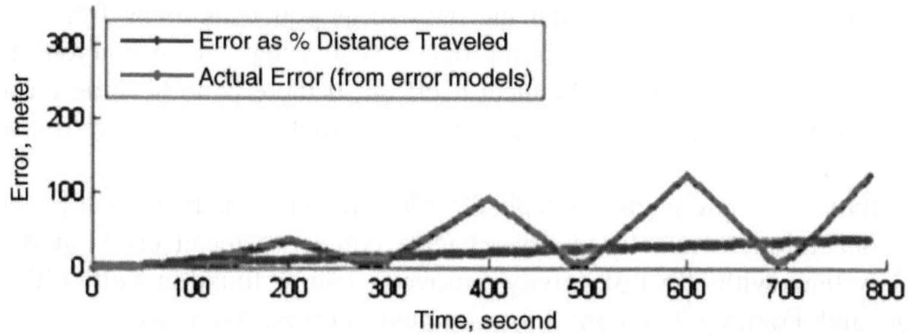

Fig. 8.23 Error growth models—error modeled as a percentage of distance travelled will often underestimate the possible error in poor magnetic environments where both heading and scaling errors occur

tracking error when considering indoor tracking applications. Percentage of distance travelled estimates give the best case performance for systems; that is, the data is taken walking at constant pace in a straight path, in an area where the compass is reliable, and so, zero heading error is assumed. Thus, the percentage distance travelled error ONLY accounts for errors introduced by improper scaling. Scaling is NOT typically the major source of error in inertial systems when tracking indoors! As we described above, the gyro errors can contribute significantly to errors in position if not compensated. Errors expressed as a percentage of distance travelled tend to significantly underestimate potential errors in GPS denied environments, which is where many users will want to use the system.

To simplify the issues, let us focus on the 2D problem and assume that the person is walking forward. There are two major sources of X–Y error: heading error and scaling error. If the compass works relatively well, then the heading error can be discounted. If that is not the case and the compass is significantly degraded, which happens often in environments where GPS denied tracking is needed; the error will change in a complicated manner. The accuracy will not only will be a function of distance travelled but also a function of time as well as the shape of the path taken. Even with useful compass data, the error depends on the shape of the path taken and the accuracy can both decrease and increase over time as shown in Fig. 8.23.

Consider the example of a person walking back and forth on a straight line for 200 m (at 1 m per second—equating metrics of time and distance travelled in this example), then making a 180° turn and coming back to the starting point then repeating the same path one more time. In Fig. 8.23 below, the straight line shows the computed error growth as a percentage of distance travelled (5 % for this example) and the zig-zag curve shows the computed worst case error dynamics calculated accounting for both heading (gyro) and scaling errors using values that are typical for MEMS inertial sensors. Notice that as the person turns at the 200 s point (equivalently 200 m in this example) to return to their initial location, the error begins to decrease. This is because we assume the scaling error is approximately constant throughout the path so as you walk away from the start point, the error increases with each improperly scaled step, and as you return to the start point the scaling error will cancel itself out if there is no heading error. The heading error, however, continues to grow and quickly overshadows the scaling error.

The above example is quite simplistic. More detailed mathematical analysis of error propagation in *accelerometry*—a term coined to mean error analysis for inertial systems with effect of gravity removed—can be found in Kelly (2010) and in Wan and Foxlin (2010) for foot-mounted sensors. Wan and Foxlin (2010) provide a rule of thumb error growth for unaided inertial systems as percentage of the bounding diameter of the course per minute.

The values for gyro drift and scaling error that should be used to estimate heading and scaling error will depend on the quality of sensor being used. It is important to understand that if only scaling errors are considered, the error will be underestimated significantly. Higher level navigation algorithms are required to mitigate error growth in situations where magnetic sensor data is often unreliable. Algorithms that fuse together map information (see Chap. 9) along with multiple sensors with complimentary properties are needed to achieve robust navigation solutions. Sensors which are degraded in a specific situation can be de-emphasized or eliminated from the navigation calculations. In many environments where GPS denied tracking is needed, accuracy of the even base sensors cannot be well represented as a percent of distance travelled; hence, it is not useful for the integrated system to be characterized in this manner.

Accuracy can be improved in one of two distinct ways: (1) higher performance (and higher cost) hardware or (2) signal processing algorithms that incorporate redundancy and other external information.

8.6 Summary

This chapter was intended to provide an introduction to the use of inertial and other body worn sensors as part of a solution for GPS denied navigation system. The availability of low cost MEMS inertial sensors in commercial smartphones and gaming devices enables easy access to sensor data for research and product development. The key to making these low accuracy MEMS inertial sensors part of a precision positioning system is developing methods to both *minimize* free inertial position error growth and *bound* accumulated inertial position errors.

It is well accepted that a high accuracy navigation solution requires the ability to fuse input from multiple sensors making use of all available navigation information. In this chapter, we discussed the use of velocity sensors (optical and Doppler radar), zero velocity updates and magnetic sensors to control inertial error growth. RF ranging for location trilateration (Chaps. 2, 3, 5, 6, 7), and Wi-Fi finger printing (Chap. 4) can also provide information that can be used to bound inertial error growth or provide system initialization/reinitialization. In Chap. 9, we discuss other methods for using feature mapping to control error growth.

References

N. Al-Sheimy, The promise of MEMS to the navigation and mobile mapping community. ION (2009)

L. Bao, S.S. Intille, Activity Recognition from User-Annotated Acceleration Data. Pervasive Computing (2004)

J.L. Barron, D.J. Fleet et al., Performance of Optical Flow Techniques. Int. J. Comput. Vision **12**, 43–77 (1994)

G. Barrows (2011) centeye.com/technology-and-research/optical-flow/

H. Bay, A. Ess et al., SURF: speeded up robust features. Comput. Vis. Image. Underst. **110**(3), 346–359 (2008)

U. Blanke, B. Schiele, Sensing location in the pocket. in *Proceedings UbiComp* (2008)

J. Bouchaud, Winners emerge in the consumer electronics and cell phone MEMS segments in 2010. Market watch (2011)

C.V. Bouten, K.T. Koekkoek et al., A triaxial accelerometer and portable data processing unit for the assessment of daily physical activity. IEEE Trans Bio Med Eng **44**(3), 136–147 (1997)

T.J Brand, R.E. Phillips, Foot-to-foot range measurement as an aid to personal navigation. ION 59th Annual Meeting. Albuquerque, NM: 113–125 (2003)

R.G. Brown, Integrated navigation systems and kalman filtering: a perspective. J Inst Navig **19**(4), 355–362 (1972)

T. Chau, A review of analytical techniques for gait data. part 1: fuzzy, statistical and fractal methods. Gait Posture **13**, 49–66 (2001a)

T. Chau, A review of analytical techniques for gait data. part 2: neural network and wavelet methods. Gait Posture **13**, 102–120 (2001b)

Y. Cui, J. Chipchase et al., *A Cross Culture Study on Phone Carrying and Physical Personalization* (Springer, Berlin, 2007). Usability and Internationalization

L. Fang, J. Antsaklis et al., Design of a wireless assisted pedestrian dead reckoning system-the navmote experience. IEEE Trans. Instrum. Meas. **54**(6), 2342–2358 (2005)

M.A. Fischler, R.C. Bolles, Random sample consensus: a paradigm for model fitting with applications to image analysis and automated cartography. Comm. ACM. **24**(6), 381–395 (1981)

E. Foxlin, Inertial Head-Tracker Sensor Fusion by a Complementary Separate-Bias Kalman Filter. VRAIS, IEEE (1996)

E. Foxlin, Pedestrian tracking with shoe-mounted inertial sensors. IEEE Comput Graph Appl (2005)

Frost and Sullivan, Motion detection sensors: MEMS inertial sensors. Tech. Insights (2007)

B. Funk, A. Bandyopadhyay et al., Method and System for Locating and Monitoring First Responders. USPTO. US, TRX Systems. 0077326 (2007)

R. Ghodssi P.Y. Lin, eds. MEMS Materials and Processes Handbook, Springer (2011)

S. Godha, G. Lachapelle et al., Integrated GPS/INS System for Pedestrian Navigation in a Signal Degraded Environment. ION GNSS, Fort Worth, ION (2006)

F.R. Hampel, E.M. Ronchetti et al., *Robust Statistics: The Approach Based on Influence Functions* (Wiley-Interscience, New York, 1986)

C.Harris, M. Stephens, A Combined Corner and Edge Detector. *Proceedings of the* 4th *Alvey Vision Conference* (1988)

C.G. Harris, *Geometry from visual motion* (MIT Press, Cambridge, 1992). MA

R. Herren, A. Sparti et al., The prediction of speed and incline in outdoor running in humans using accelerometry. Med. Sci. Sports Exerc. **31**(7) (1999)

W. Higgins, A Comparison of Complementary and Kalman Filtering. IEEE. Thans. Aerosp. Electron. Syst. **11**(3), 321–325 (1975)

R E. Hopkins, N.M. Barbour et al., Miniature inertial and augmentation sensors for integrated inertial/gps based navigation applications. Low-Cost Navigation Sensors and Integration Technology. NATO Research and Technology Organisation Sensors and Electronics Technology Panel (RTO-EN-SET-116). Madrid, Spain (2009)

T. Judd, *A Personal Dead Reckoning Module* (ION GPS, Kansas City, 1997)

J. Karvounis, Theory, design, and implementation of landmark promotion cooperative simultaneous localization and mapping. Electr. Comput. Eng. College Park, University of Maryland **Ph.D** (2011)

A. Kelly, Error Propagation in Aided Discrete 2D Accelerometry, The Robotics Institute Carnegie Mellon University (2010)

J. Kolodko,L. Vlacic, Motion Vision: Design of Compact Motion Sensing Solutions for Navigation of Autonomous Systems (2005)

J.B. Kuipers, Quaternions and Rotation Sequences: A Primer with Applications to Orbits, Aerospace and Virtual Reality Princeton University Press (1999)

Q. Ladetto, On foot navigation: continuous step calibration using both complementary recursive prediction and adaptive Kalman filtering. ION. GPS. (2000)

Q.Ladetto, J.v. Seeters et al., Digital Magnetic Compass and Gyroscope for Dismounted Solider Position and Navigation. Military Capabilities enabled by Advances in Navigation Sensors, Sensors and Electronics Technology Panel, NATO-RTO meetings, Istanbul, Turkey (2002)

M.Laverne, M. George et al., Velocity- and Range-Aided Micro Inertial Measurement for Dismounted Soldier Navigation. ION Joint Navigation Conference, Colorado Springs (2011)

S. Lee, K. Mase, Recognition of walking behaviors for pedestrian navigation. IEEE Conference on Control Applications (CCA01), Mexico City, Mexico (2001)

B.D. Lucas, T. Kanade, An iterative image registration technique with an application to stereo vision. DARPA Image Understanding Workshop (1981)

S. Maus, S. Macmillan et al., The US/UK World Magnetic Model for 20010–2015, NOAA Technical Report NESDIS/NGDC (2010)

R. McCroskey, P. Samanant et al., GLANSER: an emergency responder locator system for indoor and gps-denied applications. ION. GNSS. Portland, Oregon, ION: 2901–2909. (2010)

S. Onyshko, D. Winter, A mathematical model for the dynamics of human locomotion. J. Biomech **13**, 361–368 (1980)

J. Shi, C. Tomasi, Good features to track. IEEE. Conf. Comput. Vis. Pattern Recogn. 593–600 (1994)

W. Soehren, W. Hawkinson, Prototype Personal Navigation System. IEEE A&E SYSTEMS MAGAZINE (April) (2008)

U. Steinhoff, B. Schiele, Dead reckoning from the pocket: an experimental study. IEEE. Int. Conf. Pervasive Comput. Commun. (2009)

D. Strelow, S. Singh, Optimal motion estimation from visual and inertial measurements. in *Proceedings of the Workshop on Applications of Computer Vision* (2002)

D. Titterton, J. Weston, *Strapdown Inertial Navigation Technology* (United Kingdom, IEE and AIAA, 2004)

M.J. Veth, Navigation using images, a survey of techniques. J. Inst. Navig. **58**(2), 127–139 (2011)

S. Wan, E. Foxlin, Improved Pedestrian Navigation Based on Drift- Reduced MEMS IMU Chip. ION 2010 International Technical Meeting. San Diego (2010)

F. Weimann, G. Abwerzger et al., *A Pedestrian Navigation System for Urban and Indoor Environments* (ION GNSS, Fort Worth, 2007)

O.J. Woodman, *An introduction to inertial navigation* (University of Cambridge, Cambridge, 2007)

F. Zajac, R. Neptune et al., Biomechanics and muscle coordination of human walking Part I: Introduction to concepts, power transfer, dynamics and simulations. Gait Posture **16**, 215–232 (2002)

F. Zajac, R. Neptune et al., Biomechanics and muscle coordination of human walking Part II: Lessons from dynamical simulations and clinical implications. Gait Posture **16**, 1–17 (2003)

Chapter 9
Localization and Mapping Corrections

In the last chapter, the main focus was on inertial body reference, *idiothetic* sensors, which provide *internal information* about the subject's movements. In this chapter, we add information from local reference, or *allothetic* sensors, which provide *external information* about the environment. In Chaps. 2–7, there has been much discussion of RF ranging sensors which are a type of allothetic sensor providing ranging to fixed beacons or other tracked personnel/platforms. Another example of a common local reference sensor is an image sensor. Even inertial sensors, which are typically used as body reference sensors, can provide local map reference data by inferring the location of terrain features based on the sensor data (Funk et al. 2007; Bandyopadhyay et al. 2008). These allothetic sensors allow us to create a feature map of what is around us and to locate ourselves within that map—localization and mapping.

In this chapter, we review some different allothetic sensors and the types of features that can be extracted for localization. The ability to extract unique features that can be recognized when "seen" again is the basis for creating feature maps that can be used to aid in localization. Next, the theoretical formulation and common solution approaches for the localization and mapping problem are reviewed. Finally, an example is given that addresses some of the practical issues for implementing localization and mapping solutions.

9.1 Localization and Mapping Overview

The goal of localization and mapping is to compute the most probable observer location within the discovered map given the past sensor and control values (if available). Called *simultaneous localization and mapping* (SLAM), SLAM requires the use of sensors to construct a geometric or topological map of the environment and then use that map for localization (Smith and Cheeseman 1986; Durrant–Whyte 1988; Smith et al. 1990; Dissanayake et al. 2001; Guivant and Nebot 2001; Montemerlo et al. 2003a; Montemerlo and Thrun 2003b; Thrun et al. 2006).

C. Gentile et al., *Geolocation Techniques*, DOI: 10.1007/978-1-4614-1836-8_9,

The map information also enables us to constrain the growth of errors in body reference sensor systems. The ability to constrain the errors is dependent on the quality the idiothetic and allothetic sensors.

In SLAM, both the trajectory of the observer—positions, velocities, and headings (etc.)—together with features of the map are estimated online without the requirement for any a priori knowledge of location. Although, navigation and mapping systems may have access to pre-existing map data. This map data might consist of GIS (geographic information system) shape files (including building outlines, roads, etc.), satellite imagery, elevation maps, and building maps (CAD files, floor plans, etc.). This existing map information can be used for to refine SLAM algorithms results where map data exists while still allowing new features that are discovered to be included in the global map.

Work by Meyer and Filliat (Filliat and Meyer 2003; Meyer and Filliat 2003) provides a useful summary of map-based navigation, which involves three processes:

- **Map-learning**—the process of transforming the data acquired during exploration to a suitable representation and structure constituting a map.
- **Localization**—the process of deriving the current position within the map.
- **Path-planning**—the process of choosing a course of actions to reach a goal, given the current position and map.

Localization and map-learning are interdependent processes; the positions of tracked entities and discovered features/landmarks are estimated relative to the currently known map. On the other hand, path-planning is a somewhat independent process that takes place once the map has been built and the subject's position estimated.

These three processes may rely on both idiothetic and allothetic sensor data. Idiothetic information may include speed, acceleration, leg movement for dismounts, wheel rotation for vehicles, etc. Through dead reckoning, these data provide position estimates of the subject in a metric space. Idiothetic sensors can also provide local map reference data by inferring the location of terrain features based on how the subject moves through the environment. For example, they have been effectively used to locate features in structured environments such as stairways and elevators in buildings (Funk et al. 2007; Bandyopadhyay et al. 2008).

Allothetic information can be used to directly recognize a place or a situation; in this case, any cue such as image features, sonar time-of-flight, color, etc., may be used. Allothetic information can also be used to derive subject motion from measurements of the environment. That is accomplished by converting information expressed in the space related to the idiothetic data based on metric models of the associated sensors. With such a metric model, it is possible to infer the relative positions of two places in which allothetic information has been gathered (Filliat and Meyer 2003). For example, frame-to-frame stereo camera feature tracking can be used to solve for six degrees of freedom motion of the camera (see Chap. 8).

The limitations and advantages of these two sources of information are complementary. Indeed, the main problem associated with the derived metric motion information is that, because it involves a dead reckoning process, it is subject to ***cumulative error*** (for example, heading error in an inertial system). This leads to a continuous decrease in quality; therefore, such information cannot be trusted over long periods of time. On the contrary, the quality of feature based map information is constant over time, but it suffers from the ***perceptual aliasing problem***, e.g., for a given sensor system, two distinct places (landmarks) in the environment may appear the same, for example, doors or light fixtures.

Consequently, to build reliable maps and to navigate for long periods of time, the user track and map information must be combined. In other words, map information must compensate for sensor information drift while user motion/track information must allow perceptually aliased allothetic information to be disambiguated. When both allothetic and idiothetic sources of information are available, there are many ways to integrate them in a representation useful for navigation. Classically, the corresponding representations are referred to as metric maps or topological maps (Filliat and Meyer 2003).

In metric maps, geometric properties of the environment such as the positions of objects are stored in a common reference frame. A metric map can be represented as a 2D floor plan or a 3D architectural map. The quality of the synthetic metric map is dependent on the quality of the idiothetic and allothetic sensors. For example, the scale and shape of the metric map are affected by the quality of the position estimated by idiothetic sensors. The drift of the position estimate is difficult to correct without making assumptions about particular properties of the environment, such as orthogonal hallways; or alternatively, without closing the loop, that is, revisiting a feature with previously recorded location and using that knowledge to estimate biases and correct computed position errors. Converting raw allothetic information such as range to a feature into a metric space is dependent on the properties of the sensor, such as measurement accuracy, and also on the local properties of the environment, for example, optical features are difficult to extract from blank walls or dimly lit areas.

In topological maps, it is the allothetic characterizations of places (features/landmarks) that the subject can reach that are stored, along with some information about their relative positions, for example, a list of discovered features/landmarks with connections to other features that can be directly accessed from the given feature. This type of high-level connection diagram of the environment is valuable in path planning. Additional details about the advantages and drawbacks of these representations can be found in Filliat and Meyer (2003).

9.2 Map Features

For each sensor type, extracting reference information from sensor measurements that can be used for navigation requires finding "unique" information, *a feature*, in the sensor data that is suitable for tracking. This means a feature that can be recognized by the sensor algorithms when encountered again. The sensor features (also referred to as landmarks) can be saved to form a map of the environment which is used to aid navigation. Mapped features can be used to provide navigation corrections when a feature is revisited.[1]

When we think of a map for navigation, several types of maps may come to mind, for example, GIS maps like Open Street Maps or elevation contour maps. These are maps that humans can interpret to aid in navigation. In SLAM, as the subjects traverse the world, they collect map landmarks or features to be used by navigation algorithms. The types of features collected can be quite different. In this section, we review a few types of map features that might be used by a navigation system.

9.2.1 Optical Features

The easiest setting to think about SLAM is in the context of an optical navigation system. The system "sees" a landmark and its relative location and logs it. Then, when the subject revisits the landmark, if any errors in position have accumulated, the subject's location can be updated based on the landmark's prior location estimate. The human brain is quite adept at selecting and matching landmarks in varying conditions, but this is a difficult problem for a machine vision system.

One of the classic challenges for computer vision systems is to make object identification reliable when the same object is viewed from different perspectives and distances, and in different lighting conditions. The premise of many vision algorithms is that interesting features on an object can be extracted together with their relative spatial locations to provide a feature based description of the object that is robust to changes in these parameters.

Another classic challenge for computer vision is to detect objects and structures that are partially blocked. Feature based approaches are well suited to tackle these problems because they treat an object as the sum of its parts rather than the precise match of the whole.

Optical landmarks also suffer from perceptual aliasing, for example, in an office building, many doors look the same. There has been significant research in computer vision system to address these issues and algorithms have been developed with varying degrees of robustness. Algorithms trade off computational complexity to achieve better object recognition performance.

In computer vision research, feature extraction methods have been developed in an attempt to overcome these issues. Corner based features are useful for detecting, characterizing and identifying man-made objects. A well-known algorithm is Harris corner detector (Harris and Stephens 1988). Selected features must be sufficiently distinct so there is low probability of mismatch. Identifying distinctive landmarks is not always simple. For example, viewed from varying distances the

---

[1] Feature tracking can also be used to directly solve for the resulting motion of a sensor if enough information is gathered to infer the relative movement of features in a metric map as a result of the subject motion, for example, stereo camera feature tracking.

objects will have different scales. Lindeberg introduced the concept of automatic scale selection. He showed that for feature detectors expressed in terms of Gaussian derivatives, when estimating image deformations, such as in image matching computations, scale levels with associated deformation estimates can be selected from the scales at which normalized measures of uncertainty assume local minima with respect to scales (Lindeberg 1998).

Another common image processing algorithm used for object identification is *Scale Invariant Feature Transform* (SIFT). (Lowe 1999) The algorithm is designed to detect and describe *local* features in images. Its basic premise is that objects can most reliably be recognized based on local features and their relative spatial locations. The SIFT algorithm identifies "key points" based on contrast gradients. The SIFT key points allow one to efficiently match small portions of cluttered images under rotations, scaling, change of brightness and contrast, and other transformations (Lowe 1999, 2004).

SIFT gets mixed reviews when used for SLAM applications. One major complaint is that the algorithm is computationally intensive (Lemaire and Lacroix 2007), which hinders real-time implementation. When Lowe used SIFT as a means for conducting stereo vision SLAM, the system ran at 2 Hz on a Pentium III 700 MHz processor (Se et al. 2005), a very slow computer by today's standards. The positive aspect of SIFT is its ability to produce distinctive features from natural landmarks (Miro et al. 2005; Se et al. 2005; Sim et al. 2005; Elinas et al. 2006). The distinctiveness of SIFT allows SLAM algorithms to perform global localization more easily and allows closing-the-loop approaches to work robustly (Se et al. 2005; Elinas et al. 2006).

Speeded Up Robust Features (SURF) Bay et al. (2008) was developed to address some of the computational issues of SIFT. It is loosely based on SIFT, but it uses integral images for image convolutions which is computationally faster. An integral image is an image where the value at any point (x, y) in the image is the sum of all the pixels from the origin of the original image up to and including (x, y) (Bay et al. 2006). SURF approximated, and even outperformed, SIFT and select variants (PCA–SIFT and GLOH) with respect to repeatability, distinctiveness, robustness; it also computed and compared features much faster (Bay et al. 2006).

In work for TRX, Karvounis implemented a SURF demonstration running at 30 Hz on a desktop computer—I7 Quad-Core 2.4 GHz processor (Karvounis 2011a). For these tests, a Logitech 9000 webcam at 320 × 240 resolutions was used to capture images. A database of known landmarks was created manually containing images of several "landmarks" in an office setting.

A SURF visualization was created that displays real-time updates of the camera image in the bottom panel (see Fig. 9.1). Then, as the camera is moved, each captured camera frame from the bottom box is compared with all the landmark images stored in the database. The top image displays a black box until the bottom frame matches one of the images stored in the database. Once a match is found, see Fig. 9.1, the top image shows the matched landmark from the database. The green lines indicate the feature matches and the cyan frame indicates the relative position of the captured image with respect to the database landmark image.

Fig. 9.1 Surf feature match

Color is not used in these algorithms. Color is an important property used by humans for object recognition; however, color perception in machine vision is very complex. A person is able to perceive color as relatively constant in differing lighting conditions. On the other hand, machine vision systems are generally not so sophisticated. For example, the color histogram derived from a digital image may vary markedly for the same object under differing lighting conditions. But as long as the illumination is held fairly constant, color histograms can be a very effective feature for object identification (Abdel-Hakim and Farag 2006; Sande et al. 2010). Frame-to-frame lighting is more likely to be nearly constant but over longer periods lighting is likely to change.

The discussion of the algorithms and software in this section is centered on image recognition. An in-depth discussion of hardware is beyond the scope of this chapter; however, since a camera may serve as the "eye" of the navigation system, its characteristics can greatly affect the functionality of the system. The quality of the images produced by the camera directly affects the processing speed as well as the ability to identify objects. The properties of the lens directly affect the field of view and the ability to carry out optical ranging. At greater distances, the resolution of the camera can be the limiting factor for feature recognition and ranging.

9.2.2 Inference-Based Features

The desire for improved localization using only the sensors available on a cell phone is driving researchers to focus on developing methods that leverage only the cell phone's embedded sensor information to its maximum benefit for pedestrian navigation. A useful source of environmental information can be derived from a tracked subject's motion. A standard approach to tracking is to use an inertial navigation unit (INU) in a dead reckoning mode making use of only the idiothetic dead reckoning information provided by the INU sensors. Inertial sensors can provide allothetic map reference data by inferring the location of terrain features based on how the subject moves through the environment. In making use of this additional information, the capability of the INU is improved to function as a smart, standalone positioning device providing a rich set of inputs for SLAM algorithms. For example, just as a stereo-optical sensor might provide SURF features and descriptors with range information for each selected feature detected in an optical frame, a "smart" navigation unit can provide inertial building and shape features and signal-based features (e.g., magnetic or signal strength when these sensors are available in the navigation unit) for input to SLAM algorithms.

For example, in buildings, floor plans represent a specific partition of a 2D space into spatially characteristic areas such as hallways, rooms, points of entry or exit including stairwells, elevators, and escalators. The existence of a hallway might be inferred if a subject moves for a long period in a confined straight area. Climbing stairs indicates the presence of a stairwell and an elevation change without climbing stairs might imply an elevator. The location and orientation of each inferred feature are known based on the idiothetic information. Several researchers have now shown that these inferred features can be used to mitigate the accumulation of inertial dead reckoning errors (Funk et al. 2007; Bandyopadhyay et al. 2008; Robertson et al. 2009a, b, 2010; Borenstein 2010; Wang et al. 2012). For example, inferred knowledge of hallways and other building grid constraints may be enforced on the navigation solution to yield an effective angular drift correction.

TRX has developed algorithms that detect such building features from track histories (Funk et al. 2007; Bandyopadhyay et al. 2008). These types of algorithms have been tested and evaluated in realistic scenarios with inertial sensors alone and found to markedly improve position accuracy. For example, in one 25 min long test, the error from pure inertial-based location estimate was reduced from 48 m to less than 3 m using the mapped-based constraint algorithms; see Fig. 9.2. Adding other sensor signature data can be used to improve uniqueness of inferred features. Investigators from Duke University and EJUST have begun to pick up on these ideas for recognizing and associating inertial signal features with fixed building features (Wang et al. 2012).

Investigators at the German Aerospace Center have developed a similar pedestrian 2D map inference system called FootSLAM (Robertson et al. 2009a, b, 2010). The algorithms builds on occupancy grid methods developed for robotic SLAM that use odometry based path data to develop a 2D map of open areas based on where the robot travelled. Instead of odometry, FootSLAM uses inertial-based dead reckoning as the input to FAST-SLAM algorithms (see section on Particle Filter based SLAM). Similar to the work at TRX, no visual or ranging sensors are used; instead the 2D is inferred based on the path data.

GPS and INUs are baseline metric sensors but they can provide inferred allothetic information. They should be distinguished from cameras, thermal imagers,

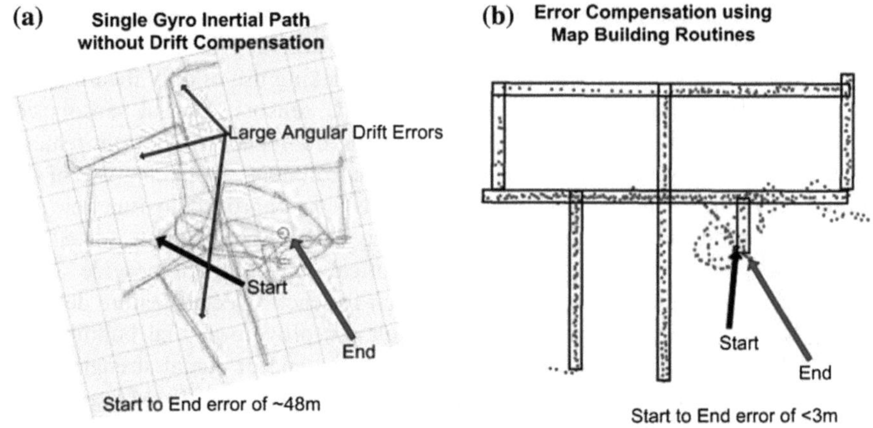

Fig. 9.2 **a** Uncompensated inertial path. **b** Generated map and compensated inertial path

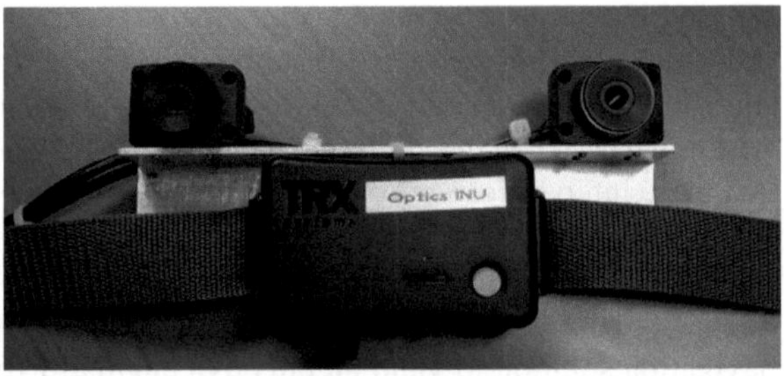

Fig. 9.3 A composite sensor—an optical INU integrating stereo vision for feature extraction with an INU

etc. that can produce "pure" topological measurements of relative range, range rate or bearing to a landmark. *Composite sensors*, for example combining vision and inertial measurements (Fig. 9.3) can combine metric and topological data to build hybrid maps that enable long term navigation. Together the combined sensors produce a "composite data array" consisting of a vector-valued path of INU position, velocity, heading, etc., together with the time-space paths of environmental features extracted from the cameras and inferred from subject motion.

For example, one might infer a hallway in a building by walking down it with only an inertial sensor; however, combining the inertial data with optical (or other) information, one may be able to estimate the length and width of the hallway as well. Figure 9.4 shows the stereo left and right camera images from the optical INU. The blue lines indicate algorithm detected hallway features. The red blocks show features that had a stereo match and the yellow lines link to matched features in the left and right image. The hallway width estimation results based on stereo line detection and matching is 1.61 m. The actual width of the hallway is about 1.52 m so the estimate is off by 0.09 m (3.5 inches).

In buildings, rigid assumptions can be made on the architecture of buildings to aid in identifying building features and the underlying map. These same assumptions do not necessarily hold in natural structures, such as caves.

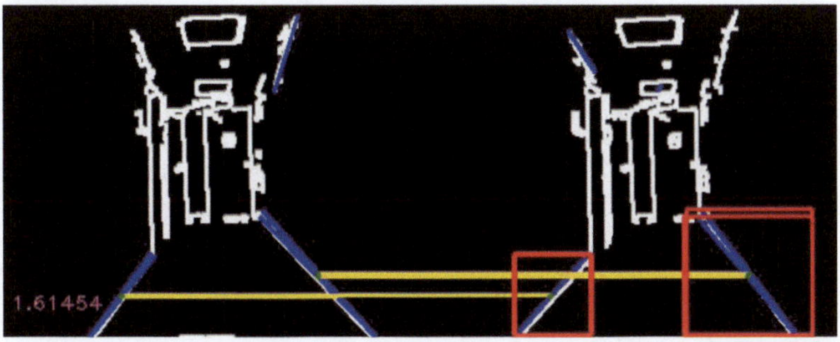

Fig. 9.4 Stereo images displaying feature matched hallway width computation

Nevertheless, it is reasonable to assume that natural trackable features will exist. Assumptions on the types of features can be adapted to allow the identification of key natural features in the environment. For example, different regions can have unique magnetic signatures, which can be measured by three-axis magnetic field sensors found in most smartphones, or received signal strength signatures (RSS), which can be accessed from most radios including, for example, Wi-Fi and Bluetooth. *Fingerprinting* methods for radio and other signals was discussed in detail in Chap. 4. In these techniques, the facility signatures are mapped a priori and then the signature map is used for localization.

9.2.3 Magnetic Features

Figure 9.5 shows an example of results from one of a sequence of magnetic signature experiments collected using a YAS529—MS-3C 3[2] axis magnetic field sensor while the tracked subject traversed the hallways of the AV Williams Building at the University of Maryland. Each corridor was found to display a consistent magnetic signature when the corridor was traversed multiple times. These signatures were recorded for three corridors as shown in Fig. 9.5. In each of the plots, the total magnetic field magnitude is plotted (y-axis) versus the sample number (x-axis) for two different traversals of each hallway. Note, there is some small variation between the two traversals for each hallway, but the hallways are clearly distinguishable.

To further test the uniqueness of the signatures, once the magnetic signatures for each hallway were recorded, tests were conducted where a small segment of one of the hallways was traversed resulting in a magnetic path signature. These magnetic signatures were tested against the three corridor database and the segments could be correctly identified in the part of the corridor where they were

2 http://pdf1.alldatasheet.com/datasheet-pdf/view/205144/YAMAHA/YAS529.html.

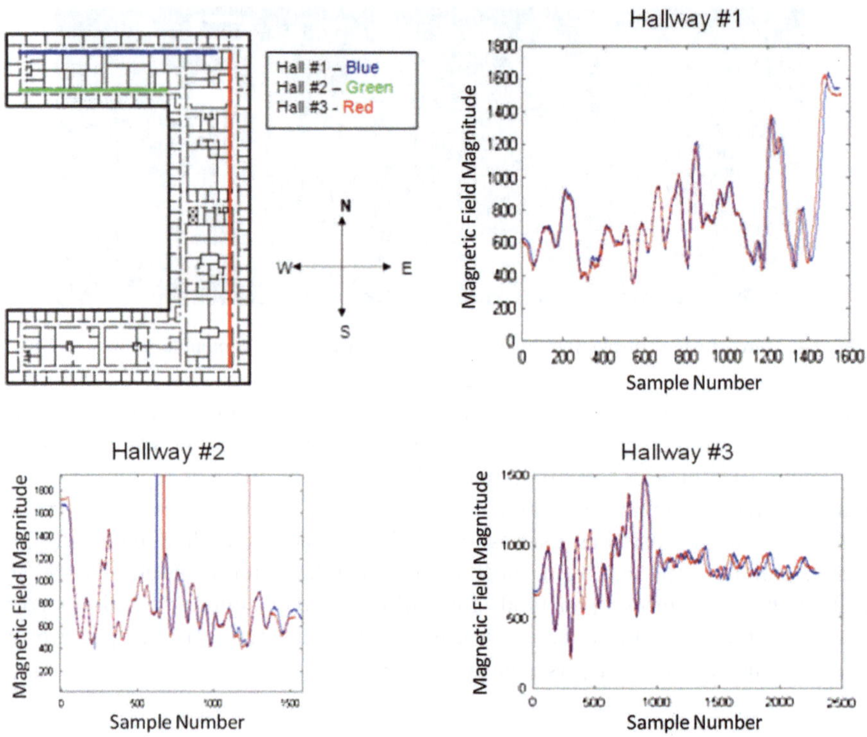

Fig. 9.5 Unique magnetic signatures of hallways

recorded.[3] These results clearly demonstrate the promise of magnetic signature fingerprints in aiding indoor localization when the signatures are available a priori.

A limitation of recording sampled signature data is that the data is speed and direction dependent. A computational method such as dynamic time warping is needed to account for variations in walking speed during the data collections. Dynamic time warping is a well-known technique for finding optimal alignment between two time dependent sequences and it is often is used in video and audio processing (Sakoe and Chiba 1978; Muller 2007).

Continuously matching path segments (in a large dataset) is computationally costly. Additionally, one may not have an a priori map as we did in the above experiment. Building a map of magnetic or signal features as the subject traverses an area, and using them for corrections in a SLAM implementation is an alternative to the fingerprinting techniques from Chap. 4.

Selecting only *interesting* features will minimize computation. Careful consideration of feature selection is critical for robustness. For example, an

[3] The subjects walked close to the center of the hallways during these tests at constant speed.

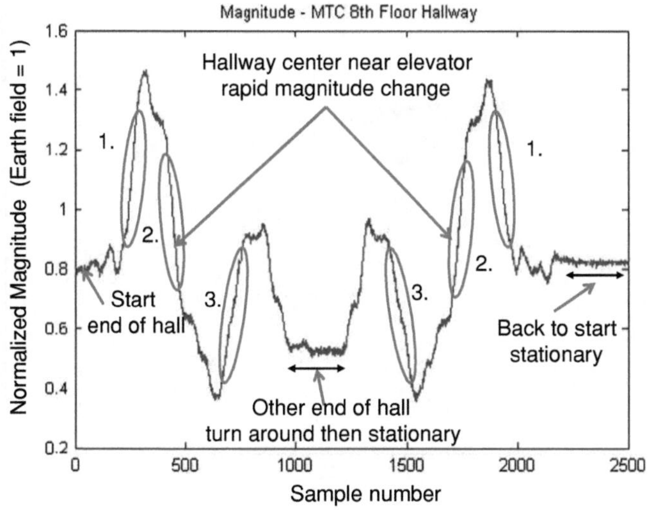

Fig. 9.6 Magnetic field Magnitude variation over time

approximately constant field may be fairly easy to match. Indoor environments typically provide a rich set of features for magnetic signatures. In outdoor environments, magnetic features may be sparse or indistinguishable. Once a feature is confirmed it can be deemed a landmark with an associated position. Recognized revisits to the landmark would subsequently provide a mechanism for mitigating accumulated dead reckoning errors.

To simplify computation, consider a well-localized magnetic feature, for example, an extreme or a *sharp transition in magnetic magnitude*. Sharp transitions are common in manmade structures with power systems and other metal causing magnetic disturbances. Figure 9.6 shows the magnitude of the magnetic field vector as a subject traverses back and forth in the hallway in an office building demonstrating the consistency of the signature. From Fig. 9.6, three sharp transition features are selected from the hallway traversal. These same three transition features are easily seen in each traversal.

The magnetic features are shown superimposed on a plot of the inertial path data of a user traversing back and forth in this hallway 10 times (Fig. 9.7). The inertial path shows clear scaling and drift errors. The path color represents the magnetic field magnitude. For each of the three magnetic features a minimum (triangle) and maximum (square) value are marked on the path.

Figure 9.8 shows a zoomed view of going back and forth in the hallway once. It is clear from these figures that the features may offer some scaling and drift correction if the features can be recognized and matched.

For signal based features such as magnetic fields, high sample rate data or other derived parameters can be saved as a feature descriptor detailing the unique aspects of the feature which can be used for matching if the features are observed at a later time. While the example above focuses on magnetic data, signature

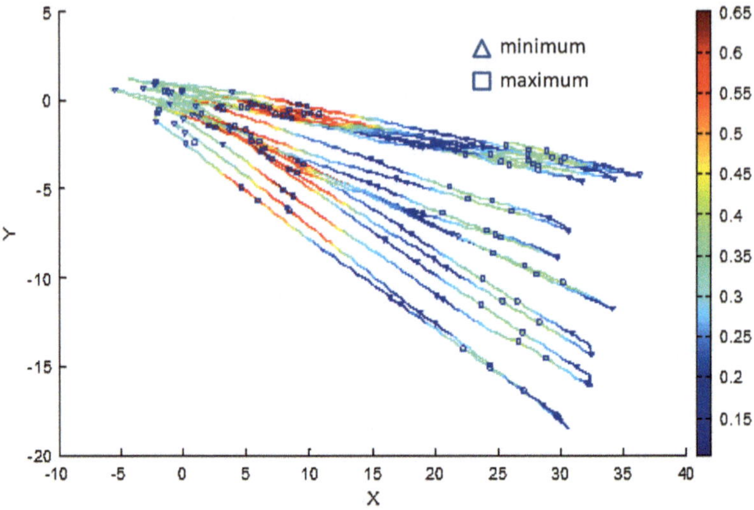

Fig. 9.7 Magnetic features superimposed on inertial track (X, Y position in meters from start point)

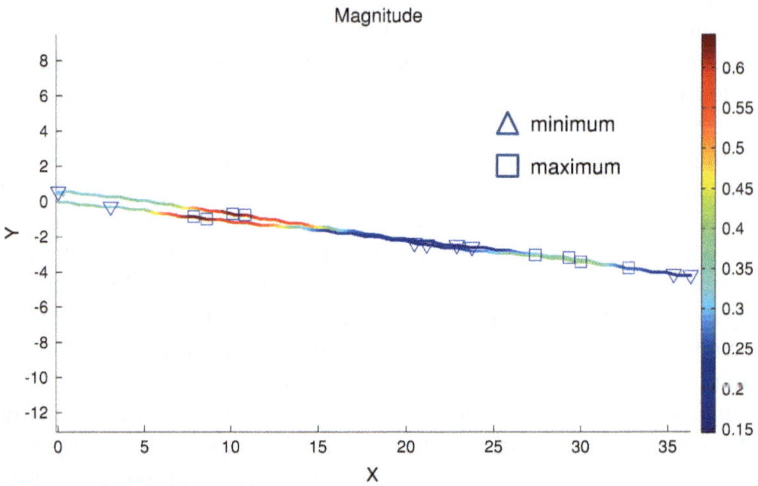

Fig. 9.8 Magnetic features superimposed on inertial track—zoom in view (X, Y position in meters from start point)

features are also valuable from other types of sensor data, such as radiation measurements or received signal strength.

We have highlighted a few allothetic sensors that provide map features that can be used for SLAM formulations. There are others we have not touched on such as LIDAR and SONAR. In Thrun et al. (2006) provide models for these and other sensors in the context of SLAM application. In the next sections, the formulation and solution of the SLAM problem, as well as issues with real-time implementation are discussed.

9.3 Simultaneous Localization and Mapping Formulation

In robotics, the SLAM problem is considered "solved." This theoretical solution has been one of the notable successes of the robotics community (Smith and Cheeseman 1986; Durrant-Whyte 1988; Smith et al. 1990; Durrant-Whyte and Bailey 2006a, b). Because position estimates and measurements are imperfect, the solution to the SLAM problem required the development of a way to update an uncertain geometric or topological environment model based on new observations that maintained consistent interpretation of relations between all of the uncertain features (Smith and Cheeseman 1986; Durrant-Whyte 1988). Work by Smith and Cheesman and Durrant-Whyte (Smith and Cheeseman 1986; Durrant-Whyte 1988) established a statistical basis for describing relationships between fixed landmarks with geometric uncertainty. A key contribution of this work was to show that, due to the common error in estimated observer location between landmarks, there must be a high degree of correlation between estimates of the location of different landmarks in a map. In fact, these correlations grow with successive observations of the landmarks. *Practically, this means that the relative location between any two landmarks may be known with high accuracy, even when the absolute location of a specific landmark is quite uncertain.* The combined mapping and localization problem, once formulated as a single estimation problem, is convergent—that is, the estimated map converges monotonically to a relative map with zero uncertainty. Additionally, the absolute accuracy of the map and subject location reaches a lower bound defined only by the uncertainty in the initialization (Smith and Cheeseman 1986; Durrant-Whyte 1988). The correlations between landmarks are the critical part of the problem and the stronger the correlations grow, the better the solution (Smith and Cheeseman 1986; Durrant-Whyte 1988; Smith et al. 1990; Durrant-Whyte and Bailey 2006a, b).

The SLAM problem can be broken into two pieces. The *observation model* (*or sensor model*) $p(z_t|x_t)$ describes the probability of making an observation z_t of selected landmarks when the observer location and landmark locations are known. In SLAM, the system state x_t includes the observer pose as well as the map. It is reasonable to assume that once the observer location and map are defined, observations are conditionally independent given the map and the current observer state. The *motion model* $p(x_t|u_t, x_{t-1})$ for the observer is assumed to be a Markov process in which the next state depends only on the immediately preceding state x_{t-1} and the applied control u_t (which may be unknown as is the case in personnel tracking) and is independent of both the observations and the map. The SLAM algorithm is then solved by a Bayes filter in a standard two-step time update, measurement update form.

1. **Time Update**: prediction of the state given the previous state and the control input

$$p(x_t|z_{1:t-1}, u_{1:t}) = \int p(x_t|u_t, x_{t-1})p(x_{t-1}|z_{1:t-1}, u_{1:t-1})dx_{t-1},$$

   and
2. **Measurement Update**: update of the predicted value given the most recent sensor data

$$p(x_t|z_{1:t}, u_{1:t}) = \eta p(z_t|x_t)p(x_t|z_{1:t-1}, u_{1:t})$$

where η is a normalization constant (Thrun et al. 2006).

The derivation of this and similarly all the popular recursive state estimation filters rely on the Markov assumption, which postulates that past and future data are independent given the current state. The Bayes filter is not practically implementable at this level of abstraction. Approximations are often made to control computational complexity, e.g., linearity of the state dynamics, Gaussian noise, etc. The resulting unmodeled dynamics or other model inaccuracies can cause violations of this assumption. In practice, the filters are surprisingly robust to such violations (Thrun et al. 2006).

In probabilistic form, the SLAM problem requires that the *joint* posterior probability density of the landmark locations and tracked subject's state (at time t), given the recorded observations and control inputs up to and including time t together with the initial state of the tracked subject, be computed for all times t. Solutions to the probabilistic SLAM problem involve finding an appropriate representation for both the observation model and the motion model, preferably recursive, which allows efficient and consistent computation of the prior and posterior distributions.

The SLAM problem has been formulated and solved as a theoretical problem in a number of different forms. However, issues remain in realizing general SLAM solutions in practice and notably in building and using perceptually rich maps as part of a SLAM algorithm. By far, the most common representation is in the form of a state-space model with additive Gaussian noise, leading to the use of the extended Kalman filter (EKF) to solve the SLAM problem.

The popularity stems from the fact that the EKF provides a recursive solution to the navigation problem and a means of computing consistent estimates for the uncertainty in subject and map landmark locations. This is despite the fact that many sensor noise models are not well represented by additive Gaussian noise.

9.3.1 Kalman Filter

Here, we take a short diversion to briefly discuss one of the most popular Bayesian filters, the Kalman Filter and a couple of its extensions, and to highlight some of the properties of the Kalman filter that drive its popularity. More detailed discussions and complete derivations can be found in Kailath (1980), Thrun et al. (2006).

A Kalman filter is a computationally tractable mechanism to incorporate

1. imprecise knowledge about a system—system dynamic models, noise models
2. system observations—measurements, sensor models

to yield an estimate of the current state. Under the assumptions that the system is linear and the model and observation errors are independent Gaussian random variables, the Kalman state estimate is an ***optimal*** estimate. There are several possible definitions for optimality

$$\text{Minimum Mean Square Error } \hat{\theta} = \underset{\theta}{\text{argmin}} \|y - f(\theta)\|^2$$

$$\text{Maximum Likelihood } \hat{\theta} = \underset{\theta}{\text{argmax}} \; p(\theta|y)$$

$$\text{Minimum Variance } \hat{\theta} = \underset{\breve{\theta}}{\text{argmin}} \; E(\theta - \breve{\theta})^2$$

$$\text{Maximum a Posteriori (MAP) } \hat{\theta} = \underset{\theta}{\text{argmax}} \; p(y|\theta)$$

It turns out because of the nice properties for linear systems and Gaussian noise, the Kalman state estimate satisfies all of these optimality criteria. In addition, the estimate is

- Unbiased[4]: the expected value of the estimate is the same as the parameter, and
- Consistent: the variance decreases to 0 with further observations

Because computer realizations of the algorithm are necessarily implemented in discrete time, here we summarize the Kalman filter for a discrete linear system. The linear system state is $x_k \in \mathbb{R}^n$, the control $u_k \in \mathbb{R}^p$, the measurements $y_k \in \mathbb{R}^m$, and additive, independent, zero mean, state noise $w_k \in \mathbb{R}^n$ and measurement noise $v_k \in \mathbb{R}^m$:

$$\begin{aligned} x_k &= Ax_k + Bu_k + w_k \\ y_k &= Cx_k + v_k \end{aligned}, \quad \begin{pmatrix} w_k \\ v_k \end{pmatrix} \sim N\left(\begin{bmatrix} 0 \\ 0 \end{bmatrix}, \begin{bmatrix} Q & 0 \\ 0 & R \end{bmatrix} \right)$$

Figure 9.9 shows the system model and Fig. 9.10 shows the standard form of the recursive estimator. Starting with an estimate of the initial state, $\hat{x}_0 \in \mathbb{R}^n$ and given a control input $u_1 \in \mathbb{R}^p$ the next state, $\hat{x}_{2|1}$, is predicted. The observations at time 2 are then used to update the state $\hat{x}_{2|2}$ and so on. It would be a good guess that the best prediction of the state given the control inputs can be obtained by

[4] The Cramer Rao Lower Bound (CRLB) gives smallest variance achievable by an unbiased estimate.

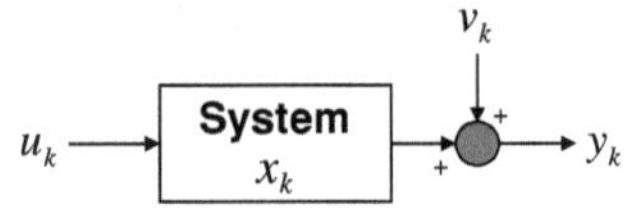

Fig. 9.9 System model

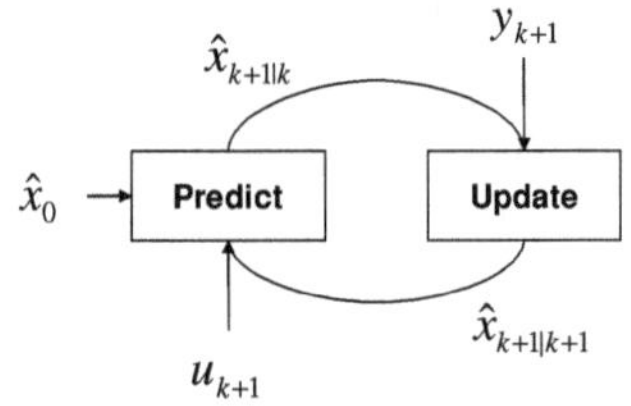

Fig. 9.10 Recursive estimator

simply applying the system model, the difficult piece is to decide how to optimally update the state given the observations. The Kalman filter provides the optimal update and additionally provides an error covariance that provides information on how good the estimate is.

Figure 9.11 shows the discrete time Kalman Filter algorithm.

The Kalman filter prediction step uses the system model to update the state given the control $\hat{x}_{k+1|k} = A\hat{x}_{k|k} + Bu_k$ and the state estimate given observations is $\hat{x}_{k+1|k+1} = \hat{x}_{k|k+1} + K(y_{k+1} - C\hat{x}_{k+1|k})$. The Kalman gain, K, is chosen to minimize the error covariance, P.

The derivations of the update equations for the Kalman gain and error covariance are made under the assumptions that the system is linear and the model and observation errors are independent Gaussian random variables. Given a complete system model, these assumptions imply a Markov property that past and future data are independent given the current state (Kailath 1980; Thrun et al. 2006).

The mathematical model introduced above is similar to the Markov model introduced in Chap. 4 for robot localization. A key difference is that each saved map feature is added to the system state and also tracked. This can cause a large increase in computational complexity over methods that assume a known map. A method for overcoming some of the practical implementation issues associated with the added computational complexity is discussed in the section SLAM Implementation.

If the system and or measurement model is nonlinear,

$$x_k = f(x_k, u_k, w_k)$$
$$y_k = g(x_k, v_k)$$

an extension of the Kalman Filter (the Extended Kalman Filter EKF) is made by substituting a linearized version of the system model, $\bar{A}_k = \frac{\partial f}{\partial x}\Big|_{(\hat{x}_{k|k}, u_k)}$, $\bar{B}_k = \frac{\partial f}{\partial u}\Big|_{(\hat{x}_{k|k}, u_k)}$ and measurement model $\bar{C}_k = \frac{\partial g}{\partial x}\Big|_{(x_{k|k})}$, into the computation of the prediction and update of error covariance, and computation of Kalman gain.

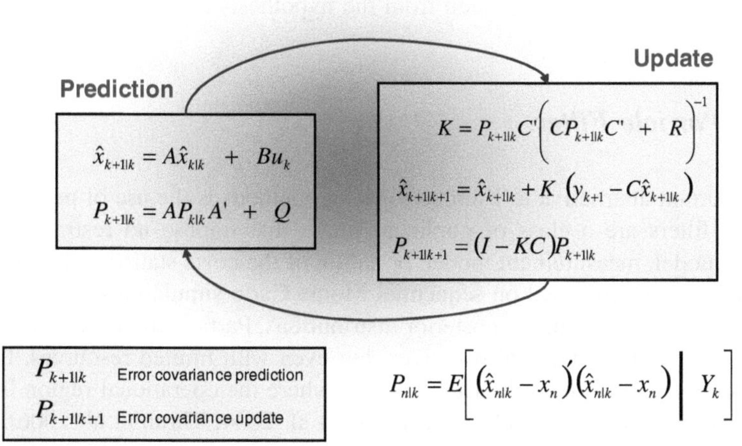

Fig. 9.11 Discrete time Kalman filter

The model prediction and update use the full nonlinear system model. The resulting unmodeled dynamics or other model inaccuracies induce violations of the original assumptions of the Kalman Filter derivation and so the EKF is no longer an optimal solution. Despite this, in practice the EKF often provides a useful solution when the linearization offers a good estimate over the current operating range of the system. Unfortunately, this is not always the case. This is because the linearization does not preserve the true mean and true covariance of the posterior distributions (Thrun et al. 2006).

A popular method that often performs better is the unscented Kalman Filter (UKF). The UKF performs a stochastic linearization through the use of a weighted statistical linear regression process, see Thrun et al. (2006) for more details. While the EKF is accurate to the first-order Taylor series expansion, the UKF is accurate to the first two terms in the expansion (Thrun et al. 2006).

The standard formulation of both the EKF-SLAM and UKF-SLAM solution is especially vulnerable to incorrect association of observations to landmarks. A single incorrect data association can induce divergence into the algorithm for map estimation, often causing catastrophic failure of the localization algorithm. (Durrant-Whyte and Bailey 2006a, b) One way to handle uncertain association of observations to landmarks is to generate a separate track estimate for each association hypothesis, creating over time an everbranching tree of tracks. This *multi-hypothesis data association* is important for robust SLAM implementation. Multihypothesis data association is especially important in loop closure, allowing a separate hypothesis for suspected loops and also a "no-loop" hypothesis for cases where the perceived environment is structurally similar.

A major hurdle in multihypothesis data association is the computational overhead of maintaining separate map estimates for each hypothesis. The number of tracks is typically limited by the available computational resources, and low-likelihood tracks must be pruned from the hypothesis tree.

9.3.2 *Particle Filter*

An important alternative to Kalman Filtering methods is the use of particle filters. Particle filters are a class of nonlinear filters that impose no restriction on the system model, measurement model, or nature of the noise statistics. Particle filters compute a solution based on sequential Monte Carlo simulations of particles that are selected to represent the posterior distributions. Particle filters are only optimal given infinite computational resources, but even with limited resources, they can give better solutions than the EKF in cases where the operational region is highly nonlinear. (Gustafssson et al. 2002; Ristic et al. 2004; Thrun et al. 2006).

One thing to be very cautious about is that computational complexity for nonlinear filters generally grows exponentially with the dimension of the system, whereas for the Kalman filter computational complexity grows as the cube of the dimension. While there are ways to keep the computational complexity under control, it is something that cannot be overlooked. The particle filter approach to modeling uncertainty is only possible because of the availability of fast, low-cost computers with large memories.

FAST-SLAM, with its basis in recursive Monte Carlo sampling, or particle filtering, was the first method to directly represent the nonlinear process model and nonGaussian pose distribution (Montemerlo et al. 2003a; Montemerlo and Thrun 2003b). Prior to the development of FAST-SLAM, the large state-space dimension in SLAM due to the number of map states made direct application of particle filters computationally infeasible. This issue is solved in FAST-SLAM by using a Rao-Blackwellized particle filter where the joint subject and map state is factored into a subject component, and a map component that is conditioned on the subject trajectory:

$$p(x_{0:t}, m|z_{0:t}, u_{0:t}, x_0) = p(x_{0:t}|z_{0:t}, u_{0:t}, x_0)p(m|x_{0:t}, z_{0:t}).$$

Note that, the probability distribution of the subject is on the entire trajectory rather than the single state as it is in EKF. When conditioned on the trajectory, the map landmarks become independent. This follows since given the exact pose states from which the observations are made, the observations are independent and therefore the map states are also independent.

The independence of map states is an important difference and the reason behind the speed improvements of FAST-SLAM over EKF algorithms. Because of the independence of the map states, updating the map, for a given pose trajectory particle (a single realization of the subject trajectory) is very fast. The map can be represented as a set of independent Gaussians. Each observed landmark can be processed individually as an EKF measurement update from a known pose.

FAST-SLAM linearizes the observation model, which is typically a reasonable approximation for range-bearing measurements when the subject's pose is known. Unobserved landmarks are independent and so unchanged.

Propagating the pose states is performed by particle filtering. The essential structure of FAST-SLAM, then, is a trajectory represented by weighted samples (particles) and a map is computed by EKF updates. The map accompanying each particle is composed of independent Gaussian distributions.

The FAST-SLAM algorithm is inherently a multihypothesis solution, with each particle having its own map estimate. A significant advantage of the FAST-SLAM algorithm is its ability to perform per particle data association (Montemerlo and Thrun 2003b).

Many types of recursive probabilistic state estimate algorithms have been developed to solve the SLAM problem in an approximate, computationally tractable way. While EKF-SLAM and FAST-SLAM are the two most important solution methods, newer alternatives have been proposed (Durrant-Whyte and Bailey 2006a, b; Karvounis 2011a). Information Filters and their extensions are of particular interest. Information Filters are duals of the Kalman Filter that have both computational and representation advantages when applied to location and mapping problems (Thrun et al. 2006).

9.3.3 Graph SLAM

GraphSLAM algorithms are also important SLAM implementations but the solution is typically not computed in real-time so we will not cover them here. For more information on GraphSLAM methods refer to Thurn and Montemerlo (2005); Thrun et al. (2006); Koller and Friedman (2009).

One particular GraphSLAM algorithm that supports real-time implementation is based on Factor Graphs (Loeliger 2004). Factor graphs provide a unified approach for modeling complex systems and to deriving practical message passing algorithms for the associated detection and estimation problems. Factor graphs allow most well-known signal processing techniques including Kalman and particle filtering to be used as components of such algorithms (Loeliger 2004).

Researchers at Georgia Tech and MIT have applied factor graph methods for incremental smoothing in inertial navigation systems (Indelman et al. 2012; Kaess et al. 2012). The system navigation states are nodes in the graph and each IMU measurement introduces a new factor to the graph connecting to the navigation state nodes. This factor may also be connected to other nodes used for parameterizing errors in the IMU measurements such as bias and scale factor. These nodes can be added at a lower frequency than the navigation state nodes. Other aiding sensors are simply additional sources of factors that get added to the graph asynchronously whenever their measurements are available. In this way, the factor graph formulation allows multirate, asynchronous measurements to be incorporated in a natural way (Indelman et al. 2012). The nonlinear optimization problem

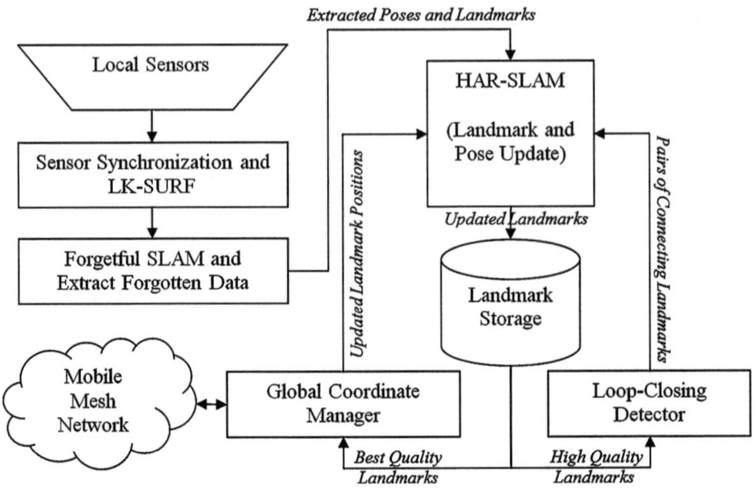

Fig. 9.12 HAR-SLAM Algorithm flow chart

encoded by the factor graph is solved by repeated linearization within a standard Gauss–Newton style nonlinear optimizer. The optimization can proceed incrementally because most of the calculations are the same as in the previous step and can be reused. As long as only sequential IMU measurements are processed, the resulting graph will have a chain-like structure. By maintaining all information within a single graph, the filter and smoother can operate asynchronously. This allows the problem to be split into a high speed navigation component and a higher latency loop closure component (Kaess et al. 2012).

9.4 SLAM Implementation

While theoretically the SLAM problem has been solved, a major issue that is faced in developing real-time implementations of SLAM is that as the number of tracked features/landmarks increases, the computation required at each step increases as a square of the number of landmarks. Required map storage also increases as a square of the number of landmarks (Dissanayake et al. 2001). Many people have developed SLAM implementations to address this issue (Montemerlo et al. 2003a; Montemerlo and Thrun 2003b; Kim and Sukkarieh 2004; Veth 2011; Karvounis 2011a). For example, computational complexity can be reduced by subdividing the map and by making the covariance matrix more sparse.

Here, we consider a particular implementation done at TRX Systems that attempts to address the computational issues in tracking an increasing number of features. Karvounis implemented an extreme version of this approach called Hierarchical SLAM or HAR-SLAM. Figure 9.12 gives a flow chart overview of how the system works. Full details of the algorithms are described in Karvounis

(2011a, c). The HAR-SLAM algorithm has similarities to the factor graph approach (Loeliger 2004; Kaess et al. 2012) in that both result in a chain-like structure of system states.

In this approach, the lowest level SLAM algorithm (Forgetful-SLAM: Fig. 9.12 left hand side) maintains active tracking of only the landmarks that can currently be seen or have been seen in the last N minutes (up to some max number of tracked landmarks). By limiting the set of landmarks tracked, the computational complexity remains bounded. Note that in Forgetful-SLAM landmarks are only matched to the landmarks currently seen by the camera; they are not matched to landmarks from the global map. This is a purely local SLAM layer and it will not offer the capability of correcting based on a previously known landmark (often referred to as "closing the loop"). That type of correction is handled by the higher level algorithm, HAR-SLAM.

The landmarks/features that are dropped from the Forgetful-SLAM algorithm are not actually forgotten; instead they are promoted and tracked within the global map by the HAR-SLAM algorithm, if they are determined to be "good", meaning that their covariance matrix $\mathbf{P}_{\text{landmark}}$ is small and "relevant", meaning that changes in the landmark location will affect the pose. To determine how much a good landmark can affect a pose, a metric combining the cross-covariance matrix between the landmark and pose, $\mathbf{P}_{\text{cross}}$, with the inverse of the landmark covariance matrix, $\mathbf{P}_{\text{landmark}}$, is used:

$$\max \text{Eigen value}\left(\mathbf{P}_{\text{cross}}^{T}(\mathbf{P}_{\text{landmark}})^{-1}\mathbf{P}_{\text{cross}}\right)$$

Landmarks are promoted when the max Eigenvalue is greater than a threshold. As landmarks are removed from Forgetful-SLAM and promoted, their correlations are tied to the last pose (historical position and orientation). There is a state vector and covariance matrix per pose, a state vector and covariance matrix per landmark, and a cross-covariance matrix per link.

As new poses are promoted from Forgetful-SLAM, any updates ripple back through the chain of historical pose estimates. Each pose is updated through a correlating Kalman Filter, and each landmark is updated through its own Kalman Filter. This directional update procedure defines the global level update (HAR-SLAM). A key advantage of this method is that both storage and computations grow only linearly with the number of landmarks and poses. (Karvounis 2011a, c).

The global coordinate manager is secondary loop that is run to manage the coordinate transforms for merging map data from other tracked subjects when matching features are detected in their respective maps. This is discussed in more detail later in this chapter. This property of remembering all poses and linking landmarks only to a single pose allows multiple tracked subjects to link maps together and asynchronously update portions of the map.

Figure 9.13 shows a high level diagram of HAR-SLAM. Each landmark and pose has a state vector x_i and associated covariance matrix P_i. In the Forgetful-SLAM section, features are fully linked to each other and the tracked subject's pose by cross-covariances $P_{i,j}$. Features no longer in view may be selected for

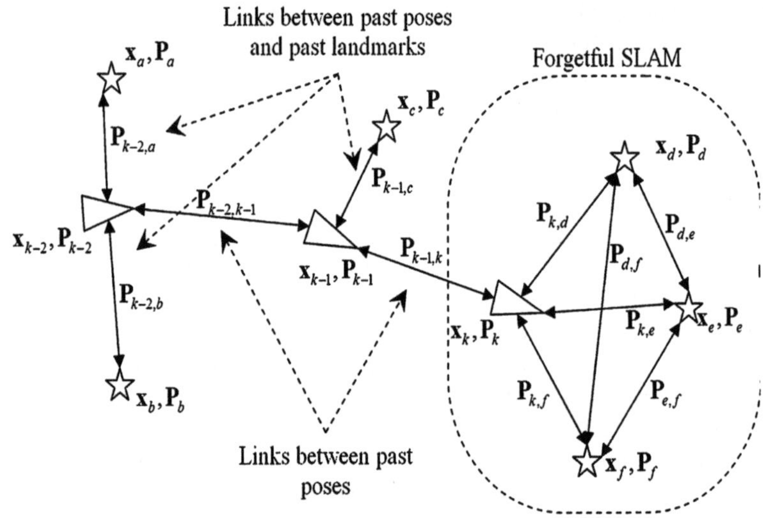

Fig. 9.13 Overview diagram of HAR-SLAM

promotion. Once features are promoted cross-correlations tie them to only the last pose from which the landmark was seen. All other links are broken. In the HAR-SLAM update, each pose and any associated landmark(s) are updated depending only on their direct links using Kalman gains.

Remembering the entire path instead of only the best current estimate allows HAR-SLAM to more quickly recover from errors by adjusting the entire historical path and all associated landmarks. This property of remembering historical poses and linking landmarks only to a single pose allows multiple tracked subjects to link maps together and asynchronously update portions of the map.

Feature management and promotion are the key differences between Forgetful-SLAM and the standard EKF-SLAM. In Forgetful-SLAM, features that are no longer seen are removed and considered for promotion to the global map. In order for a high level SLAM algorithm to assemble and track the "forgotten" features, the features need to be recoverable. A cross-correlation matrix is generated per lost feature that relates the feature to the previous pose (the last pose from which the feature was observed). Only features/landmarks where the max eigenvalue metric is greater than a threshold are promoted.

Karvounis developed HAR-SLAM primarily with the goal of limiting computation. As such, a main advantage of HAR-SLAM is its low computational cost. The cost grows linearly with the number of states and landmarks, while typical Kalman based SLAM algorithms are quadratic in cost (Fig. 9.14). FAST-SLAM is the closest to HAR-SLAM in computational cost, with the same linear growth with poses and number of landmarks; however, FAST-SLAM is based on particle filtering and so it maintains several particles, each with its own map, whereas a single map is maintained in HAR-SLAM.

Fig. 9.14 EKF vs forgetful-SLAM

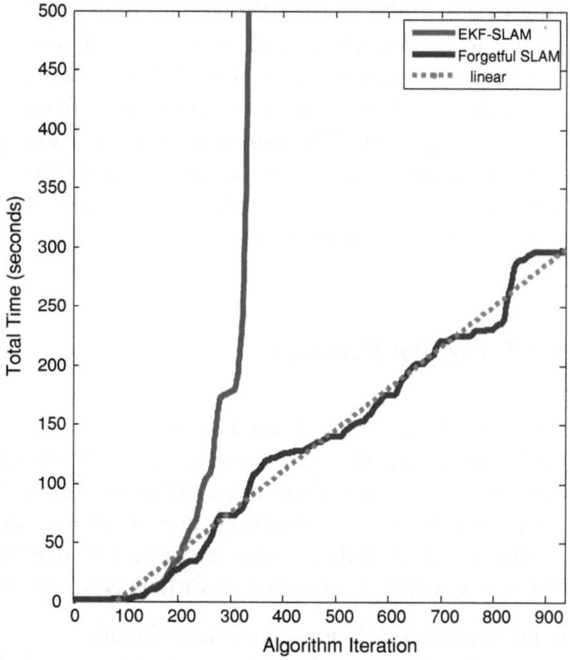

The Markov assumptions or complete state assumption that underlies all Bayesian Filters would imply that knowledge of prior states is not needed. However, unmodeled dynamics, other model inaccuracies or approximations, and correlations in inputs to the filters are all common in implementations and cause violations of the assumption. As it turns out, there are other benefits to maintaining prior pose history.

An added benefit that comes from linking landmarks through poses is that the pose history provides a directed approach for actively correcting the map and the entire path history. Because the historical corrections do not affect the current pose estimate, the rippling changes can have some delay, if necessary, to manage computational resources. Another advantage of keeping the entire pose history is that it facilitates closing the loop when matching features. A simple shortest path algorithm can find the chain of connecting poses between two landmarks, and this provides a directed path for updating the entire system (and computing needed cross correlations). Breaking the update into a chain reduces computation complexity to a point where the lower level SLAM and feature extraction algorithms are where the majority of computational resources are spent.

A key contribution in the development of the first SLAM algorithms was to show that, due to the common error in estimated observer location, there must be a high degree of correlation between estimates of the location of different landmarks in a map (Durrant-Whyte 1988; Smith et al. 1990; Durrant-Whyte and Bailey 2006a, b). The correlations between landmarks are a critical part of the problem and the more the correlations grow, the better the solution (Durrant-Whyte 1988;

Smith et al. 1990; Durrant-Whyte and Bailey 2006a, b). In Forgetful-SLAM, all feature to feature and feature to pose correlations are tracked and only the best features are promoted. Once promoted, in HAR-SLAM the features are extracted into a chain and tied to only to the last pose from which the feature was seen (as shown in Fig. 9.13). This eliminates cross-correlation links between features and between all but one pose. This change from the theoretical fully connected solution was made to improve computational speed and it is been demonstrated to be an effective approach in practice.

9.4.1 Outlier Removal

Kalman Filters are the method of choice for many navigation problems because the Kalman filter offers a computationally efficient optimal solution in the case that the underlying system has linear dynamics and the noise is Gaussian additive. Unfortunately, these assumptions do not hold for many navigation systems.

The standard Kalman filter algorithm is unable to handle the nonGaussian errors frequently encountered in various types of ranging systems, for example:

- incorrectly matching stereo image features,
- missed or incorrect detections caused by poor lighting
- ranging to unexpected people/objects moving in the field of the sensor.

Failure to recognize and reject these disturbances can cause non recoverable navigation errors in Kalman filter based navigation systems.

One option is to estimate the nonGaussian error probability and then apply a particle filter which can handle nonGaussian disturbances (Ristic et al. 2004). Particle filters have been used successfully in this way but at some computational cost. Another option is to develop a robust method for recognizing and rejecting outliers before allowing them to enter the Kalman filter. To minimize computational burden, Karvounis developed a Robust Kalman Filter that is able to recognize and reject the disturbances based on expected motion (Karvounis 2011a, b, c).

Typically, robust filters remove outliers before entering the Kalman Filter stage. What is novel and interesting about Karvounis' approach is that the median filter is inserted between the prediction and measurement update step of the Kalman Filter (Karvounis 2011b). Principle Component Analysis (PCA) is used to map the multidimensional observed features into a 1D space. Error vectors are computed by multiplying the error between the measured and the predicted observation values by the Kalman gain to find the effect of individual observation errors on the state. PCA is used to compute the principal vector in the state error space that causes the projected errors to be maximally distributed, making it sensitive to outliers. This technique is agnostic to the number of dimensions and the number of measurements. Including the Kalman gain scaling is important because it provides a weighting of the observation errors based on the how much the state is affected by the error, not just on the quality of the measurement. Combining PCA with a median filter provides a robust way to remove outliers.

Consider a nonlinear system model

$$x_k = f(x_k, u_k, w_k)$$
$$y_k = g(x_k, v_k)$$

where the system state is $x_k \in \mathbb{R}^n$, the control $u_k \in \mathbb{R}^p$, the measurements $y_k \in \mathbb{R}^m$, with the assumption of Gaussian zero mean state noise $w_k \in \mathbb{R}^n$ with variance Q_k and Gaussian zero mean measurement noise $v_k \in \mathbb{R}^m$ variance R_k. The linearized system is given by

$$\begin{aligned} x_k &= A_k x_k + B_k u_k + w_k \\ y_k &= C_k x_k + v_k \end{aligned}, \begin{pmatrix} w_k \\ v_k \end{pmatrix} \sim N\left(\begin{bmatrix} 0 \\ 0 \end{bmatrix}, \begin{bmatrix} Q_k & 0 \\ 0 & R_k \end{bmatrix} \right)$$

where, $A_k = \frac{\partial f}{\partial x}\Big|_{(\hat{x}_{k|k}, u_k)}$, $B_k = \frac{\partial f}{\partial u}\Big|_{(\hat{x}_{k|k}, u_k)}$ and $C_k = \frac{\partial g}{\partial x}\Big|_{(x_{k|k})}$.

The Kalman filter prediction step is the same as in a standard EKF. It uses the system model to update the state given the control $\hat{x}_{k+1|k} = A_{k+1}\hat{x}_{k|k} + B_k u_k$ and compute a predicted estimate covariance $P_{k+1|k} = A_{k+1} P_{k|k} A'_{k+1} + Q_k$.

In order to accommodate the fact that the measurement noise is not actually Gaussian in practice, the outlier removal function is inserted at this stage into the Kalman Filter. Assuming that each of the N observed features is independent from other features, each feature's covariance can be extracted from the block diagonal measurement covariance matrix R_k. The weights are determined by considering the effect each feature has on the state if the Kalman gain is applied. So for each observed feature,

$$K = P_{k+1|k} C^{(i)'}_{k+1} \left(C^{(i)}_{k+1} P_{k+1|k} C^{(i)'}_{k+1} + R^{(i)}_{k+1} \right)^{-1}$$
$$\tilde{x}^{(i)}_{k+1} = K \left(y^{(i)}_{k+1} - g^{(i)}(\hat{x}_{k+1|k}) \right)$$

For measurement related functions, $C^{(i)}_{k+1}, R^{(i)}_{k+1}, g^{(i)}, y^{(i)}_{k+1}$, the superscript (i) indicates the portion related to the selected feature. Note that each of $K, P_{k+1|k}, \hat{x}_{k+1|k}, \tilde{x}^{(i)}_{k+1}$, are full size. For $\tilde{x}^{(i)}_{k+1}$, the superscript (i) indicates that this is the state correction that is indicated due to the variation of that observed feature from what was predicted.

The mean and variance of the state corrections is then computed over the set of all features.

$$\bar{x}_{k+1} = \frac{1}{n}\sum_{i=1}^{N} \tilde{x}^{(i)}_{k+1} \quad \tilde{X}_{k+1} = \sum_{i=1}^{N} (\tilde{x}^{(i)}_{k+1} - \bar{x}_{k+1})(\tilde{x}^{(i)}_{k+1} - \bar{x}_{k+1})'$$

The largest eigenvector v of $\tilde{X}_{k+1}$ is the principal vector in the state space that causes the projected error corrections to be maximally distributed. Each feature's weight is then determined by projecting the state correction for that feature onto that principle vector: $w_i = v' x^{(i)}_{k+1}$. Outliers in this space are then eliminated using a median filter (see Chap. 8). Note that by using the state correction as a common metric for selecting outliers, measurements of different dimensions can be compared.

Next, for each of measurement related functions, $C^{(i)}_{k+1}, R^{(i)}_{k+1}, g^{(i)}, y^{(i)}_{k+1}$, for all i in the set of features that were not eliminated by the median filter, the matrices and functions must be reformed (now having reduced observation dimension). To make clear the reduction in dimension, we indicate them by $\tilde{C}_{k+1}, \tilde{R}_{k+1}, \tilde{g}, \tilde{y}_{k+1}$ in the update equations for the EKF.

$$K = P_{k+1|k}\tilde{C}'_{k+1}\left(\tilde{C}_{k+1}P_{k+1|k}\tilde{C}'_{k+1} + \tilde{R}_{k+1}\right)^{-1}$$

$$\hat{x}^{(i)}_{k+1|k+1} = \hat{x}^{(i)}_{k+1|k} + K\left(\tilde{y}^{(i)}_{k+1} - \tilde{g}(\hat{x}^{(i)}_{k+1|k})\right)$$

$$P_{k+1|k+1} = (I - K\tilde{C}'_{k+1})P_{k+1|k}$$

In the next section, we review some experiments that show the performance benefit of the Robust Kalman Filter when using stereo-optical measurements as part of an optical SLAM algorithm.

9.4.2 Experimental Results

Experiments to compare the performance of selected SLAM algorithms were conducted at TRX. The test was conducted using a robot from the University of Maryland's Autonomous Systems Lab (ASL). The ASL robot has the capability to report location via encoders, which provides a position estimate with roughly a 0.1 % error over distance travelled,[5] however, the robot uses gyros for heading, and these can have a drift in the heading estimate. Figure 9.15, the right image, shows in the center a PC that controls the robot and processes all data. The rear of the robot has a router that is setup to network with other robots but this capability was not used in the experiments reported here. Sonar sensors surrounding the robot were also available but not used in this experiment. The PC is powered by lithium ion batteries and the robot is powered by nickel metal hydride batteries.

The ASL robot was equipped with a TRX INU containing six-axis inertial, three-axis magnetic and barometric pressure sensors and enhanced with stereo Firefly cameras from Point Gray, as circled in the left image of Fig. 9.15. The Firefly cameras have a global shutter, which minimizes image blur, and a trigger that allow us to sync the images with the inertial measurements from the TRX INU. To selected and track stereo-optical features, LK-SURF, a hybrid feature

[5] This assumes traveling in a straight line track without wheel slip.

Fig. 9.15 University of Maryland automatic systems lab robot with TRX INU and machine vision cameras

Fig. 9.16 CAD drawing of the test location

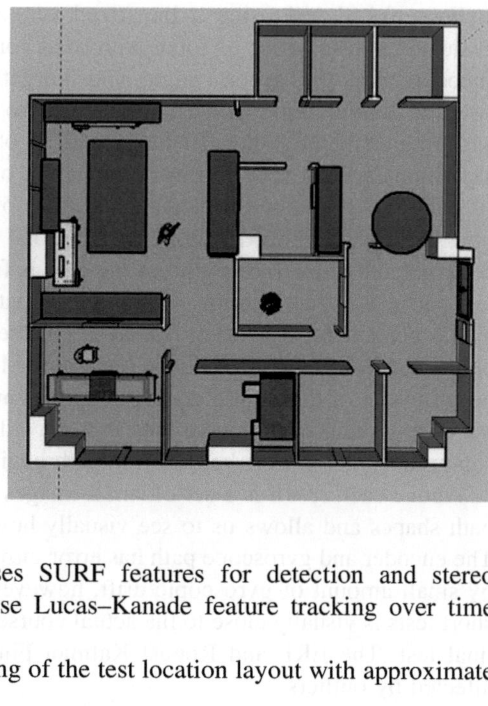

tracker, was implemented that uses SURF features for detection and stereo matching then modifies them to use Lucas–Kanade feature tracking over time (Karvounis 2011a, c).

Figure 9.16 shows a CAD drawing of the test location layout with approximate location of furniture.

9.4.2.1 Robust Kalman Filter Versus Standard Extended Kalman Filter

Tests were first run to show the performance of the Robust Kalman Filter versus the standard Extended Kalman Filter for integrating optical and inertial measurements. While the robots have capabilities for autonomous operations, data was collected by remotely controlling the robot in

1. a short path around the lab table,
2. a long path around the center offices,
3. a figure-eight path around the center offices and lab table, and
4. four laps around the center offices.

Images were collected at 20 frames per second and extracted/matched features were logged at each time step to enable a comparison of exactly the same path/features for each algorithm.

On these simple paths traversed by a robot in a lab, it was demonstrated that the Robust Kalman Filter did not suffer from poorly matched or tracked features and produced a path within about a meter of the true path as long as 50 % of the observed features align with the expected model (Hampel et al. 1986; Karvounis 2011a). The standard Kalman Filter (with no pre-filtering), on the other hand, drastically altered the estimated position of the robot inducing an error of over 60 m in location.

Forgetful-SLAM using a linearized version of a Robust Kalman Filter was evaluated over a series of robot paths and compared with a path based on wheel encoders and gyroscopic angle and Forgetful-SLAM using a standard EKF (without outlier removal). The encoder/gyro path is included as a baseline for performance if the controller has access to other vehicle sensors; optical SLAM algorithms are not used in this computation of this path.

Table 9.1 gives an estimate percent error over distance travelled for each path/filter. The wheel encoders measure the total distance travelled, and the error is determined by how far the end of the path is from the true location. In each of the test paths, the true location of the end of path is the same as the start location. While percent of distance travelled is not the best metric in for tracking system performance (Chap. 8, section Accuracy Metrics), it allows a comparison of performance of different algorithms on the same base data set when a system for measure ground truth course data is not available. Note that, scaling error is not captured by this metric because the path begins and ends at the same point.

Above each result is a set of small images of the path. This shows the various path shapes and allows us to see visually how each filter performed on the path. The encoder and gyroscope path has error introduced by deviations in distance and by small amount of gyroscopic drift, however the path shape over the first three short tests is visually close to the actual course and the effect of drift is clear in the final test. The EKF and Robust Kalman Filter are able to reduce drift but are affected by outliers.

The Robust Kalman Filter consistently matches shape to the baseline encoder and gyroscopic path. The EKF path suffers significantly from outliers, causing the path to be distorted. The Robust Kalman Filter performed the best by a significant margin in some cases but was out performed by the encoder/gyro path in one of the shorter paths. As the path length/complexity increased, the Robust Kalman Filter showed more consistently good performance. The standard EKF always performed worse because it was unable to reject the outliers.

Table 9.1 Path snapshot and filter performance comparison using percent error over distance travelled. Eight paths used to compare inertial paths, Robust Kalman Filter paths, and Extended Kalman Filter paths

Path	Wheel Encoder and Gyroscope	Forgetful SLAM using an EKF	Forgetful SLAM using RKF
Short Clockwise Path	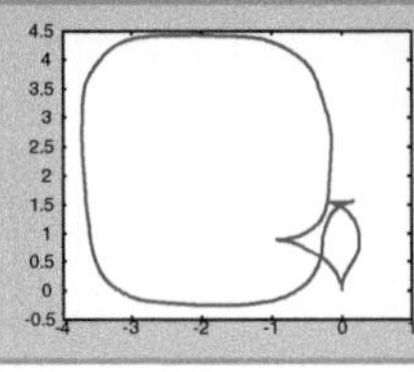	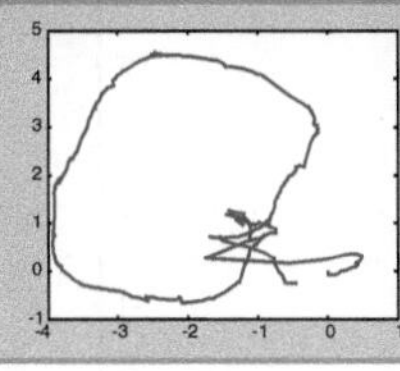	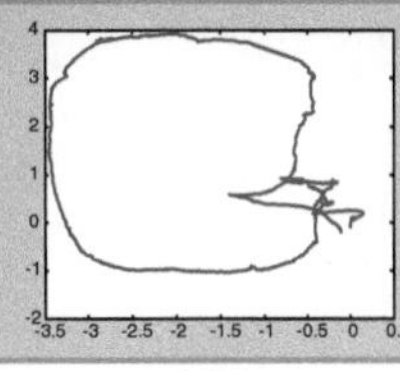
Short Clockwise	0.85%	2.75%	1.06%
Long Clockwise Path	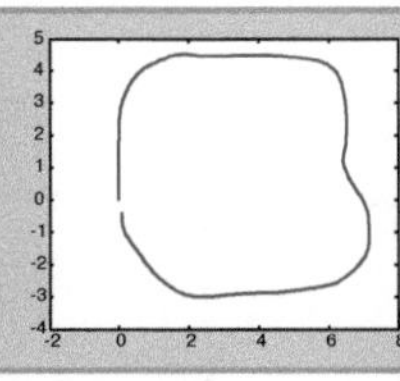	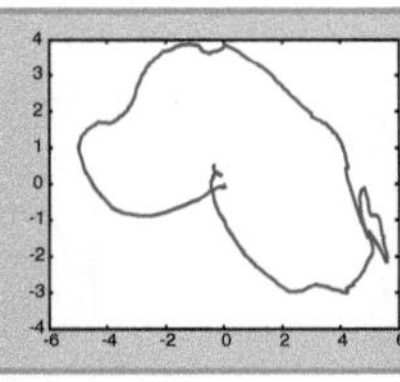	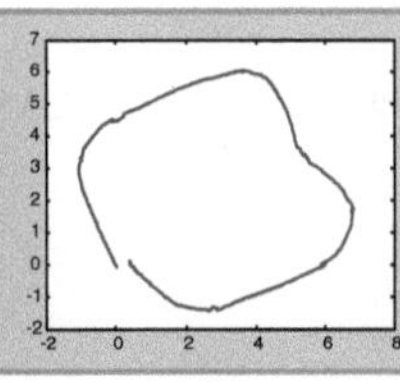
Long Clockwise	2.44%	2.48%	1.76%
Figure-Eight Path	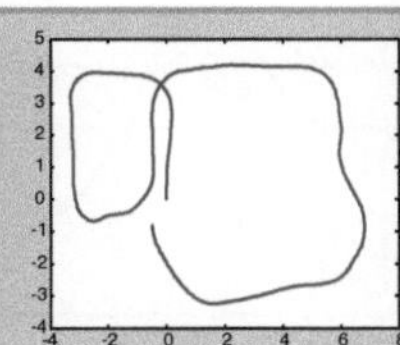	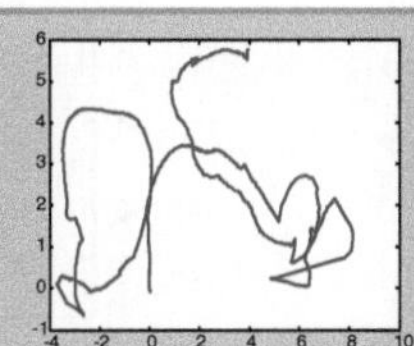	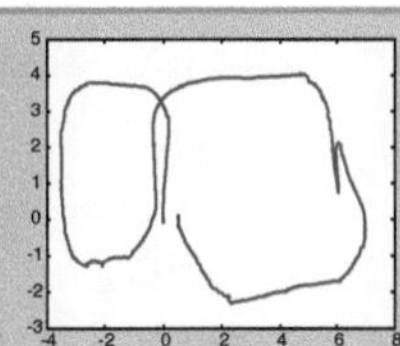
Figure-Eight	2.75%	18.41%	1.35%
Four Loop Path	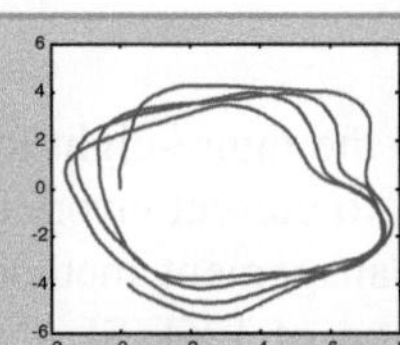	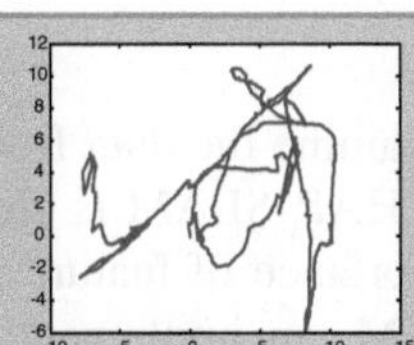	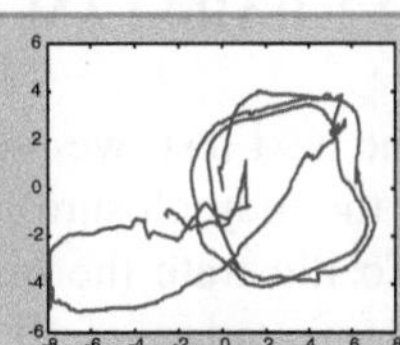
Four Loop	3.85%	8.50%	2.65%

While the Robust Kalman Filter appears to remove most outliers, it is not entirely immune to outliers; the four loop path shows anomalies indicating outliers are present. The median filter outlier rejection rule only works when outliers make up less than 50 % of the samples. If, for example, a moving object covers the entire image frame, there is no guarantee that there are any correctly tracked features. These anomalies may be able to be corrected if the higher level HAR SLAM algorithm were performed to allow a global map to be created. Without the global map, loop closure is not performed.

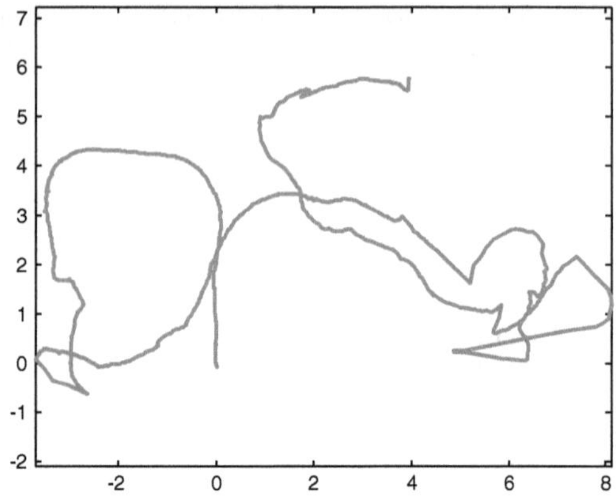

Fig. 9.17 EKF-SLAM

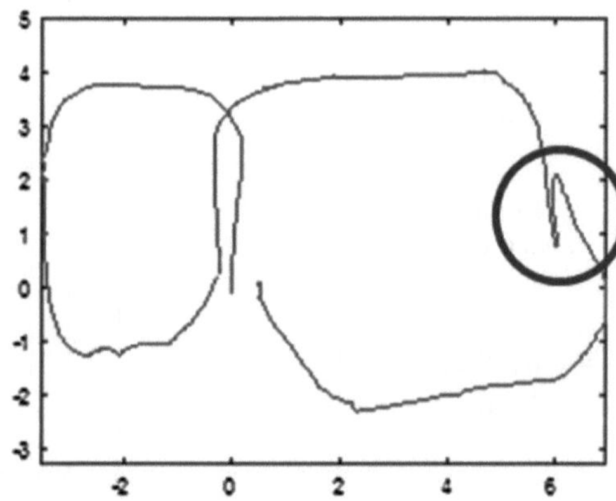

Fig. 9.18 Forgetful-SLAM

9.4.2.2 HAR-SLAM

In the next test, we reexamine the data from the figure-eight loop, to demonstrate that the loop closure in HAR-SLAM is able to correct errors caused by outliers.

To illustrate the importance of feature management methods in any real-time implementation of SLAM, an implementation of EKF-SLAM was run with no attempt to prune selected features, neither to remove outliers nor to reduce computation. The EKF-SLAM implementation (with every single feature ever seen saved!) took approximately 16 h to compute the path estimate for a 2 min path (Karvounis 2011a, c). Even with the extensive time taken, the result has many errors induced by outliers (Fig. 9.17). Simple feature management methods can improve this considerably.

In Forgetful-SLAM, a Robust Kalman Filter is used (Karvounis 2011a), which significantly reduces, but does not eliminate, the affect of outliers on the solution. An error caused by an outlier can be seen at (1, 6) in Fig. 9.18. On the same computer, Forgetful-SLAM took about 5 min to run the same path. While not yet real-time, this is a huge improvement over the EKF-SLAM running time.

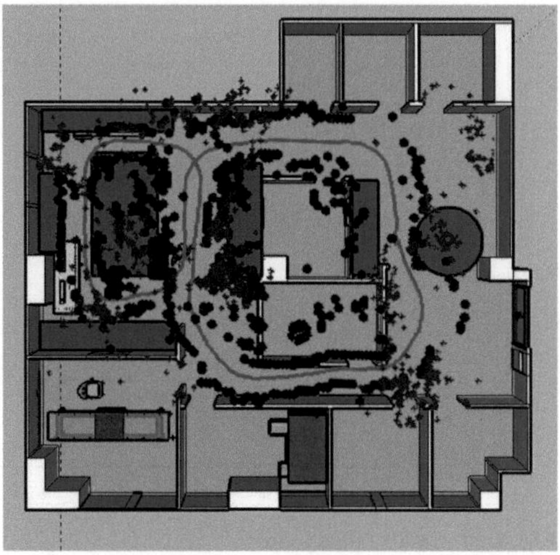

Fig. 9.19 HAR-SLAM map and path drawn on *top* of the CAD drawing

Closing the loop in HAR-SLAM by recognizing a feature and correcting the location enables the errors caused by outliers to be almost completely eliminated. HAR-SLAM took an additional minute to run, making it 6 min total. Figure 9.19 shows the HAR-SLAM map features (gray), filtered sonar data (black) and the HAR-SLAM path (red) overlaid on the CAD drawing. The HAR-SLAM path is very close to the true path.

9.4.3 Map-Joining

HAR-SLAM allows multiple robots to join maps on the fly in near real-time; whereas, other algorithms such as SEIF (Sparse Extended Information Filter) proposed joining maps in batch mode (Dissanayake et al. 2001). As depicted in the HAR-SLAM flow diagram in Fig. 9.12, as landmarks are promoted, a function is run to check for loop closures and another to compare features to map data from other tracked subjects in order to merge the features into a joint map. The promotion criteria is potentially different for the individual subject's global map which used to determine loop closure and the joint map, higher confidence being required to be promoted to the joint map. Landmarks promoted to the joint map are and shared among all tracked subjects on the network. Each landmark promoted to the joint map is check for matching by each individual. The same landmark matching technique that is used in loop-closing is used to determine matches between maps.

In order to join maps, a coordinate transform from each tracked subject's local coordinate system into the common coordinate system must be estimated. The transform consists of a translation to move the origin of the local coordinate

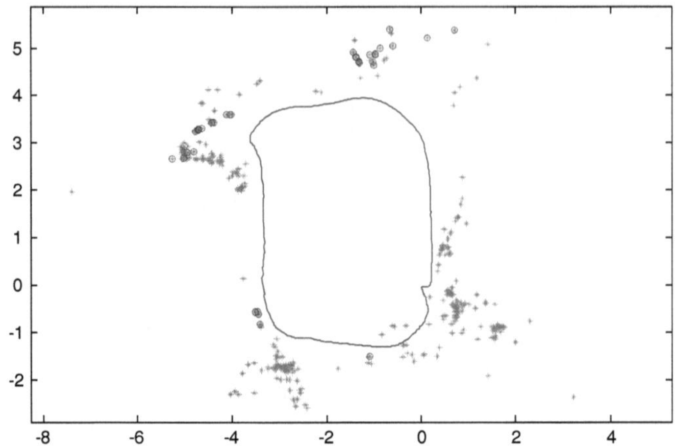

Fig. 9.20 Small office loop

system to that of the common coordinate system, and rotation to align the axes. Because of the uncertainty in map features, the global coordinate manager uses an EKF to estimate the coordinate transform for each track subject. Details of the update can be found in Karvounis (2011a, c) along with some discussion of a method for improving robustness.

The global coordinate manager can update the coordinate transform estimates as often as each time the coordinate manger detects that two or more tracked subjects have seen the same feature, but this is not necessary and computationally it may be too expensive. Each update can affect large portions of the joint map, and as those features are moved, in turn, each individual subject's map and historical track must also be updated. To conserve resources the update might be operated at some fixed interval or after some fixed number of feature matches are detected.

Figures 9.20, 9.21, and 9.22 show the HAR-SLAM results from three independent paths taken in the same test location. The associated features for each path are indicated by '+'. In one path the robot loops around a lab table (small loop), in another the robot loops around the lab table and the center offices (small and big loop), and in a third the robot loops around the center offices (big loop). In addition to path corrections, HAR-SLAM maintains a map of landmarks/promoted features. In each of the figures, the promoted features are circled.

This ability to merge the maps relies on selecting robust optical features. In this experiment only a few features are selected for promotion. The joint map is created by matching promoted features and then performing a global coordinate transform.

Figure 9.23 shows the three paths and the features in the joint map. Many of the selected landmarks are brought within close range of each other.

This example shows the promise of near-real-time joint map discovery using optical features but more work is needed in variable environments (including natural features, variable lighting, moving objects, etc.).

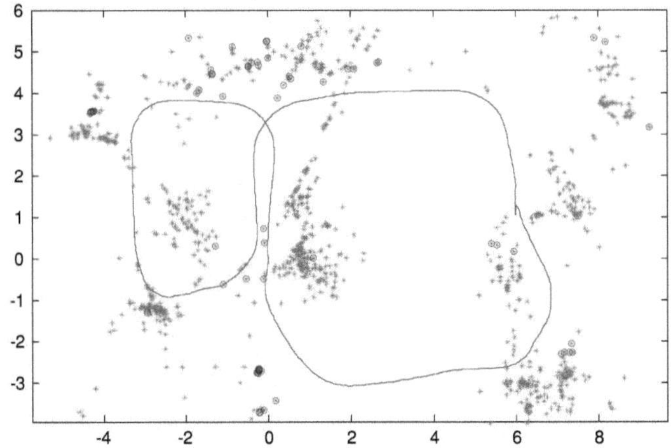

Fig. 9.21 Large and small loops

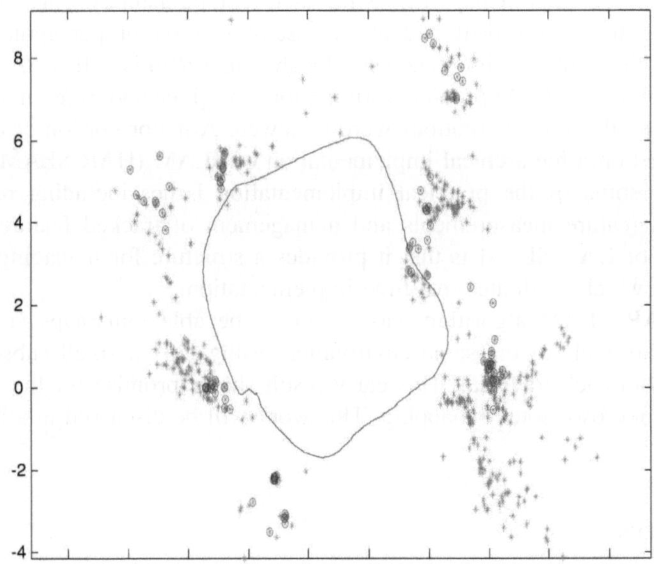

Fig. 9.22 Large office loop

9.5 Summary

This chapter gave an overview of localization and mapping with a focus on near real-time implementation. Methods for feature detection using selected allothetic sensors were reviewed. It is possible to create maps that enable long-term tracking with good accuracy by combining allothetic feature information with the idiothetic

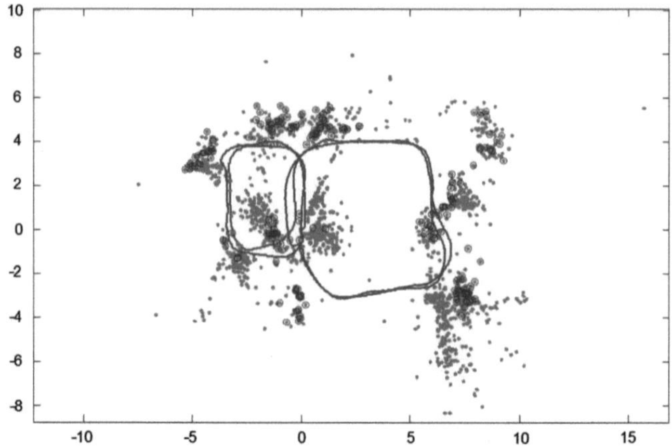

Fig. 9.23 Joint map with optical features

inertial tracking data (Chap. 8), using their complementary characteristics to compensate for sensor drift and allow disambiguation of perceptually aliased features. This ability to simultaneously localize and map is called SLAM.

A probabilistic SLAM problem formulation was given and different approaches for obtaining theoretical solutions were reviewed. A major portion of the chapter was focused on a hierarchical implementation of SLAM (HAR SLAM) designed to address some of the practical implementation issues including rejection of anomalous feature measurements and management of tracked features. A main advantage of HAR-SLAM is that it provides a structure for managing computational cost which facilitates real-time implementation.

The HAR SLAM algorithm was shown to be able join maps created from different traversals of the same environment using only a small subset of good features from each traversal. This early result shows promise for the use of this algorithm in crowd source mapping. This work will be discussed in a later paper.

References

A.E. Abdel-Hakim, A.A. Farag, in *CSIFT: A SIFT Descriptor with color invariant characteristics*. IEEE Computer Society Conference on Computer Vision and Pattern Recognition (2006)

A. Bandyopadhyay, D. Hakim et al., System and method for determining location of personnel and/or assets both indoors and outdoors via map generation and/or map matching techniques. USPTO. US, TRX Systems. Utility (2008)

H. Bay, A. Ess et al., SURF: Speeded up robust features. Comput. Vis. Image Underst. (CVIU) **110**(3), 346–359 (2008)

H. Bay, T. Tuytelaars et al., *SURF: Speeded Up Robust Features*. ECCV (2006)

J. Borenstein, Heading Error Removal System for Tracking Devices USPTO, University of Michigan. US 2010/0256939 Al (2010)

G. Dissanayake, P. Newman et al., A solution to the simultaneous localization and map building (SLAM) problem. IEEE Trans. Robot. Autom. **17**(3), 229–241 (2001)

H. Durrant-Whyte, Uncertain geometry in robotics. IEEE J. Robot. Autom. **4**(1), 23–31 (1988)

H. Durrant-Whyte, T. Bailey, Simultaneous Localization and Mapping: Part I. IEEE Robotics & Automation Magazine (June): 99–108 (2006a)

H. Durrant-Whyte, T. Bailey Simultaneous Localization and Mapping (SLAM): Part II. IEEE Robotics & Automation Magazine (September): 108–117 (2006b)

P. Elinas, R. Sim et al., in *σSLAM: Stereo vision SLAM using the Rao-Blackwellised particle filter and a novel mixture proposal distribution*. International Conference on Robotics and Automation, Orlando, Florida, IEEE (2006)

D. Filliat, J.-A. Meyer, Map-based navigation in mobile robots: I. a review of localization strategies. Cog. Sys. Res. **4**(4), 243–282 (2003)

B. Funk, A. Bandyopadhyay et al., Method and system for locating and monitoring first responders. USPTO. US, TRX Systems. 0077326, (2007)

J.E. Guivant, E.M. Nebot, Optimization of the simultaneous localization and map-building algorithm for real-time implementation. IEEE Trans. Robot. Automat. **17**(3), 242–257 (2001)

F. Gustafsson, F. Gunnarsson et al., Particle filters for positioning, navigation and tracking. IEEE Trans. Signal Process. **50**(2), 425–437 (2002)

F.R. Hampel, E.M. Ronchetti et al., *Robust Statistics: The Approach Based on Influence Functions* (Wiley, New York, 1986)

C. Harris, M. Stephens, in *A Combined Corner and Edge Detector*. Proceedings of the 4th Alvey Vision Conference (1988)

V. Indelman, S. Williams et al., in *Factor graph based incremental smoothing in inertial navigation systems*. International Conferences on Information Fusion (2012)

M. Kaess, S. Williams et al., in *Concurrent Filtering and Smoothing*. International Conference on Information Fusion (2012)

T. Kailath, *Linear Systems* (Prentice Hall, Englewood Cliffs, 1980)

J. Karvounis, Theory, Design, and Implementation of Landmark Promotion Cooperative Simultaneous Localization and Mapping. Electrical and Computer Engineering. College Park, University of Maryland. Ph.D (2011a)

J. Karvounis, Robust Kalman Filter. Joint Navigation Conference. Colorado Springs, CO, ION, (2011b)

J. Karvounis, Theory, Design, and Implementation of Landmark Promotion Cooperative Simultaneous Localization and Mapping. US—Provisional Patent, TRX. Provisional (2011c)

J. Kim, S. Sukkarieh, in *Improving the Real-Time Efficiency of Inertial SLAM and Understanding its Observability*. International Conference on Intelligent Robots and Systems, Sendai, Japan, (2004)

D. Koller, N. Friedman, Probabilistic Graphical Models: Principles and Techniques (Adaptive Computation and Machine Learning series), (MIT Press, Cambridge 2009)

T. Lemaire, S. Lacroix, SLAM with panoramic vision. J. Field Robot. **24**(1–2), 91–111 (2007)

T. Lindeberg, Feature detection with automatic scale selection. IJCV **30**(2), 79–116 (1998)

H.-A. Loeliger, An Introduction to Factor Graphs. IEEE Signal Processing Magazine (2004)

D.G. Lowe, in *Object recognition from local scale-invariant features*. International Conference on Computer Vision, (1999)

D.G. Lowe, Distinctive image features from scale-invariant keypoints. Int. J. Comput. Vision **60**(2), 91–110 (2004)

J.-A. Meyer, D. Filliat, Map-based navigation in mobile robots: II. A review of map-learning and path-planning strategies. Cog. Sys. Res. **4**(4), 283–317 (2003)

J.V. Miro, G. Dissanayake, et al., *Vision-based SLAM using natural features in indoor environments*. Intelligent Sensors, Sensor Networks and Information Processing Conference, IEEE, (2005)

M. Montemerlo, S. Thrun, in *Simultaneous Localization and Mapping with Unknown Data Association using Fast SLAM*. Proceedings of the IEEE International Joint Conference on Robotics and Automation, (2003b)

M. Montemerlo, S. Thrun, et al., in *Fast SLAM 2.0: An Improved Particle Filtering Algorithm for Simultaneous Localization and Mapping that Provably Converges.* International Joint Conference on Artificial Intelligence (2003a)

M. Muller, *Dynamic Time Warping. Information Retrieval for Music and Motion* (Springer, Berlin, 2007)

B. Ristic, S. Arunlampalam et al., Beyond the Kalman Filter Particle Filters for Tracking Applications, Artech House, (2004)

P. Robertson, M. Angermann, et al., Simultaneous Localization and Mapping for Pedestrians using only Foot-Mounted Inertial Sensors Ubicomp Orlando, Florida, (2009a)

P. Robertson, M. Angermann et al., in *Inertial Systems Based Joint Mapping and Positioning for Pedestrian Navigation.* ION GNSS. (Savannah, Georgia, 2009b)

P. Robertson, M. Angermann et al., in *SLAM Dance: Inertial-Based Joint Mapping and Positioning for Pedestrian Navigation.* Inside GNSS (2010)

H. Sakoe, S. Chiba, Dynamic programming algorithm optimization for spoken word recognition. IEEE Trans. Acoust. Speech Signal Process **26**(1), 159–165 (1978)

K. van de Sande, T. Gevers et al., Evaluating color descriptors for object and scene recognition. IEEE Trans. Pattern Anal. Mach. Intell. **32**(9), 1582–1596 (2010)

S. Se, D.G. Lowe et al., Vision-based global localization and mapping for mobile robots. Transactions on Robotics, IEEE, (2005)

R. Sim, P. Elinas et al., Vision-based SLAM using the Rao-Blackwellised Particle Filter. IJCAI Workshop on Reasoning with Uncertainty in Robotics, IJCAI, (2005)

R. Smith, P. Cheeseman, On the representation of spatail uncertainty. Int. J. Robot. Res. **5**(4), 56–68 (1986)

R. Smith, M. Self et al., *Estimating uncertain spatial relationships in robotics.* ed. by I.J. Cox, G.T. Wilfon. Autonomous Robot Vehicles. (Springer, New York, 1990), pp. 167–193

S. Thrun, W. Burgard et al., *Probablistic Robotics* (MIT Press, Cambrige, 2006)

S. Thurn, M. Montemerlo, The graph SLAM algorithm with applications to large-scale mapping of urban structures. Int. J. Robot. Res. **5**(6), 403–429 (2005)

M.J. Veth, Navigation using images, a survey of techniques. J. Inst. Navig. **58**(2), 127–139 (2011)

H. Wang, S. Sen et al., Unsupervised Indoor Location. MobiSys, (2012)

Index

C. Gentile et al., *Geolocation Techniques*, DOI: 10.1007/978-1-4614-1836-8,

MIX
Papier aus verantwortungsvollen Quellen
Paper from responsible sources
FSC® C105338

If you have any concerns about our products,
you can contact us on
ProductSafety@springernature.com

In case Publisher is established outside the EU,
the EU authorized representative is:
Springer Nature Customer Service Center GmbH
Europaplatz 3, 69115 Heidelberg, Germany

Printed by Libri Plureos GmbH
in Hamburg, Germany